12th Topical on Refinery Processing 2009

Topical Conference at the 2009 AIChE Spring National Meeting

Tampa, Florida
26 – 30 April 2009

ISBN: 978-1-61567-259-2

Printed from e-media with permission by:

Curran Associates, Inc.
57 Morehouse Lane
Red Hook, NY 12571

Printed by Curran Associates, Inc. (2009)

For permission requests, please contact AIChE
at the address below.

AIChE
3 Park Avenue
New York, NY 10016-5991

Phone: (203) 702-7660
Fax: (203) 775-5177

www.aiche.org

Additional copies of this publication are available from:

Curran Associates, Inc.
57 Morehouse Lane
Red Hook, NY 12571 USA
Phone: 845-758-0400
Fax: 845-758-2634
Email: curran@proceedings.com
Web: www.proceedings.com

TABLE OF CONTENTS

Paper Abstract Submitted for
AIChE National Spring Meeting
Tampa, USA, April 16-30, 2009

Topical Conference: Refining Processing (T7)
Session (T7001): CO2 Reduction Tutorial

CO2 Capture Options for Power Production Facilities

G. Protopapas, O. Aguilar
Shell Global Solutions US, Inc.
3333 Highway 6 South, Houston, TX, 77082
georgios.protopapas@shell.com, oscar.aguilar@shell.com

EXTENDED ABSTRACT

Environmental concerns have led to increased interest in processes that aim to capture CO2. The produced CO2 stream can be stored in depleted oil fields or aquifers. Post-combustion carbon capture is an attractive option, since this technology can easily be implemented in existing facilities without significant modifications. However, as CO2 capture is an energy intensive process, there is a significant efficiency penalty with the addition of the capture plant. This effect can be minimized with appropriate integration between the power plant and the CO2 capture plants. A study on the potential benefits of various integration options on the overall efficiency of a Combined Cycle Gas Turbine power plant and a post-combustion CO2 capture plant system was conducted to determine the optimum integration configuration between plants. The CCGT power plant considered consists of 2 GT's and 1 ST with a design capacity of approximately 800MWe at a net electrical LHV efficiency of approximately 58%. Addition of an amine based CO2 capture plant, including drying and compression, and designed to produce high purity CO2 at a pressure level suitable for aquifer storage, led to a net electrical LHV efficiency reduction to approximately 45%. A number of integration options as well as their combinations were simulated using a steady state, thermodynamic cycle modeling software tool for the power plant and an in-house developed spreadsheet tool for the capture plant. The best performance achieved resulted in an overall efficiency of approximately 50%. State of the art technology, which is readily available in the market at the time of the study was considered throughout the evaluation.

Keywords: Carbon capture, energy, CO2 reduction

Preliminary Findings on OSHA's Refinery NEP

James Lay, P.E.
Lisa A. Long (Presenter)
Michael Marshall, P.E. (Presenter)

U.S. DOL – OSHA
200 Constitution Ave, NW
Washington, DC 20210
202-693-2409
Long.lisa@dol.gov

Prepared for Presentation at
American Institute of Chemical Engineers
2009 Spring National Meeting
5th Global Congress on Process Safety
43rd Annual Loss Prevention Symposium
Tampa, Florida
April 26–30, 2009

UNPUBLISHED

Abstract
On June 7, 2007, U.S. DOL - OSHA (OSHA) initiated a Petroleum Refinery Process Safety Management National Emphasis Program (NEP) in response to a large number of fatal or catastrophic incidents in the petroleum refining industry. The NEP describes OSHA's intent to inspect all refineries (SIC 2911, NAICS 32411) in federal jurisdiction within 2 years, and provides a new protocol for conducting these inspections. OSHA state plan states were also encouraged to inspect refineries within their jurisdiction using the same protocol, and many are doing so.

This paper will briefly describe the inspection protocol, update OSHA's progress in completing these inspections, and review findings of the NEP and other PSM inspections conducted since 1992, including the elements of the PSM standards that were most frequently cited.

Background
Following a major incident at BP's Texas City Refinery in April 2005, OSHA compiled and evaluated data from PSM incidents and found that petroleum refining had more serious PSM incidents than any other industry segment. The data revealed that between 1992 and 2005, fatal (one or more employees killed) and catastrophic (three or more employees hospitalized) PSM incidents in all industries resulted in 138 fatalities and 553 injuries. During this period, 36 of the 240 total serious PSM incidents occurred in petroleum refineries, resulting in 52 employee deaths (38% of total) and 250 employee injuries (45%). As a result of the incident history in petroleum refining, OSHA developed a Petroleum Refinery National Emphasis Program (NEP) that was issued on June 7, 2007. The Refinery NEP outlines policies and procedures for OSHA Compliance Officers to use in conducting inspections of petroleum refineries, and mandates that OSHA inspect all petroleum refineries, excluding those in the Voluntary Protection Program (VPP), within federal jurisdiction over two years. Although VPP sites are excluded from NEP inspections, their PSM programs are evaluated as part of the VPP process. OSHA strongly encouraged State Plan States to adopt the NEP inspection strategy, and all states with refineries have either implemented the Refinery NEP or a similar program.

The Refinery NEP emphasizes *implementation* of PSM programs. Compliance officers use a list of static questions, which is published in the Refinery NEP, and a smaller list of dynamic questions, which is unpublished and periodically changed. The questions focus on issues identified as contributing to previous incidents and contain compliance guidance to assist Compliance Officers in evaluating PSM implementation in a unit selected by the inspection team. Contractors working in or adjacent to the selected unit are also inspected.

Inspection Numbers and Fines
Inspection data[1] demonstrates that OSHA has found a significant number of PSM violations in the refining industry. Between June 7, 2007 and February 12, 2009, OSHA has opened inspections of 42 refineries and 81 contractors. Citations have been issued in 30 of these refinery inspections; the remainder are ongoing. Table 1 illustrates the proposed penalties associated with the violations, broken down by classification.

[1] Data presented is from inspections conducted by Federal OSHA. Data does not include inspections conducted by State-Plan States.

Table 1: Refinery NEP Penalties (June 7, 2007 to February 12, 2009)

	Refineries		Contractors	
Violation Classification	**Number of Violations**	**Total Penalties**	**Number of Violations**	**Total Penalties**
Willful	6	$379,000	0	$0
Repeat	17	$447,500	0	$0
Serious	462	$1,661,705	18	$39,225
Unclassified	9	$429,000	0	$0
Other	29	$31,450	13	$7,487
Total	**523**	**$2,948,655**	**31**	**$46,172**

On average, Refinery NEP inspections at refineries have resulted in 17.4 violations with penalties of $98,300. Refinery NEP inspections of contractors at refineries have resulted in an average of 0.4 citations with penalties of $658.

Violations broken down by Element and Sub-element of the PSM Standard

OSHA inspection data was analyzed to which element and sub-element of the PSM standard were most frequently violated. The following table shows the top 10 elements and sub-elements.

Table 2 – Refinery NEP Most Frequently Cited PSM Elements

Element	Description	Number of Violations
J	Mechanical Integrity (MI)	404
D	Process Safety Information (PSI)	270
F	Operating Procedures (OP)	224
E	Process Hazard Analysis (PHA)	195
M	Incident Investigation (II)	159
L	Management of Change (MOC)	94
O	Compliance Audits (CA)	78
H	Contractors	35
C	Employee Participation	16
G	Training	14

Table 3 - Refinery NEP Most Frequently Cited PSM Sub-elements

Sub-element	Description	Number of Violations
d(3)	PSI – information pertaining to the equipment in the process	246
j(4)	MI – inspection and Testing	198
f(1)	OP – develop and implement written operating procedures	183
e(3)	PHA – requirements of PHA	101
j(5)	MI – equipment deficiencies	95
j(2)	MI – written procedures	77
e(5)	PHA – system to promptly address recommendations	74

m(1)	II – investigate each incident and near miss	63
m(4)	II – prepare an investigation report	48
o(4)	OA – promptly determine and document response to findings	47

As an aid in understanding the nature of the violations, each of the top 10 most frequently cited Sub-elements will be discussed and examples of language from actual citations presented.

Discussion of Most Frequently Cited Sub-elements

d(3) PSI – information pertaining to the equipment in the process	246 Instances

Requirement:

1910.119 d(3) – Information pertaining to the equipment in the process.

(i) Information pertaining to equipment in the process shall include:

(A) Materials of construction;
(B) Piping and instrumentation diagrams (P&ID's);
(C) Electrical classification;
(D) Relief system design and design basis;
(E) Ventilation system design;
(F) Design codes and standards employed;
(G) Material and energy balances for processes built after May 26, 1992; and,
(H) Safety systems (e.g. interlocks, detection or suppression systems).

(ii) The employer shall document that equipment complies with recognized and generally accepted good engineering practices.

(iii) For existing equipment designed and constructed in accordance with codes, standards or practices no longer in general use, the employer shall determine and document that the equipment is designed, maintained, inspected, tested, and operating in a safe manner.

Discussion:

The most frequent instance of violation within the d(3) sub-element was failure to have accurate P&ID's in the PSI, followed by a failure to document that equipment follows RAGAGEP, and failure to document information for relief system basis and design.

When P&ID's contain errors, they typically contain more than one error, which accounts for the large number of instances of P&ID violations. These P&ID errors were for various types of equipment and controls.

There have been numerous citations for failure to follow RAGAGEP. The citation history shows that this violation has been found in many areas including facility siting, relief valve design, safety instrumented systems, fireproofing, control of intervening valves, and electrical classification. Instances of violations for failure to document relief system design and design basis included many instances of not completely documenting design codes and basis. Some examples of paraphrased citation language are given below.

Example Citation Language:

- P&ID's are inaccurate: temperature indicator in field is not shown on P&ID, valves on the P&ID were not found in the field, valves in were not on the P&ID, Car sealed valves not shown as such on the P&ID, wrong valve sizes on P&ID.
- The employer failed to ensure that recognized and generally accepted good engineering practices (RAGAGEP) were followed with respect to the set pressures of relief devices. The relief devices on an exchanger were set at 540 and 600 psig while the MAWP for the exchanger was 480 psig.
- The employer did not comply with RAGAGEP when it failed to protect employees inside structures exposed to explosion, fire, and high pressure hazards as a result of a release of highly flammable materials. A facility siting study found deficiencies that needed correction in the Auxiliary Control Room, Central Control Room, and Laboratory.
- The employer did not document that it complied with RAGAGEP such as:
  - API 753 Section 3 when it failed to protect employees inside a wooden trailer being used as an operator shelter that was placed within the 1 psi overpressure zone;
  - ASME Boiler and Pressure Vessel Code (BPVC), Division 1 Section VIII, UG-135(d) when it failed to provide adequate controls to ensure that the isolation valve for a safety valve was in the open position;
  - API 521 where a number of relief valves were evaluated to be undersized in regards to the external fire case scenario.
  - ISA-S84.01, for calibrating, inspecting, testing, and maintaining Safety Instrumentation Systems (SIS) for emergency shutdown and interlocks.
  - NFPA 496-2003 Section 7.4.7 in that failure of the positive pressure air system inside the control building would not activate an alarm at a constantly attended location.
  - API 752, a knock-out drum was evaluated to not be able to provide adequate retention time to prevent liquid carryover to the flare.
  - NFPA 101 Section 7.8.1 when Illumination of the means of egress was not adequately maintained throughout the unit.
- PSI did not include the relief system design and design basis for: relief valves, removal of fractionator PSV, crude unit blowdown system, coker unit blowdown system, cat unit blowdown system.

j(4) MI – inspection and Testing	198 Instances

Requirement:

1910.119 j(4) – Inspection and testing.

(i) Inspection and tests shall be performed on process equipment.

(ii) Inspection and testing procedures shall follow recognized and generally accepted good engineering practices.

(iii) The frequency of inspections and tests of process equipment shall be consistent with applicable manufacturers' recommendations and good engineering practices, and more frequently if determined to be necessary by prior operating experience.

(iv) The employer shall document each inspection and test that has been performed on process equipment. The documentation shall identify the date of the inspection or test, the serial number or other identifier of the equipment on which the inspection or test was performed, a description of the inspection or test performed, and the results of the inspection or test.

Discussion:
Within the j(4) sub-element, failure to perform inspection and testing was most frequently cited. This was followed by failure to follow RAGAGEP for inspection and testing, inappropriate frequency of inspection and testing, and, finally, failure to properly document inspection and testing. Failures in the specific requirements for the inspection and testing occurred for many different types of equipment and controls. While there were a variety of aspects of inspection and testing failures, there were numerous violations regarding thickness measurements.

Example Citation Language:

- The employer does not ensure that thickness monitoring for process equipment including: a piping circuit, is conducted in accordance with the MI inspection program.
- Inspections and testing procedures did not follow RAGAGEP such as API 510 Section 6.6.2.3 when a bench tested safety valve failed to pop at its set pressure due to fouling. The valve was repaired and returned to service with the same 6 year inspection frequency. RAGAGEP requires that the inspection interval shall be reduced when a pressure relieving device is heavily fouled.
- The frequency of inspections and tests of process equipment was not conducted as specified by the employer's master plan to maintain mechanical integrity; such as, but not limited to, a cathodic protection system, and process analyzers.
- The employer did not follow RAGAGEP, such as API 570 when it failed to resolve anomalous data for pressure vessels and piping, such as increasing thickness measurements over previous readings.
- A critical piping circuit did not have any inspection data established since the line was installed in 1972.
- The employer did not ensure that thickness readings were documented for an FCCU.

f(1) OP – develop and implement written operating procedures	183 Instances

Requirement:
1910.119 f(1) – The employer shall develop and implement written operating procedures that provide clear instructions for safely conducting activities involved in each covered process consistent with the process safety information and shall address at least the following elements.

(i) Steps for each operating phase:

(A) Initial startup;
(B) Normal operations;
(C) Temporary operations;
(D) Emergency shutdown including the conditions under which emergency shutdown is required, and the assignment of shutdown responsibility to qualified operators to ensure that emergency shutdown is executed in a safe and timely manner.
(E) Emergency Operations;
(F) Normal shutdown; and,
(G) Startup following a turnaround, or after an emergency shutdown.

(ii) Operation limits:

(A) Consequences of deviation; and
(B) Steps required to correct or avoid deviation.

(iii) Safety and health considerations:

(A) Properties of, and hazards presented by, the chemicals used in the process;

(B) Precautions necessary to prevent exposure, including engineering controls, administrative controls, and personal protective equipment;
(C) Control measures to be taken if physical contact or airborne exposure occurs;
(D) Quality control for raw materials and control of hazardous chemical inventory levels, and,
(E) Any special or unique hazards.

Discussion:
The most frequent citations for developing and written operating procedures involved developing procedures for emergency shutdown. Other commonly cited violations included issues related to safe operating limits and failure to develop operating procedures for all operations.

Example Citation Language:

- The employer did not ensure that the written operation procedures addressed the conditions under which emergency shutdown is required, and the assignment of shutdown responsibilities to qualified operators to ensure that emergency shutdown is executed in a safe and timely manner.
- The employer did not develop and implement written procedures that provide clear instruction consistent with process safety information that address operating limits. The procedure for variable limits was not consistent with information lists on the control board / DCS system. The consequences of deviation list of operating variables did not address the safety and health consequences.
- The employer did not develop or implement written procedures that provided clear instructions for safely conducting activities in the covered process including but not limited to: the use of chains or car seals to ensure that block valves on pressure relief lines were kept in the open position, the Crude Unit blowdown system in all operating procedures of that unit, the Coker Unit blowdown system in all operating procedures of that unit, the Cat Unit blowdown system in all operating procedures of that unit.
- Written operating procedures for Main Flare operations did not contain steps for transition from normal operations to emergency operations when the Main Flare loses flame.
- Procedures were not developed for switching lube oil pumps on a compressor, resulting in the shutdown of the compressor and causing a mechanical integrity incident.

e(3) PHA – requirements of PHA	101 Instances

Requirement:
1910.119 e(3) – The process hazard analysis shall address:

(i) The hazards of the process;
(ii) The identification of any previous incident which had a likely potential for catastrophic consequences in the workplace;
(iii) Engineering and administrative controls applicable to the hazards and their interrelationships such as appropriate application of detection methodologies to provide early warning of releases. (Acceptable detection methods might include process monitoring and control instrumentation with alarms, and detection hardware such as hydrocarbon sensors.);
(v) Consequences of failure of engineering and administrative controls;
(vi) Facility siting;
(vii) Human factors; and

(viii) A qualitative evaluation of a range of the possible safety and health effects of failure of controls on employees in the workplace.

Discussion:
Over half of the citations issued for violations of e(3) have been related to failures to address facility siting or human factors. A substantial number of violations have also been issued regarding failure to address the consequences of failure of the engineering and administrative controls, and failure to identify previous incidents which could have had catastrophic consequences.

Example Citation Language:

- The process hazard analysis did not address facility siting. The PHA did not address the location of a piperack containing hydrogen, benzene, naptha, and butane found to be within 22 ft of the operator shelter. The PHA did not address the discharge of flammable gasses including but not limited to butane, propane, and naptha from the stack of an atmospheric blowdown drum which could impact a furnace and an operator shelter located 43 ft and 91 ft away, respectively.
- The employer did not ensure that the PHA addressed human factors such as: the number of alarms and warning that board operators can handle in an emergency or the hazards associated with process operators isolating critical valves in an emergency.
- The employer did not identify, evaluate, address the consequences of all scenarios that might result in the discharge of hot, heavier-than-air, or liquid hydrocarbons when it conducted the process hazard analysis including but not limited to: crude unit blowdown system, coker unit blowdown system, cat unit blowdown system, or closed valve on a pressure relief line.
- The PHA did not address the lack of fireproofing on pipe rack supports located in close proximity to process equipment such as but not limited to pumps.
- The employer's PHA did not control hazards related to heat tracing and insulation of Condenser outlet piping to reduce the likelihood of low spots freezing, which could result in overpressure of a Solvent Regenerator.
- The PHA revalidation of the Topper unit failed to evaluate and control the hazards identified on the human factor checklist.
- The PHA failed to consider unmarked eqiupment on the unit to include exchangers and safety relief valves which could lead to confusion by operators and/or contractors.
- The process hazard analysis did not identify previous incidents which had a likely potential for catastrophic consequences in the workplace. Sewers beneath a UPS battery room having an explosive atmosphere due to a water seal failure in the sewer system.

j(5) MI – equipment deficiencies	95 instances

Requirement:
1910.119 j(5) – Equipment deficiencies. The employer shall correct deficiencies in equipment that are outside acceptable limits (defined by the process safety information in paragraph (d) of this section) before further use or in a safe and timely manner when necessary means are taken to assure safe operation.

Discussion:
The frequency of violation in the j(5) sub-element is particularly noteworthy because there are no further divisions of this sub-element. All of the violations in this sub-element concerned running equipment outside of acceptable limits. The instances of these violations were for various types of equipment.

Example Citation Language:

- The employer did not correct deficiencies in equipment that were outside acceptable limits, including piping thickness reading below the minimum thickness (3 instances).
- The employer did not correct deficiencies in equipment to ensure safety in operation and maintenance before further use or in a safe and timely manner. Three pressurized enclosures for electrical equipment did not have an alarm or did not have an alarm monitored at a constantly attended location.
- Broken bolts in a flange connection on the splitter reboiler exchanger were not repaired.
- The employer did not correct deficiencies in the equipment that were outside acceptable limits including fireproofing insulation removed from conduits powering EIVs was never replaced, and continued to fail critical instrument inspections.
- A relief valve was evaluated to be undersized for protection against the loss of cooling and power failure relief scenarios
- Gage glasses were not maintained in a clean/readable condition.
- Toxic and flammable gas detectors located in duct work to the control room for H2S, and LEL were not in working order.

j(2) MI – written Procedures	77 instances

Requirement:
1910.119 j(2) - Written procedures. The employer shall establish and implement written procedures to maintain the on-going integrity of the process equipment.

Discussion:
From the data, it appears that most facilities did have procedures for most aspects of mechanical integrity. However, there were many instances of facilities not having procedures to handle unique situations such as on equipment that is insulated, has integrally bonded liners or injection points. There were also a few instances of not having procedures to deal with deficiencies or anomalous data.

Example Citation Language:

- The employer did not develop an inspection procedure to inspect for "corrosion-under-insulation" on pressure vessels.
- The employer did not establish and implement written procedures to maintain the on-going mechanical integrity of equipment, including inspections for non-metallic linings of pressure vessels.
- The employer did not write procedures to maintain the on-going mechanical integrity of process equipment in the refinery such as, but not limited to procedures for anomalous data related to thickness growths recorded in piping with criteria as to what would be considered an increase from a previous thickness measurement.
- The employer did not ensure that mechanical integrity procedures for relief devices were implemented at the facility. Multiple instances of relief valves were found that were past their individual inspection intervals for inspection and testing in the Hydrocracker and Alkylation Units.
- MI procedures did not include Information on alterations of vessels including welding qualifications and certifications for adding a nozzle to a heat exchanger
- MI procedures did not include inspection frequency for pressure vessels

- MI program procedures did not address how the employer determines and establishes a representative number of thickness measurements locations (TML) to satisfy requirements for conducting adequate internal and on-steam inspections.

e(5) PHA – system to promptly address recommendations	74 instances

Requirement:
1910.119 e(5) – The employer shall establish a system to promptly address the team's findings and recommendations; assure that the recommendations are resolved in a timely manner and that the resolution is documented; document what actions are to be taken; complete actions as soon as possible; develop a written schedule of when these actions are to be completed; communicate the actions to operation, maintenance and other employees whose work assignments are in the process and who may be affected by the recommendations or actions.

Discussion:
The majority of violations of e(5) involved instances of failure to resolve recommendation either at all or in a timely manner. A large number of these PHA recommendations could be considered to be "low-hanging–fruit" for employers, and often are relatively easy to resolve, e.g. car sealing isolation valves of relief devices. However, the violation history suggests that many recommendations that do not necessarily require large capital outlays have not been resolved.

Example Citation Language:
- The employer did not establish a system to promptly address the process hazard analysis team's findings and recommendations, as shown, for example, in an audit conducted in 2002 in which twenty three findings from a 1996 PHA were identified as not resolved.
- The most recent PHA listed items that have not been completed since the previous PHA, possibly resulting in loss of containment or the formation of a vapor cloud resulting in an explosion. These included: adding a charge pump suction temperature to the DCS with alarm and local indication to avoid cavitation and pump failure, installing an independent low level alarm on a storage tank to avoid hot vapors reaching a storage tank, converting a local temperature indicator on a propane vaporizer to local indication plus board indication with alarm and control, and addition to the critical process variable list to avoid liquid propane feed to heater burners.
- The employer has not resolved an action item from the 1997 to 2005 facility siting analysis in that occupied facilities, such as but not limited to the unit control room, have not been addressed.
- The employer did not assure that recommendations identified in PHAs for the refinery were resolved in a timely manner and the employer did not document the actions to be taken for resolution of PHA recommendations. Subsequent requirements, such as developing a written schedule for completion of actions and communication of the action to affected employees, were also not completed as required.
- The employer had not completed or resolved actions recommended by the its process hazards analysis teams in an expeditious manner including the recommendation related to the review of the relief valve sizing basis for the Propane Treater to ensure protection during a chemical reaction with HP acid.
- Did not resolve the PHA recommendation on adding a temperature sensor on the external surface of the Iso Treater.

- Actions items developed from a 2002 PHA revalidation were not completed in a timely manner in that an action item to re-rate and/or protect a steam coil had not been completed as of November, 2007.

m(1) - Investigate each incident and near miss	63 instances

Requirement:
1910.119 m(1) – The employer shall investigate each incident which resulted in , or could reasonably have resulted in a catastrophic release of highly hazardous chemical in the workplace.

Discussion:
Citations under this sub-element all addressed failure to investigate potentially catastrophic incidents. Multiple instances were identified in many inspections, with high frequencies for pump problems, process leaks and releases, upsets lifting relief valves, and fires.

Example Citation Language:

- Employers did not ensure that incident investigations were conducted for incidents that could reasonable have resulted in a catastrophic release of highly hazardous chemicals in the workplace. Examples include, but are not limited to:
    - Pump fires
    - Pump seal fires (multiple)
    - Packing blown out of pump
    - Pump tripped
    - Hydrocarbon liquid and vapor release (multiple)
    - Lost level
    - Injection line rupture
    - Pipe rack butane release
    - OVHD receiver leak
    - Acid leaks (multiple)
    - Exchanger leaks
    - Heater leaks
    - Over pressure and release to flare
    - Product leaking from threaded area on level float chamber
    - Corrosion
    - Sulfur fire
    - Slide flange fire
    - FCCU fire
- A near-miss incident was not investigated that could reasonably have resulted in a catastrophic release of HHC when a pressure relief valve and the pipe around it developed a buildup of ice due to propane leaking out of the pressure relief valve rendering the relief valve inoperable when the process was in operation.
- An incident investigation was not conducted when a process tower experienced a rapid pressure rise causing the pressure to be unstable which resulted in propane being vented to the flare for several hours.

m(4) – prepare an investigation report	48 instances

Requirement:
1910.119 m(4) – A report shall be prepared at the conclusion of the investigation which includes at a minimum:

(i) Date of Incident;
(ii) Date investigation began;
(iii) A description of the incident;
(iv) The factors that contributed to the incident; and
(v) Any recommendations resulting from the investigation.

Discussion:
Citations have been issued for all aspects of this sub-element, including failure to prepare any incident investigation reports at all. The most common missing component was failure to identify factors that contributed to an incident. Additionally, when factors were listed, they often had a marginal, if any, relationship to the actual causal factors of the incident.

Example Citation Language:

- Reports prepared at the conclusion of investigations of incidents did not include the date the investigation began (6 instances).
- No determinations of how cross-contamination occurred or how it was corrected.
- Root causes/conditions/factors that resulted in 2 crude heater tube leak fires were not identified in the incident investigation report.
- Reports for incidents did not include the factors that contributed to the incident:
  - Incident report addressing smoke in a substation.
  - Incident report addressing operations exceeding design temperatures on a reactor and associated piping and equipment.
  - Incident report addressing fan shaft breakage.
  - Incident report addressing accidental shutdown of extractor reflux pump.
- The report prepared at the conclusion of an incident investigation did not include the factors contributing to the incident, including, but not limited to, that piping inspections, evaluations, and recommendations were not adequate or effective in detecting internal corrosion on a piping weld.
- The incident investigation report did not include the factors contributing to the incident, including, but not limited to, that the quality control procedure was not adequate or effective in assuring that contractor welding was acceptable.
- Recommendations were not established in a crude heater tube leak fire investigation report.

o(4) - responses to findings	47 instances

Requirement:
1910.119 o(4) – The employer shall promptly determine and document an appropriate response to each of the findings of the compliance audit, and document that deficiencies have been corrected.

Discussion:
These citations all involved failure to address deficiencies identified in compliance audits, including failure to address management system shortcomings to prevent recurrence of the problems identified.

Example Citation Language:

- Pre-start up safety reviews/management of change tracking and approval not completed as identified in the last two compliance audits.
- Incident investigations not promptly addressed as identified in the last two compliance audits.
- Compliance audit finding that operating procedures were not being updated were not adequately addressed.
- More than three years had elapsed since the previous compliance audit and audit findings from previous audits were not completed in a timely manner.
- Appropriate responses to compliance audit findings were not determined and documented:
    - Ventilation system designs for control rooms not available
    - Process safety system information not available
    - Equipment test and inspection frequencies were not consistent with applicable manufacturers' recommendations and good engineering practices.
- Incident investigations were not promptly addressed as identified in the last two compliance audits.
- The employer did not ensure that PSM compliance audit findings were resolved prior to closing the action item as completed. Multiple relief devices were found to be past due for inspection/testing during the 2003 and 2006 compliance audits, however, the action items were closed out without ensuring that all relief devices were properly inspected, or that the management system was adequate to ensure that relief devices would not go past their inspection due date again.
- The employer did not specifically document correction of the deficiencies in the compliance audit related to the management system for the employer's Risk Based Mechanical Integrity (RBMI) program.

Comparison of Refinery NEP Data with prior PSM Inspection Data

The Refinery NEP has been OSHA most significant PSM enforcement effort since the promulgation of the PSM standard in June 1992. Comparison of the Refinery NEP Inspection data with PSM inspection data prior to June 2007 shows several notable differences:

- **The Majority of Refinery NEP inspections have been done as planned or programmed inspections as opposed to unprogrammed inspections resulting from accidents, complaints, or referrals.** Approximately 73% of the inspections with PSM citations conducted between 1992 and June 2007 (prior to the Refinery NEP) resulted from accidents, complaints, or referrals .

- **Refinery NEP inspections are resulting in more violations and higher penalties.** Refinery NEP inspections have resulted in an average of 17.4 violations with penalties of $98,300. Prior PSM inspections (where citations were issued) in all SICs have resulted in an average of 3.6 violations with penalties of $22,865.

- **The inspection strategy used in the Refinery NEP has resulted in more violations than the PSM PQV inspection strategy.** The Refinery NEP focuses on issues identified as contributing to previous incidents and contains Inspection Priority Items (IPI) and compliance guidance to assist Compliance Officers in evaluating PSM implementation. In contrast, the PSM PQV used an audit checklist which mirrored the performance requirements of the PSM standard. In 1993, OSHA conducted 39 planned or programmed inspection under PSM Program-Quality-Verification (PQV). These inspections, conducted in 8 identified high risk SICs, resulted in an average of 2.3 violations with penalties of $2,953. Assuming equivalent experience levels and training for the OSHA compliance personnel conducting these inspections, it appears that the Refinery NEP strategy (17.4 violations per inspection) is a better tool for identifying hazards and violations than the PSM PQV strategy (2.3 violations per inspection).

- **Average PSM violations and penalties have been higher in the refining industry than in the regulated community as a whole.** Inspection data shows that PSM inspections prior to June 2007 in the petroleum refining industry (SIC 2911) resulted in an average of 4.8 violations with penalties of $91,570 compared with inspections in all SIC codes that resulted in an average of 3.6 violations with penalties of $22,865.

- **There are some similarities in the most frequently cited violation; however, the NEP's most frequently cited sub-elements show more violations related to process safety information, incident investigation, and compliance audits, and fewer related to more basic elements such as training and employee participation.** The table below shows the most frequent violations cited during the key periods compared: Refinery NEP Inspections, Inspections from 1992 until the start of the Refinery NEP for the entire covered community (for inspections where a PSM citation was issued), and Inspection from 1992 until the start of the Refinery NEP for the petroleum refining industry (for inspections where a PSM citation was issued).

- **Data for the most frequently cited sub-elements reflects OSHAs focus on PSI and RAGAGEP in the inspection priority items and training of compliance officers.** The most frequently cited sub-element, (d)(3) and more specifically (d)(3)(ii)[2] is subject to numerous IPIs in the Refinery NEP document and is a primary focus for OSHA PSM Refinery NEP training for compliance officers.

- **Focus on Incident investigation during Refinery NEP compliance officer training appears to be beneficial.** OSHA training covered requirements for incident investigation with specific emphasis on determining whether employers were conducting "near-miss" investigations (1910.119 m(1)), and adequately reporting of the factors that contributed to the incident (1910.119(m)(4)(iv)). Data indicates that element (m), Incident Investigation, is the 5th most frequently cited PSM element during the Refinery NEP where it was previously the 9th most cited PSM element. More specifically , (m)(1) and (m)(4) were the 8th and 9th most frequently cited PSM sub-element in the Refinery NEP whereas, neither of these two sub-elements were in the top 10 most frequent during the other two study periods.

- **In general, Refinery employers are not resolving PHA and audit findings and recommendations at a rate expected of large, sophisticated employers.** The Refinery NEP data shows that not resolving PHA and audit findings and recommendations were #7 and #10 most frequently cited sub-elements, respectively. Given that Refineries are typically large employers with both resources for, and knowledge

---

[2] 1910.119(d)(3)(ii) - The employer shall document that equipment complies with recognized and generally accepted good engineering practices.

of, PSM application, it is surprising that Refineries would have such a large number of violations related to addressing hazards identified by their PHA and audit teams.

Table 4 - Comparison of PSM Violations

Refinery NEP Violations		*2911 '92 to NEP Violations*		*PSM '92 to NEP Violations*	
Elements	**#**	**Elements**	**#**	**Elements**	**#**
J – MI	404	J - MI	263	F - OP	1189
D – PSI	270	F – OP	177	J – MI	1152
F - OP	224	L – MOC	138	E – PHA	991
E – PHA	195	E – PHA	110	D – PSI	744
M - II	159	D – PSI	51	G – Training	551
L – MOC	94	G – Training	46	L – MOC	518
O – CA	78	H – Contractors	43	H – Contractors	448
H - Contractors	35	M - II	42	C – Employee Participation	386
C – Employee Participation	16	C – Employee Participation	17	M – II	228
G - Training	14	N - ER	13	N – Emergency Planning and Response (ER)	177
Sub-elements		**Sub-elements**		**Sub-elements**	
d(3) PSI – information pertaining to the equipment in the process	246	j(5) MI-equipment deficiencies	113	j(2) MI – written procedures	369
j(4) MI – inspection and Testing	198	l(1) MOC - establish and implement procedures for MOC	57	e(1) II – perform an initial PHA	347
f(1) OP – develop and implement written operating procedures	183	j(4) OP – develop and implement written operating procedures	46	l(1) MOC - establish and implement procedures for MOC	287
e(3) PHA – requirements of PHA	101	l(3) MOC – affected employees must be notified and trained before changes	41	c(1) EP – develop a written plan for employee participation	243
j(5) MI-equipment deficiencies	95	j(2) MI – written Procedures	34	j(5) MI-equipment deficiencies	214
j(2) MI – written procedures	77	e(5) PHA – requirements of PHA	30	g(1)Training – train employees involved in operating a process	205

e(5) PHA – requirements of PHA findings and recommendations	74	f(4) OP – develop and implement safe work practices	30	f(1) OP – develop and implement written operating procedures	202
m(1) II – investigate each incident and near miss	63	f(3) OP – review procedures as often as necessary and certify procedures annually	17	j(4) OP – develop and implement written operating procedures	189
m(4) II – prepare an investigation report	48	g(2) Training – refresher training at least every 3 years with input from employees	17	e(5) PHA – requirements of PHA	188
o(4) OA – promptly determine and document response to findings	47	j(3) MI – Training for process maintenance activities	17	N ER – establish and implement emergency action plan including handling small releases	177

Concluding Remarks

OSHA will continue conducting Refinery NEP inspections until all refineries within federal jurisdiction are complete. It is expected that fewer violations will be found in later inspections as refineries learn from others and improve their PSM program. In 2008, OSHA announced that it is working on a new National Emphasis Program that will target other PSM covered facilities, such as chemical plants, and use a strategy similar to the Refinery NEP although abbreviated. OSHA will continue to share findings from both of these programs with stakeholders in an effort to continuously improve PSM systems and reduce the number of catastrophic incidents.

References

(1) Occupational Safety and Health Administration, *Directive Number: CPL 03-00-004, Petroleum Refinery National Emphasis Program,* June 7, 2007, http://www.osha.gov/pls/oshaweb/owadisp.show_document?p_table=DIRECTIVES&p_id=3589, Accessed January 2009.

4c Effects of CO_2 Regulation and Potential Solutions

X. X. Frank Zhu, UOP LLC, 25 East Algonquin Road, Des Plaines, IL 60071

The European Union Energy Pact (2008) contains a key target for the year 2020: a 20 % reduction in CO_2 emissions (compared with 1990). The US will follow the similar direction in setting up CO_2 regulation in the near future. It is clear that it is a big challenge for the industry to achieve 20% reduction in CO_2 emissions from the 1990 level.

This presentation will describe Honeywell Energy Management solutions for refining and petrochemical industry. Honeywell Energy management solutions involve a lot more than just modernizing utility systems, although for many sites the utilities plant is an area that has often been neglected.

Honeywell's experience has shown that a 12 to 25% energy reduction is achievable by implementing a comprehensive energy management solution with attractive returns on capital investment.

37a Design of Refinery Steam Network Process Control Philosophy by Using Dynamic Simulation

Haribabu Chittibabu[1], **Dipanjan Bhattacharya**[1], and Ramachandra P. Tekumalla[2]. (1) Bechtel Oil and Gas, New Delhi, India, (2) Advanced Simulation & Analysis, Bechtel Corporation, 3000 Post Oak Blvd., Houston, TX 77056

A dynamic simulation of a refinery steam system provided a high degree of confidence that the process control system design will handle all major plant upsets with minimal impact on operations and lesser operator intervention. Refinery steam system is highly integrated network of multi pressure level steam distribution headers, generators and consumers. Instabilities in operating steam systems can lead to major plant upsets resulting in loss of revenue. The steam system is difficult to design for stability when there are multiple steam generators and users with many differing requirements. In order to ensure smooth operation, it is essential to conduct transient analysis of the system during various operational upsets with current control design in place.

Recently, Bechtel engineers completed a comprehensive dynamic simulation of the steam utility system for a major refinery upgrader project. The steam system for this project included two steam generators (utility boiler and hydrogen plant), power and steam distribution network comprising four pressure levels of steam with steam flows to and from process units. During transient situation the steam producers and consumers ramp up or down in a non-linear fashion. This adds complexity to the dynamics and makes it virtually impossible to assess the process conditions without conducting a detail dynamic analysis. Whether the steam system and its controls would respond effectively to major operational upsets, such as power failure, hydrogen plant trip, steam turbo-generator trip and process gas compressor trips can only be determined from evaluating dynamic responses to these incidents.

The study involved developing a high-fidelity dynamic simulation model of the steam system for evaluating steam system responses to various operational upset scenarios. The responses were used to identify limitations of the "as designed" system if any. The model also proved to be an effective tool for verifying adequacy of the proposed controller set point across let down valves.

OSCILLATORY CONTROL LOOPS: THE LOST ECONOMIC BENEFITS. CAN ANYTHING BE DONE?

Srinivas Karra and M. Nazmul Karim

Department of Chemical Engineering, Texas Tech University, Lubbock, TX-79409, USA.

Abstract: If present, oscillations in control loops, the basic economic benefits that can be claimed by employing the control loops are lost. There are many reasons behind oscillatory control loops: interacting control loops, aggressive controller tuning, external disturbances and/or nonlinearities in control loop hardware. In recent years, significant research has been performed on control loop oscillation diagnosis; there are still many unresolved issues such as indecisive or wrong diagnostics for certain regions of process dynamics, failure in the case of multiple faults occurring simultaneously and, absence of quantification of root-cause. In this work, a novel methodology has been devised to bridge these gaps in oscillation diagnosis. Power spectral density (PSD) and Auto Correlation Function (ACF) based oscillation detection method followed by a model based approach for identifying and quantifying the root cause of the oscillations is proposed. The application of each component of the algorithm is described with appropriate example cases. Finally, overall methodology is validated on industrial and simulation case studies for various cases of stiction, poor controller tuning, and external disturbances.

Keywords: limit cycles, root-cause detection and quantification

1. INTRODUCTION

In refining, petrochemical and chemical processing industries (CPI), competitive pressure, prohibitively increasing costs of energy and tighter environmental regulations make it imperative to explore various avenues to identify potential areas of improvement. Identifying trends that lead to performance degradation is also important in carrying out preventative maintenance.

1.1 Prevalence of oscillations in CPI and Economic Impact

Previous research on the performance of control loops revealed that as many as 60% of all industrial controllers have some kind of performance problem (Bialkowski, 1993; Ender, 1993; Rinehart, 1997). Studies in controller performance monitoring show that many control loops in the process industry perform poorly because of bad tuning or equipment problems; however, control valve nonlinearities are the major contributors towards performance degradation (Beckman and Jury, 1997; Isermann and Raab, 1993; Ruel, 2000; Corradini and Orlando, 2002). The control valve problems account for about 32% of ''poor'' or ''failed'' control loops (Miller and Desborough, 2001). A survey (Yang and Clarke, 1999) indicated that 30% of all control loops in Canadian paper mills were oscillating because of valve problems. Limit cycles due to valve problems were identified in many refineries, causing a large variation in the final products (Ruel,2000). Oscillations in control loops increase variability in product quality, accelerate equipment wear, and may cause other issues that could potentially disrupt the operation. In more than 4000 audited loops in all industries as reported by Beckman and Jury (1997), it was found that performance for more than 50% of the control loops could be significantly improved if the control loop hardware (control valves, I/P converters) were serviced at appropriate times, regularly. Since there are several possible causes for performance degradation of control loops, it is not always easy to improve their performance by taking corrective action, as it is a challenging task to identify the root cause. Hence, detection and diagnosis of oscillatory behaviour in control loop is of paramount importance in order to take appropriate corrective action.

1.2An example of oscillatory control loops

A continuous stirred tank reactor (CSTR) system is simulated to exemplify the generation of oscillations in process variables because of stiction in one of the control valves. The process involves a mixer followed by a CSTR as shown in Figure 10. Reactant stream coming from upstream processing unit is to be diluted with solvent to the desired concentration in the mixer. Concentration of A in the mixer is controlled by manipulating the solvent addition rate (F_W). Liquid-phase reaction $A_{(l)} \rightarrow B_{(l)}$ which is of first order in A takes place in the reactor. This reaction is highly exothermic, and the temperature controller controls the temperature of the reactor by manipulating the flow rate of the coolant (F_j) flowing through the jacket. The concentration of product (C_B) in the reactor is controlled by manipulating the temperature set-point. Liquid level in both the mixer and reactor is controlled by manipulating the outlet flow rates F_0 and F respectively. The mixer, reactor and the jacket are modelled with perfectly mixed tank dynamics.

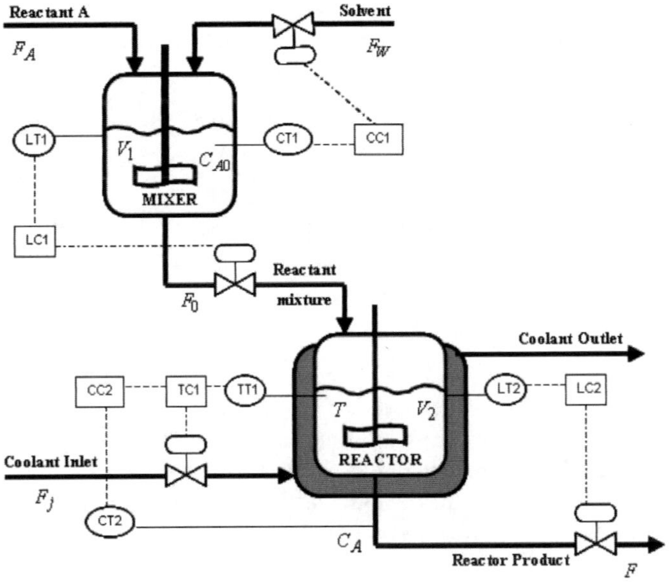

Fig. 1. Schematic of mixer-CSTR system

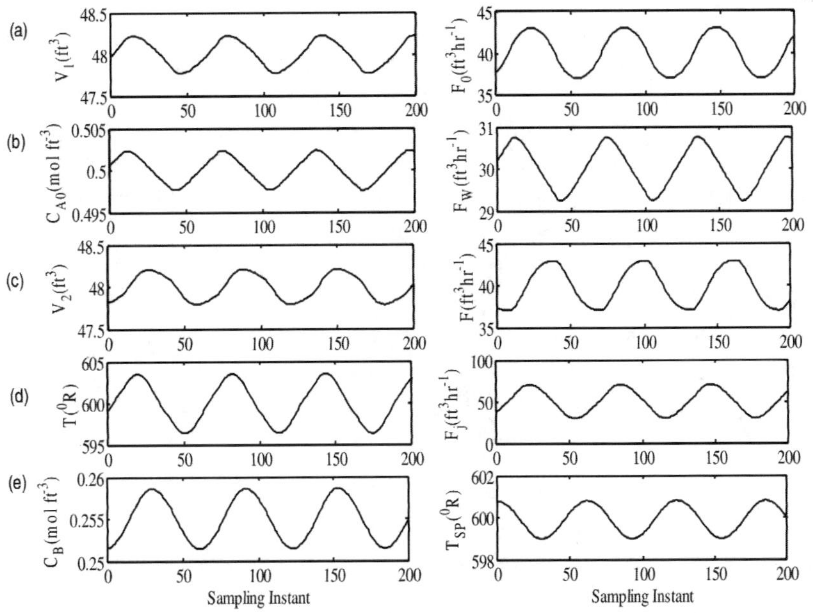

Fig. 2. Oscillations in the process variables due to the presence of stiction in mixer outlet control valve, F_0

A stiction affected control valve is simulated in the outlet line of mixer using Kano et al.(2004)'s sticky valve model. The wave forms thus resulted in the process parameters and the controller outputs are shown in Figure 2. It can be seen that the presence of stiction in a single control loop can introduce sustained oscillations in the whole downstream control system. Plot *a* shows the controller output (OP) – process variable (PV) signals of the liquid level controller of the mixer, characterized by stiction induced oscillations. Plot *b* shows the OP-PV signatures of the mixer outlet concentration controller. The root-cause for the oscillations in these signals is the oscillatory disturbances in the mixer outlet flow-rate. The oscillations in the reactor level control loop are shown in Plot *c*, caused by entering oscillatory flow into the reactor. The oscillations in reactor temperature control loop are induced by oscillatory reactor inlet and outlet flows and are shown in Plot *d*. Oscillatory trend in reactor concentration control loop is shown in Plot *e* and was a result of oscillatory behaviour in mass flux and temperature in the reactor affecting the conversions.

One can clearly see that even the origin for oscillations lie in one control loop all the neighbouring control loops are influenced and they also tend to oscillate. This results in plant-wide oscillations adversely affecting the economics.

2. COMPREHENSIVE OSCILLATION DETECTION AND ROOT-CAUSE ANALYSIS

In the context of performance assessment and troubleshooting, objectives of oscillation diagnosis of control loops can be stated as: (i) identification of sustained oscillations in control loop, and (ii) distinct detection and quantification of root-cause once the loop is confirmed as oscillatory.

2.1 Quantification of oscillatory nature

Identification of oscillatory control loops can be performed using the OP, PV and/or valve position (VP) signals. Four quantifiable properties of oscillating signals are estimated from the signal of interest: regularity, decay ratio, time period and power (Karra and Karim, 2008). One can confirm the presence of oscillations in the signal of interest if; regularity >1, decay ratio > 0.5 and power > 0.1.

2.2 Oscillation root-cause detection and quantification

The flow chart of the oscillation detection and diagnosis method is given in Figure 3. If significant oscillations are identified in control loop upon performing oscillation detection algorithm on OP or PV data, next step to be implemented is the identification of the root cause of these oscillations. The root-cause detection and quantification are based on the identified model parameters of the Hammerstein model of OP-PV mapping (refer to Figure 4). If the estimated stiction parameters, $\hat{S}$ and $\hat{J}$ contain non-zero values, there is a quantifiable dead band plus stick band and slip jump in the control valve. In that case one can conclude that root-cause for the oscillatory behaviour is stiction nonlinearity of the control valve. If $\hat{S}$ and $\hat{J}$ are close to zero, there is no stiction in the control valve. The detailed description for *S* and *J* parameters can be found in Figure 5. If there are any cyclic external disturbances entering the process causing the control loop to oscillate, it can be captured by the bias-term sequence $\{\hat{\eta}(t)\}$. The cyclic nature of $\{\hat{\eta}(t)\}$ can be quantified. If the $\{\hat{\eta}(t)\}$ is found to be oscillatory, part of the oscillations is attributed to external oscillatory disturbances. If the closed loop simulations of the identified process transfer function with the identified PID controller (whose parameters are identified from PV-OP data) yield an oscillatory control system, then the root-cause for oscillations is under-damped tuning.

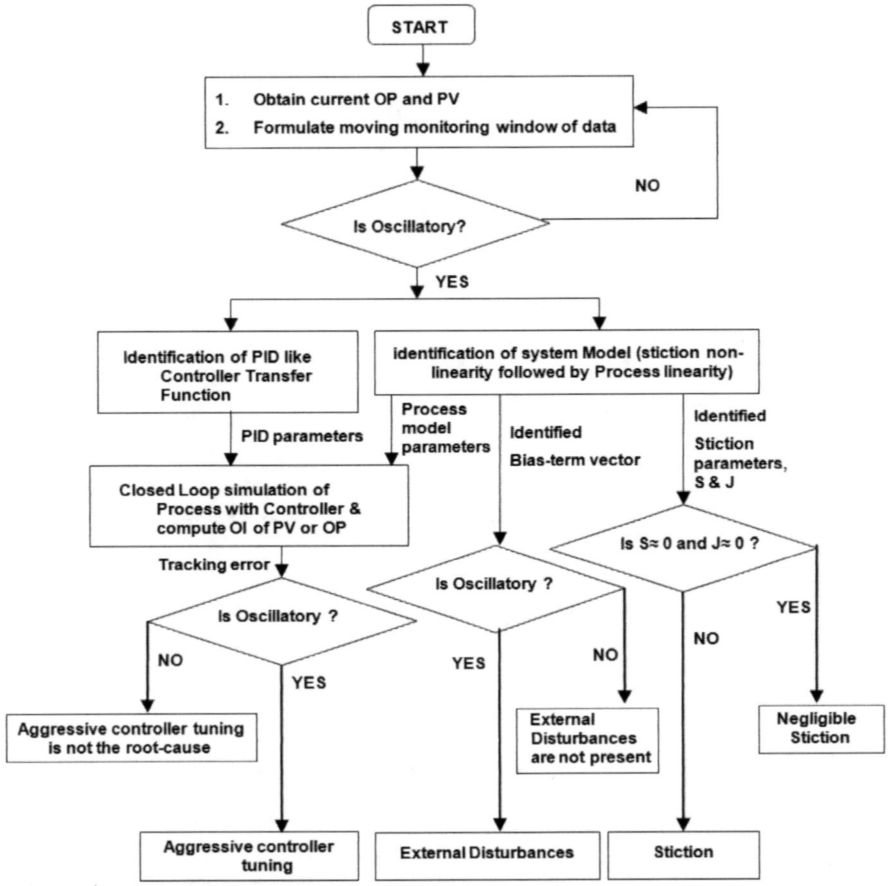

Fig. 3. Flow chart of oscillation detection and diagnosis.

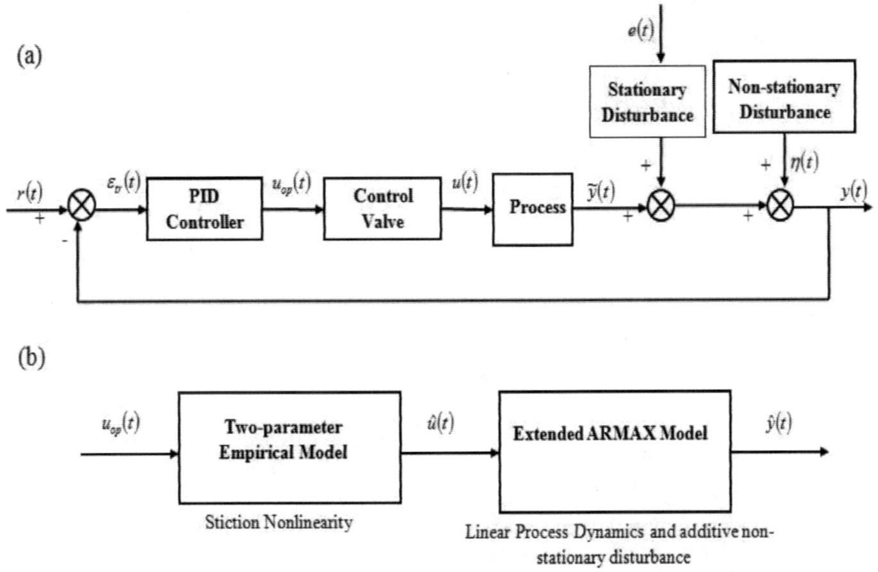

Fig. 4. Schematic of control loop; a) Block diagram of control loop involving significant valve dynamics, stationary and non-stationary disturbances entering the process, b) Hammerstein model representing control valve followed by linear process subjected to additive non-stationary disturbance. The extended ARMAX model is given by, $A(q^{-1})\ y(t) = B(q^{-1})\ u(t-t_d) + C(q^{-1})\ e(t) + \eta(t)$, where $\{\eta(t)\}$ is the nonstationary disturbance sequence.

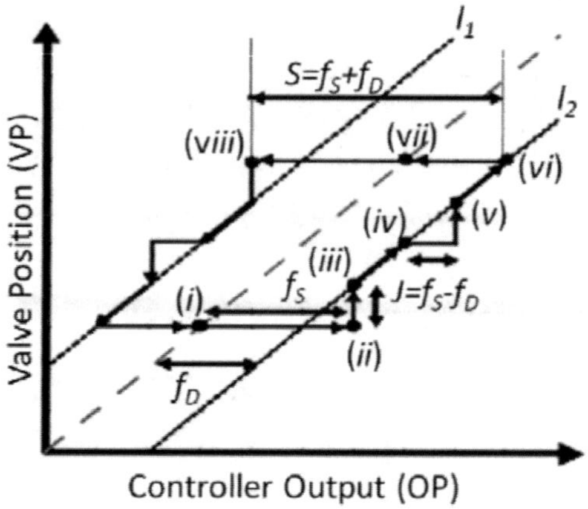

Fig. 5. Relation between controller output and valve position under valve stiction. When the valve stem is in resting position (*i*), a force that is equal to static friction (f_S) is needed to move the stem and it suddenly jumps to a new position (*iii*) (slip jump, $J = f_S - f_D$), once the force is equal to f_S at position (*ii*). While in motion the VP will lag OP by a signal equal to dynamic friction (f_D). When the stem motion is reversed in direction (*vi* to *vii*) to move the stem, a signal equal to stick band ($S = f_S + f_D$) is required.

SUMMARY AND CONCLUSIONS

In this work, a comprehensive detection and diagnosis methodology for oscillatory control loops was proposed. Oscillation detection makes use of the power spectral density and auto-correlation function of tracking error signal. A model based oscillation diagnosis method was also devised to distinctly identify, and quantify the root-causes of oscillatory control loops. Its effectiveness to identify, and quantify the root-cause even in the case of multiple faults occurring simultaneously (poor controller tuning, control valve stiction and external disturbances) was proven by simulation examples. Sensitivity analysis was performed on wide ranges of process dynamic parameters, in simulation studies. The proposed methodology was employed on many industrial control loops to identify and quantify the root-cause for their oscillatory behaviour.

REFERENCES

Beckman, D., & Jury, F. Reducing process variability with control valves, Internal Report (November), Fisher Controls International Inc., Marshalltown, Iowa, 1997.

Bialkowski, W. L. Dreams versus reality: A view from both sides of the gap. *Pulp and Paper Canada*, 1993;94(11): 19–27.

Corradini, M.L., & Orlando, G. Robust stabilization of nonlinear uncertain plants with backlash or dead zone in the actuator, *IEEE Transactions on Control Systems Technology*, 2002; 10(1); 158–166.

Ender, D.B. Process control performance: Not as good as you think. *Control Engineering*, 1993; 9: 180–190.

Isermann, R., & Raab, U. Intelligent actuators – Ways to autonomous actuating systems, Automatica, 1993; 29(5): 1315–1331.

Kano, M., Maruta, H., Kugemoto, H, & Shimizu, K. Practical model and detection algorithm for valve stiction. In *IFAC-Dycops-7th International Conference*, Cambridge (USA), 2004.

Karra, S., & Karim, M.N. Comprehensive methodology for detection and diagnosis of oscillatory control loops, submitted revised version to *Control Engineering Practice*, 2008.

Miller, R., & Desborough, L. Increasing customer value of industrial control performance – Honeywell's experience, in: CPC VI, Arizona, USA, 2001.

Rinehart, N. F. The impact of control loop performance on the successful application of advanced control. In *Aspen world conference proceedings*, Boston, MA, 1997.

Ruel, M. Stiction the hidden menace, Control Magazine, November, 2000.

Yang, J.C., & Clarke, D.W. The self-validating actuator, *Control Engineering Practice*, 1999; 7: 249–260.

EMERGENCY ISOLATION VALVE (EIV) SAFEGUARDING SURVEY

Edward M. Marszal, Kenexis, Columbus, OH

Peter G. Hereña, Kenexis, Chicago, IL

Abstract

All refiners employ engineered safeguards so they may reduce to a tolerable level the risk of process hazards. Emergency Isolation Valves (EIVs) are among the engineered safeguards commonly considered. An EIV is a special category of valve that is dedicated to the purpose of isolating large inventories of flammable or toxic material from sources or equipment whose relative likelihood of significant leakage is high.

The number, type and service of EIVs that are employed vary to a great degree from refiner to refiner and from site to site. This paper presents the results of a benchmarking study that was performed to determine the most prevalent practices in industry and to establish the degree of adoption for some of the more common EIV services. The survey reveals norms pertaining to the service, installation methodology, hardware preference and maintenance practices that are common to major refineries.

1.0 Introduction

Most refining organizations are very concerned with meeting the requirements of safety related standards and best practices in addition to their overall commitment to continuous improvement of safety and reliability. Refiners have found that there is value in providing common standards and guidelines in implementing safety functions because they can focus resources to areas of the plant where they are most needed. This reduces cost and improves safety reliability. The American Petroleum Institute (API) has released a Recommended Practice 553 for Refinery Control Valves (RP 553), which has become widely adopted by industry practitioners and regulators alike. Many refiners use RP 553 as an example for Recognized and Generally Accepted Good Engineering Practices (REGAGEP), or as a basis to generate their own policies.

Section 7 of API 553 addresses the recommended practices for the implementation of "Emergency Block Valves." Four different classes of Emergency Block Valves are listed, called Type A through Type D. Type A, B and C valves are shutoff valves that are either manually or remotely initiated. They have other adaptations, but their key feature is that they are expected to be outside of the hazard zone. In general refiners are already very familiar with how to specify, purchase and install these valves. Type D valve is what these authors refer to as

Emergency Isolation Valves (EIVs). They are different from Type A, B and C valves in that they are expected to operate within the hazard zone, or in locations that are significantly elevated with respect to grade. Most of the discussion in API 553 Section 7 refers to fire hazards, and mentions specific modifications in order to improve the reliability of the valves under severe fire conditions.

There have been a number of accidents in the past decade that underscore the need for further review; to analyze where EIVs are implemented, what common design practices are adopted, and how the functions are tested. Refiners are interested in evaluating their own practices and policies relative to their peers, then to use that information to adopt the best practices are being used throughout industry. A benchmarking survey is very effective in providing this information by allowing a high degree of information sharing while still maintaining individual confidentiality.

1.1 Benchmarking Survey Questions

The benchmarking was performed in accordance with a survey form that contained the information presented in *Table 1.* The survey form itself is provided in Appendix A.

Table 1 – Summarized Survey Questions

General Design Philosophy
What design and engineering guidelines are used
What hazards are EIVs installed to protect?
EIV Service
For which leak sources are EIVs considered?
Inlets and/or outlets for those leak sources?
Alternate isolation methods in lieu of EIVs
Minimum distance required to allow an alternate isolation method
Considerations for implementing compressor EIVs
Considerations for implementing pump EIVs
EIV Hardware
Activation
Actuator style
Signal type (ETT or DTT)
Valve failure direction
Special requirements for pneumatic actuators
Special requirements for motor-operated actuators
Testing
What is the testing frequency?
Future Implementation
What is done for existing units?

The survey was divided into five main areas, General Design Philosophy, EIV Service, EIV Hardware, Testing and Future Implementation. Some of these areas contained several follow-up questions to characterize certain details of EIV implementation. In general, the reader was asked to choose one from a group of answers that best represented their corporation's practices and attitudes. In some cases, the reader was asked to select all applicable statements within a question. In most cases a "Not Applicable" or "Other, Please Specify" was added to the question to allow some flexibility. For all questions the respondent could choose not to answer the question, in which case their null response was not included in the overall results.

General Design Philosophy Questions asked two basic questions on how EIVs are implemented. The first is what design standards or guidelines govern design, selection and integration. Additionally, a question was asked to probe whether or not EIVs are commonly installed for toxic chemical protection as well as fire protection. These questions were asked to get feedback on how established EIV design principles are currently incorporated in refineries.

The next group, EIV Service, included a number of questions that dealt with the EIV selection process. All refiners execute an analysis of some kind to judge where an EIV should be installed. The criteria to make the decision include what leak sources as considered critical, what alternate forms of isolation are acceptable and how severe is the unmitigated hazard. These questions were asked to glean information on how refiners go through that decision-making process.

Once the need for an EIV is identified, EIV Hardware questions pertain to design and installation details. There is no single correct way to install an EIV; commonly the hardware requirements reflect corporate preference, experience, risk tolerance and local policy. Examples of design details include EIV actuator type, whether or not the control signal is fail-safe, valve failure orientation and special hardware preferences.

One question was asked for each of the Testing and Future Implementation groups. The testing question asked how frequently the EIVs are tested, which is interesting because there is generally not a significant amount of guidance on this critical question. Future Implementation asked how existing units will be reviewed to evaluate the need for EIVs.

The data from the responses is presented in Section 2. Where the reader was asked to "choose one among the following answers," the results are represented by pie charts which qualitatively show the selection frequency. Null answers or "Does Not Apply" answers were not included as part of the results. Where the respondent was asked to "choose as many as applicable" among a group of possible answers, the results are represented by bar charts, indicating a percentage of the time a selection was made out of all respondents who answered the question. Summary results for implementation questions were then presented as the arithmetic mean (average) of all responses (one response per operating company).

1.2 Survey Participants

In order to understand the survey and put it in proper context, it is important to understand the demographics of the surveyed companies. Several operating companies responded to the survey (in some cases, multiple responses per company were received – in this case the responses were averaged for that company). All of these companies operate at least one refinery in the United States and are in the business as either pure-play refiners, or integrated oil and gas production, refining, and petrochemical companies. The survey team selected companies that have large market capitalizations, operate multiple refineries and have a large crude oil processing capacity. Significantly smaller companies with very low oil processing capabilities were not included in the study due to the potential that their practices may not reflect the best practices in industry.

It should be noted that the results presented in this survey are based on a somewhat limited data set. Although the survey includes companies that represent the preponderance of crude oil processing in the United States, typically only one respondent per operating company completed survey forms. Many refining organizations have recently been built up through a series of mergers and acquisitions, leading to a wide variety of safeguarding practices within a single operating company. Care was taken to send the questionnaire to highly placed individuals in the safety division of each operating company, in order to obtain results from the person who has the most exposure to existing practices.

While most refiners are working diligently to standardize their operations, that task is not complete yet. While the authors have made great efforts to provide a comprehensive analysis of operating company practices, data from every site in every organization could not be obtained. The results from the survey indicate most organizations use a common EIV standard, so the results should be an excellent barometer of current practices. However, because most EIV standards have been in effect for only a few years and EIVs have been implemented for much longer, it is possible that the usage patterns and design preferences of legacy EIVs differ from current practices.

2.0 Survey Results Summary

A summary of some of the results of the survey is shown below. The information presented below is simply a statistical compilation of the responses to the survey forms and notes containing general commentary on the results and the comments that the respondents provided in their survey forms. While the original survey analyzed a number of design scenarios, the summary results presented here only represent the application of these design features on new equipment.

2.1 General Design Philosophy

What guides EIV design and installation?

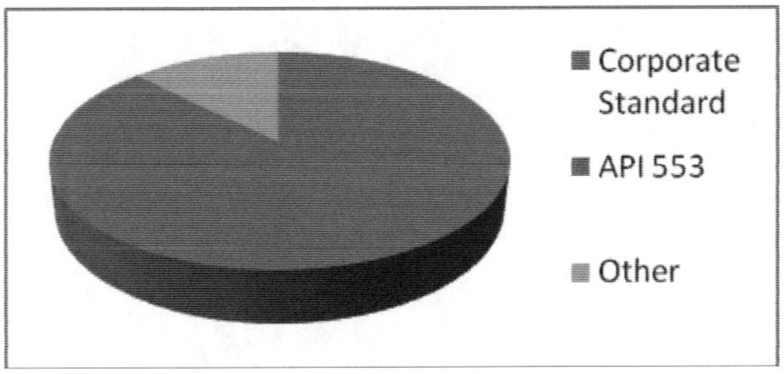

Nearly 90% of respondents use a corporate standard to determine EIV requirements. Some respondents allow the implementation of EIVs to be done on a per-site basis, or using some other undefined standard. Although no respondents use API 553 as a design standard, many used API 553 as a guideline or source material for their corporate standard.

EIV Service

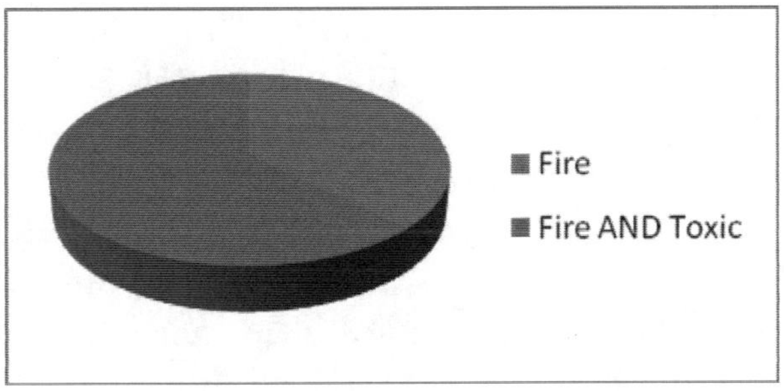

All respondents indicated EIVs are used to protect against the fire case. However, nearly 60% of respondents also use EIVs in certain toxic chemicals service. Most published literature related to EIVs concentrate on their use in fire service. EIVs are sometimes used in special service such as hydrofluoric acid (HF) containment.

2.2 *EIV Service*

High Probability Leak Sources

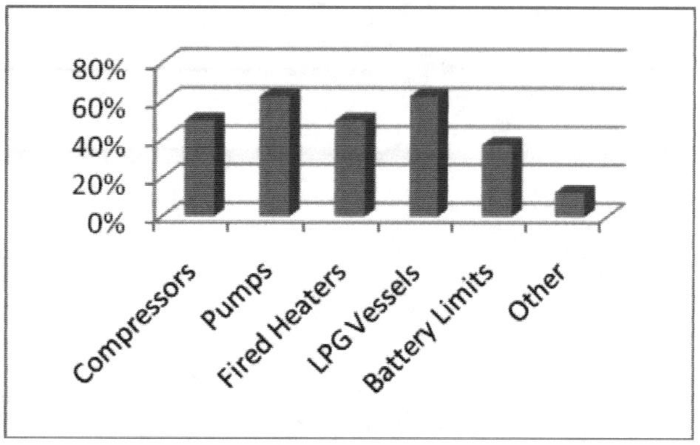

Different organizations can vary in their interpretation of a "high probability" leak source. The data show that EIVs service depends on the organization. Over 50% of respondents say they consider (though not necessarily implement) EIVs for compressors, fired heaters, pumps and compressors. Nearly 40% of respondents consider EIV isolation at battery limits.

The authors asked follow-up questions on whether "high risk" leak sources considered inlets only, outlets only, or inlets & outlets for protection. Our respondents revealed that compressor suction was always evaluated for inventory isolation and 25% also considered compressor discharge. Respondents indicated that EIVs were always considered at pump suction, and 40% of returned surveys also considered EIVs at the pump discharge. When considering fired heaters, 50% of refiners noted that only the heater inlet line was considered for inventory protection and the remainder considered implementing EIVs at both inlets and outlets.

Our respondents remarked that they were generally concerned with the pump or compressor suction vessel exacerbating a fire, and believed that there were enough various downstream protections already in place so an EIV was not necessary. Similarly, for fired heaters a major concern was the feed pump continuing to send flammable or toxic material to the fired heater, but downstream isolation was generally not needed. They noted it was also much easier to find downstream suitable fire protection that is outside of the hazard zone, which obviates the need for a special EIV.

One question was whether or not a regular control valve or a check valve could be considered to provide suitable leak protection and thus be used in lieu of an EIV. Nearly 90% of respondents indicated that check valves are not suitable as a replacement as an EIV and over 70% of respondents do not consider a control valve sufficient substitute, either because they provided insufficient leak isolation, or because they are impractical to install. For example, in the case of pumps and compressor inlets, control valves are typically not installed and a check valve to prevent forward flow to the equipment would obviously be incorrect. It should be noted that for the respondents who allow check valves to sometimes be used as an alternate isolation method, dual check valves are favored because they provide superior reliability.

Minimum Leak Source to Manual Isolation Distance

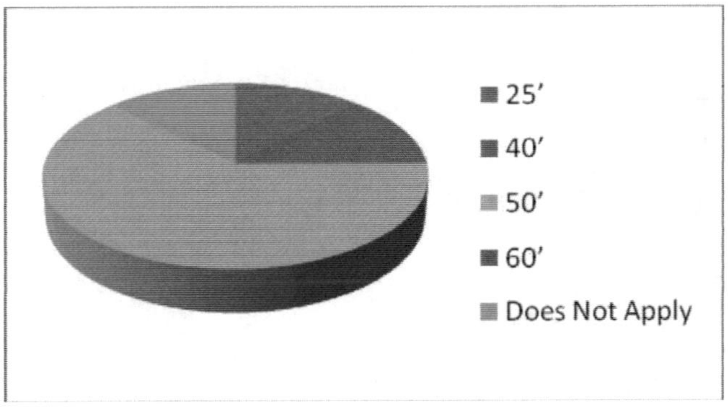

A Type "D" EIV requires special design, metallurgy and instrumentation, which can be very expensive. Therefore, many refiners would prefer to install an alternate means of isolating the leak, provided it can be sufficiently far from the leak source (and potential hazard). This minimum distance for manual isolation varies from one company to another and depends on factors such as maintenance practices, installed hardware, typical plant layout, past event history and risk tolerance. Over 60% of respondents allow a manual isolation valve in lieu of an EIV so long as the manual valve is 50 ft (15 m) from the leak source. Approximately 10% of respondents enforce a minimum distance of either 25 ft (7.5 m) or 40 ft (12 m) and the remainder did not specify. Some respondents noted that these distances could increase under certain conditions, such as elevation of the leak or material properties.

Considerations to Install Compressor EIVs

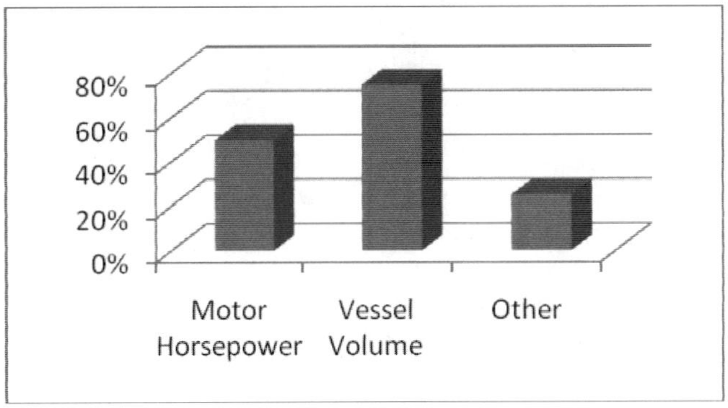

Not every compressor necessarily needs an EIV installed. The valves are complex, expensive to install and maintain and are best prioritized to the most critical equipment. Over 40% of respondents used compressor horsepower (or equivalent), and over 60% used suction vessel volume is used to judge whether or not an EIV is installed. The larger the horsepower or higher the suction or interstage vessel volume the greater hazard potential.

Considerations to Install Pump EIVs

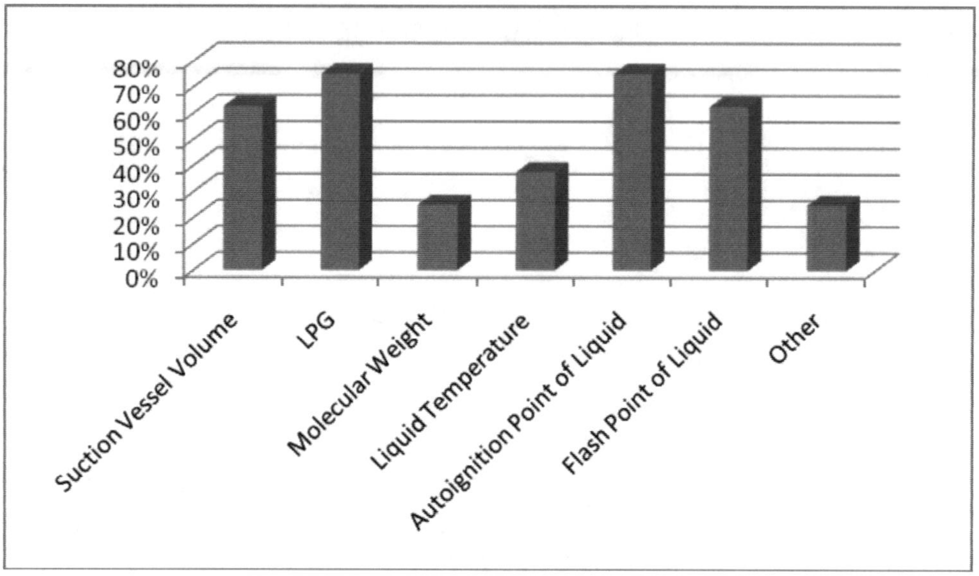

As with compressors, EIVs in pump service are often prioritized to most critical equipment, or where a loss of containment hazard is most severe. Over 40% of respondents used suction volume, LPG service, auto-ignition point or flash point in deciding whether or not an EIV is installed. Factors less frequently cited are molecular weight, liquid temperature and toxic material threshhold.

2.3 Hardware

Valve Control Location

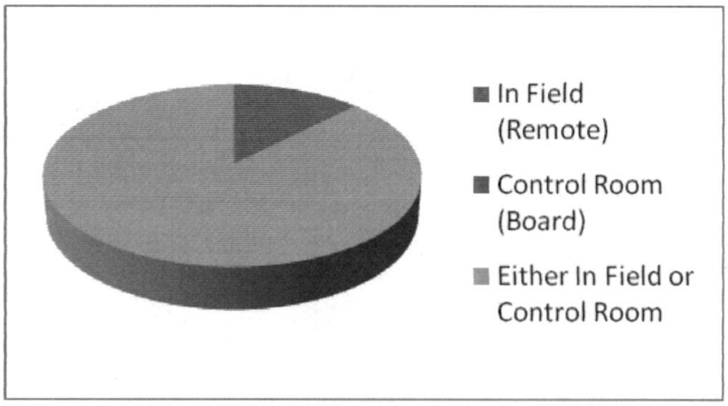

Respondents stated that they were flexible with respect to where the valve control (initiator) was located. The most common answer was to leave initiation details on a per-function basis, or to allow a board command to activate the valve in case the local command console is unavailable or inaccessible.

Valve Actuation

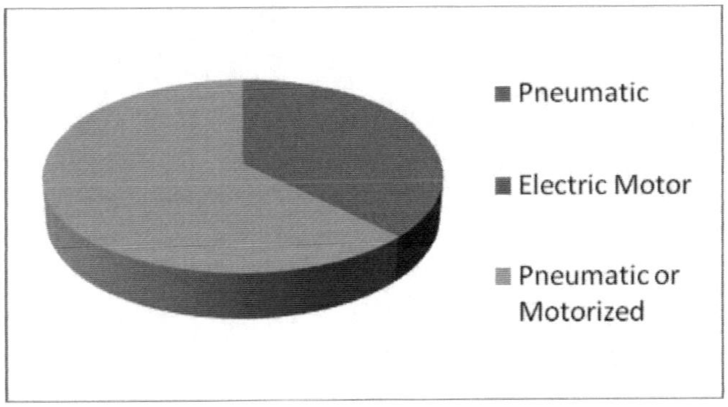

Two common methods are used to actuate an EIV; pneumatic and motor-operated. The former utilizes an air-operated actuator and the latter utilizes a motorized actuator to close the valve. Results show that nearly 40% of refiners who responded require pneumatics for valve actuation, with the most of the remainder allowing either pneumatic or motorized. Many refiners allow the individual site to decide the actuation based on site-specific experience. Some respondents use alternate means for valve actuation other than pneumatic and motorized.

Signal Philosophy

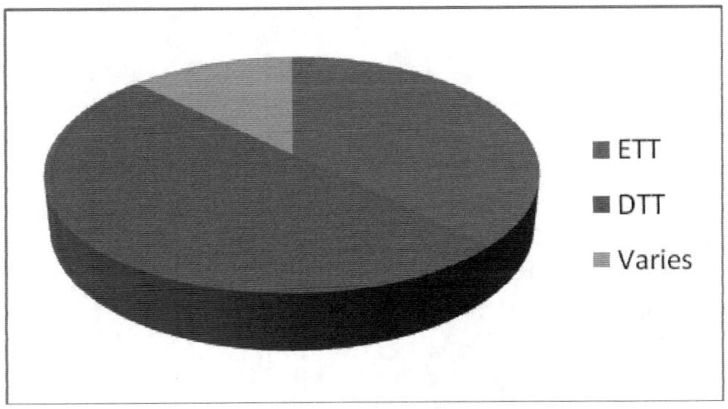

The command signal to the valve can be designed using two different philosophies; energize-to-trip (ETT) and deenergize-to-trip (DTT). ETT configuration is not fail-safe but it prevents spurious trips. DTT will activate the function in case of signal loss, but that could increase the probability of nuisance trips. Results show that 50% of respondents specify DTT functions, nearly 40% require ETT and the balance allow either.

Valve Fail Position

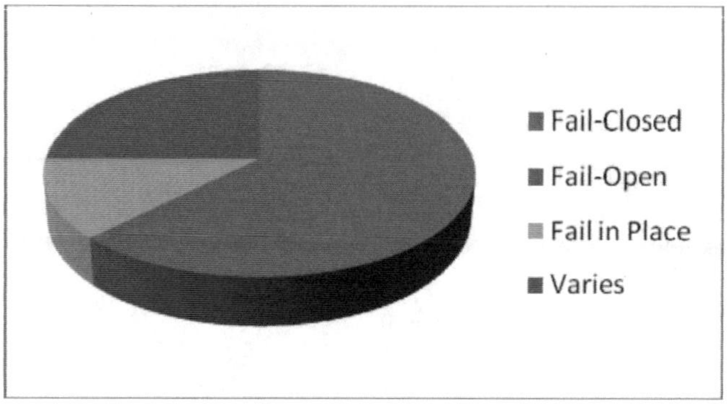

Valves can fail in three orientations; fail to the closed position, to the open position and in place. Failing to the closed position allows for the valve to take action upon failure of many internal components. For EIV service it is considered the fail-safe mode. Failing to the open position and fail-in-place reduces the chance for spurious trips but has a higher covert failure rate. Over 60% of refiners surveyed specify fail-closed valves, 25% allow a site-based or case-by-case selection and remainder specify fail-in-place valves.

Pneumatic Actuator Hardware

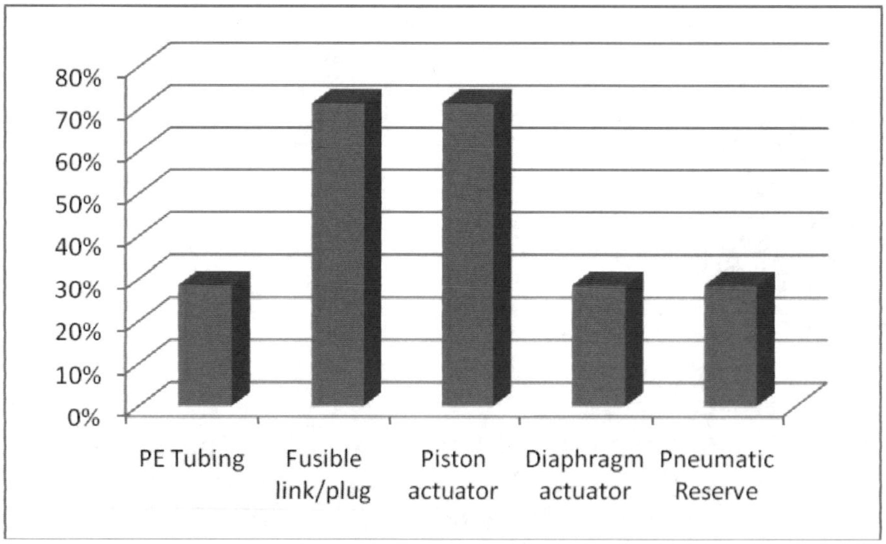

Pneumatic EIV actuators can have various specialized options added. Polyethylene (PE) tubing will melt when exposed to heat, thus allowing the valve to actuate while exposed to fire even if the signal to close does not reach the valve. A fusible link is a specially engineered metal device that melts when exposed to a specific temperature, which allows the valve to actuate in a fire. A pneumatic reserve is an air volume reserve that allows a finite number of actuations (often three) in the absence of the normal instrument air supply. Piston and diaphragm valve actuators are two different pneumatic actuators with different sealing, control and maintenance properties.

Respondents indicated that fusible link/plugs and piston actuators were the most commonly implemented accessories; required 60% or more of the time. PE tubing, diaphragm actuator and pneumatic reserve volumes were each required less than 30% of the time.

Motorized Actuator Hardware

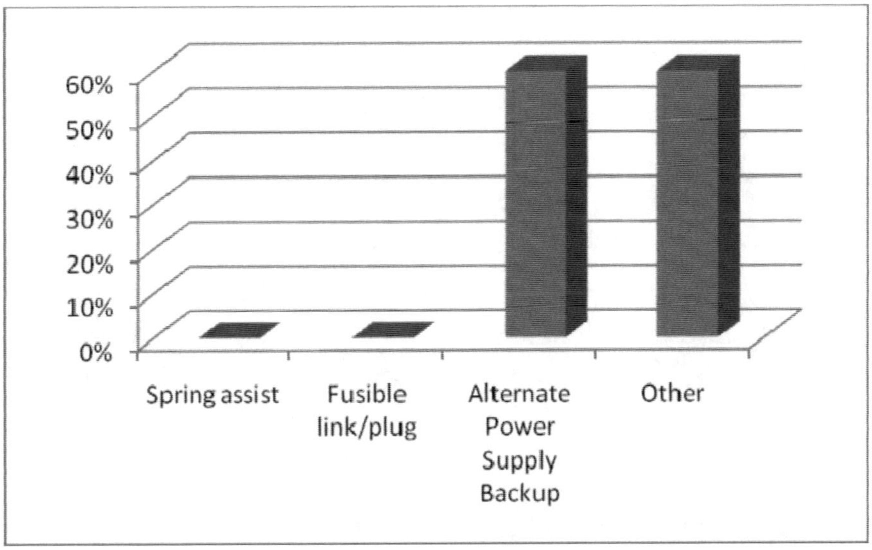

Motorized EIV actuators can also have various specialized options added. A spring assist will allow the valve to close in the case of power loss, although it is not always practical to install. A fusible link is a specially engineered metal device that melts when exposed to a specific temperature, which allows the valve to actuate in a fire. An alternate power supply (usally DC battery) can also allow the actuate in case the valve's normal power supply is unavailable.

Respondents indicated that alternate power supplies were to be specified over 50% of the time. Other requirements were specified, including special actuators hardened against high temperature, additional fireproofing, power requirements, cable length, position indication and others. However the alternate power supply backup was by far the most common single feature specified.

2.4 Testing

Testing Frequency

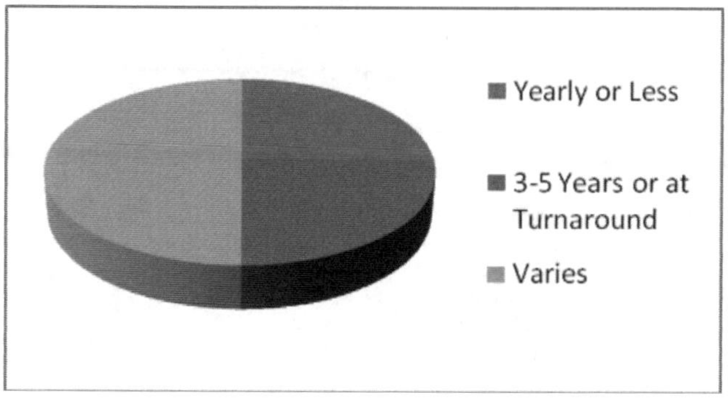

It is common for safety functions to be tested periodically in order to reveal covert failures and to confirm the overall failure rate is consistent with design. EIVs are not safety functions in the same sense because they act to mitigate an existing hazard rather than prevent the hazard. However the idea of periodic testing is beneficial because safety functions and EIVs both act on demand after long periods, perhaps years, of inactivity. Testing can reduce the probability of failure on demand. 25% of respondents specify testing period of 1 year or less, 25% specify 3-5 year interval (or at turnaround), with the remaining 50% allowing the testing interval to vary depending on the function, hazard severity or other factors.

2.5 Future Implementation

Evaluating EIVs for Existing Units

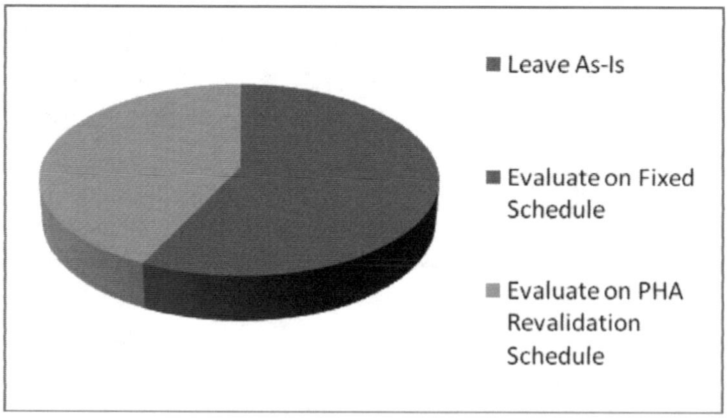

Most respondents who were surveyed indicated that they plan to evaluate the implementation of EIVs on a fixed schedule. Some respondents plan to implement EIVs on a rolling basis at the next closest PHA revalidation.

3.0 Conclusions

Benchmarking is an effective tool in assisting refiners to determine how to implement portions of standard practices where the requirements are ambiguous, open to interpretation, or multiple options are available. In many cases the practices that are implemented are remarkably consistent and form the basis for "recognized and generally accepted good engineering practice".

Appendix A: The Questionnaire

Definitions

Emergency Isolation Valve (EIV) – a valve that is employed to isolate a large inventory of flammable or toxic material from a leak source. In API RP 553 "Refinery Control Valves" an EIV is given the name of Emergency Block Valve.

Emergency Block Valve – (EBV) – the name given to an Emergency Isolation Valve in API RP 553 "Refinery Control Valves"

API RP 553 lists 4 classes of EBVs, Type A through Type D. This survey is primarily concerned with the utilization of Type D EBVs.

API 553 Type A/B/C: EBV that IS NOT expected to be in or near the fire zone of the leak source. The different types have different hardware requirements. They may or may not have remote controls and fireproofing.

API 553 Type D: An EBV that IS expected to be in or near the fire zone of the leak source, or the valve may be considerably elevated relative to grade. The valve operates with remote controls and other fireproofing requirements.

Survey

GENERAL QUESTIONS

1. For guidance on the selection and installation of EIVs, your company directs you to (select the statement that best applies):
    a. A corporate standard
    b. API 553 "Recommended Practice for Refinery Control Valves"
    c. Another guideline or Per site philosophy
2. Select the statement that best applies to your EIV philosophy:
    a. EIVs are considered primarily as fire hazard mitigation
    b. EIVs are considered for fire AND toxic material hazard mitigation

INSTALLATION GUIDELINES

3. Select the following equipment that is considered a "high probability leak source" and that may require an EIV (SELECT ALL THAT APPLY):
    a. Compressors
    b. Pumps
    c. Fired Heaters
    d. Vessels containing LPG

e. Battery Limits
f. Other (please specify) ________________

4. If you selected a Compressor as a "high probability leak source" then do you typically consider
    a. Inlets only
    b. Outlets only
    c. Inlets and Outlets
    d. Does not Apply
5. If you selected a Pump as a "high probability leak source" then do you typically consider
    a. Inlets only
    b. Outlets only
    c. Inlets and Outlets
    d. Does not Apply
6. If you selected a Fired Heater as a "high probability leak source" then do you typically consider
    a. Inlets only
    b. Outlets only
    c. Inlets and Outlets
    d. Does not Apply

ISOLATION

7. If you are considering installing an API 553 Type "D" EBV or equivalent, is a regular control valve considered an acceptable alternate isolation method?
    a. Yes, with Minor Modifications
    b. No, or Not Practical
    c. Sometimes
8. If you are considering installing an API 553 Type "D" EBV or equivalent in a pump, compressor or fired heater service, can a check valve be considered an acceptable alternate isolation method?
    a. Yes, with Minor Modifications
    b. No, or Not Practical
    c. Sometimes
9. If you answered "Yes" to the previous question, select any special design or hardware requirements that apply (SELECT ALL THAT APPLY):
    a. Double Check Valves
    b. Minimum Check Valve Distance from Leak Source
    c. Special Fireproofing
    d. Other
    e. None
10. If alternate means of isolation, such as a manually operated local block valve, are used in lieu of an API553 Type "D" EBV, what distance is required between the alternate means of isolation and the leak source?

    a. 25'
    b. 40'
    c. 50'
    d. 60'
    e. Does not Apply
    f. Other (Please specify) _______________

11. For an API 553 Type "D" EBV or equivalent, how is the remote valve control implemented?
    a. In Field (Remote)
    b. Control Room (Board)
    c. Either In Field or Control Room
    d. Does not Apply
    e. Other (Please specify) _______________

GAS SCENARIOS

12. If you are considering whether to implement an API 553 Type "D" EBV or equivalent for COMPRESSOR SERVICE, select all factors that you consider (SELECT ALL THAT APPLY):
    a. Motor (or equivalent) Horsepower
    b. Volume of interstage vessel
    c. Does not Apply
    d. Other (Please specify) _______________

LIQUID SCENARIOS

13. If you are considering whether to implement a API 553 Type "D" EBV or equivalent for PUMP SERVICE, select all factors that you consider (SELECT ALL THAT APPLY):
    a. Minimum volume or "hold up" in suction vessel
    b. LPG
    c. Molecular Weight
    d. Absolute Temperature of liquid
    e. Temperature relative to autoignition point of liquid
    f. Flash point of liquid
    g. Does not Apply
    h. Other (Please specify) _______________

EIV HARDWARE

14. When installing an API 553 Type "D" EBV or equivalent, select the one that best applies:
    a. The valve is typically a pneumatic actuator
    b. The valve is typically an electric motor actuator
    c. The valve is sometimes pneumatic and sometimes motor actuated
15. When installing an API 553 Type "D" EBV or equivalent, select the one that best applies:
    a. The valve is typically energize to trip (ETT)

b. The valve is typically de-energize to trip (DTT)
c. The valve is sometimes ETT and sometimes DTT

16. When installing an API 553 Type "D" EBV or equivalent, select the one that best applies:
    a. The valve is typically Fail-Closed
    b. The valve is typically Fail-Open
    c. The valve is typically Fail in Place
    d. The valve failure orientation varies
17. When installing an API 553 Type "D" EBV or equivalent WITH A PNEUMATIC ACTUATOR, select ANY of the following that are typically utilized (SELECT ALL THAT APPLY):
    a. Sunlight-resistant polyethylene tubing to the actuator air supply
    b. Fusible link/plug
    c. Piston actuator
    d. Diaphragm actuator
    e. Pneumatic reserve to provide multiple operations (ex, open-close-open)
    f. Does Not Apply
    g. Other (Please specify) ________________
18. When installing an API 553 Type "D" EBV or equivalent WITH AN ELECTRIC MOTOR ACTUATOR, select ANY of the following that are typically utilized (SELECT ALL THAT APPLY):
    a. Spring assist
    b. Fusible link/plug
    c. Alternate Power Supply Backup
    d. Does Not Apply
    e. Other (Please specify) ________________

EIV TESTING

19. When installing an API 553 Type "D" EBV or equivalent, select the statement that best applies:
    a. The valve is typically stroke tested yearly, or more frequently
    b. The valve is typically stroked tested every 3-5 years, or at every turnaround
    c. The valve is typically not on a periodic test cycle
    d. The test frequency varies because the valve is tested at a frequency that depends on the service

FUTURE EIV IMPLEMENTATION

20. Select the statement that best applies to your EIV implementation plans with respect to EXISTING units
    a. Leave existing units as-is
    b. EIVs will be evaluated and implemented for existing units on a fixed schedule
    c. EIVs will be evaluated and implemented for existing units on a PHA Revalidation schedule
    d. Does Not Apply
    e. Other (Please specify) ________________

Realizing the Capital and Operational Benefits of a ICSS System

Integrated Control and Safety Systems (ICSS) can provide unprecedented benefits if users balance the work process with the technology.

Chuck Miller, CFSP, Emerson Process Management

Your company is committed to installing (or replacing or upgrading) the control- and safety-systems at one of your plants. You've seen the very aggressive project schedule and recognize that speed of implementation is critical. You've heard and read about these new highly-integrated, all digital system solutions. You recall hearing that these "smart" systems are a lot easier to implement than their predecessors, but you've heard things like that before. Beside's, even if these new systems are easier to implement, isn't it likely that all that sophistication will end up costing more to operate and maintain?

Twenty-first century economics dictate that "speed of implementation" stand as a hallmark to project success, but projects still have budgets and now, more than every, shareholders additionally measure success based on their return on Capital Expenditures (CapEx).

Alongside the CapEx performance metric, and possibly more important, stands Operational Expenditures (OpEx) performance.

Improving OpEx performance in today's market requires extreme diligence beginning with the purchase of raw materials and extending all the way to how finished goods are shipped to the buyer. Simply stated, opportunities to reduce OpEx are everywhere.

When combined with the correct culture and organizational structure, new technologies can make significant contributions to improving shareholder value by improving both capital- and operational-expenditures.

Digital technologies have proven to be one of the keys to achieving CapEx and OpEx improvements. A number of studies indicate that significant CapEx and OpEx improvements can be contributed to control- and safety-systems using "Purpose Designed Integration"; meaning a system that's original design philosophy was to streamline every aspect of the systems use commencing at implementation and extending through commissioning and right into day-to-day use.

The remaining topics discussed in this paper and their order is:

- Digital Revolution
- Integrated Architecture
- Intelligent Field

- Smart Safety Instrumented System (SIS)
- The Entire Safety Loop
- Integrating the BPCS & SIS
- Reducing CapEx
- Reducing OpEx
- Culture, Organization & Competency

Digital Revolution

The digital revolution is upon us - from cameras to computers, telephones to television, music to video. Everywhere you look, digital systems are replacing their analog counterparts. The reason is increased plant availability and better performance – plus a wealth of information previously unavailable.

Distributed control systems were a major breakthrough in the mid-1970's, but these systems are based on 4-20mA signals and proprietary communications.

The industry's move into the digital revolution began with the introduction of the HART protocol more than 15 years ago.

HART is a hybrid communication protocol that enables a digital signal, primarily for increased information, to coexist with a 4-20mA control signal.

This digital bi-directional communication transforms field devices from single functional instruments to multi-functional devices that deliver multiple measured variables with single process intrusions, like:

- Device health
- Process health
- End of life predictions

Today there are more than 20 million HART devices installed in process control and safety applications. Undoubtedly you have some of these devices installed in your plants.

Foundation fieldbus is an all digital communications network that's roots are the HART technologies. The introduction of all digital communications ushered in the term "smart" plant.

Smart plants are constructed using all digital devices (i.e. transmitters, controllers, final control, etc.) In a smart plant intelligent devices work together to control, diagnose and predict.

In designing the BDO plant we want to avoid older analog communications or proprietary digital buses and instruments. These would reduce our design flexibility, inflate life-cycle costs, and complicate upgrades. – John Rezabeck, BP Chemicals

Digital technology taps the intelligence of today's microprocessor-based instruments speeding data between process, plant, and business applications. It enables systems and software to transform 1's and 0's into useful information for better process control, better asset management, and better decision-making.

You'll benefit in many areas including:

- Reduced capital and engineering costs
- Reduced operations and maintenance costs
- Increased plant availability
- Reduced process variability
- Enhanced safety and environmental compliance.

Integrated Architecture

Based upon its unique architecture, the primary litmus test for an Integrated Control and Safety System (ICSS) is that it has been designed, from the ground-up, as one system.

Typical sub-systems of a modern ICSS include (Refer to Figure 1):

- Redundant Ethernet based control network
- Distributed and scalable process controllers,
- Distributed and scalable safety controllers,
- Human machine interface (HMI),
- Engineering workstations, and
- Application servers.

Typical ICSS Architecture

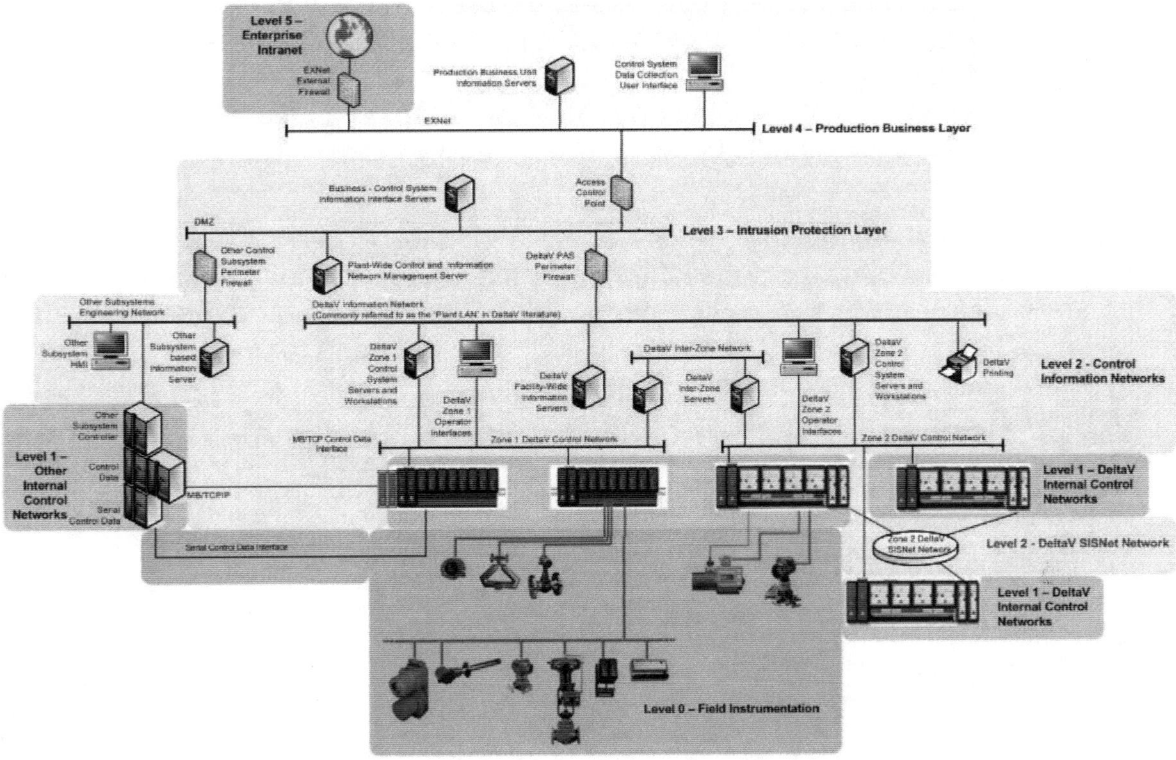

Figure 1

The first commercially available ICSS solution appeared in 1996. To some people that means that particular system is becoming obsolete and is due to be replaced. However, one of the amazing things about digital technologies is that they allow the system manufacturer to incorporate digital technology advances with minimal impact to the rest of the devices or to the system. What that means is that about the only thing that really remains of that 1996 ICCS is the name and the packaging. Theoretically a manufacturer of an ICSS could keep the system current indefinitely.

Because every implementation begins with plant design, ICSS generated CapEx savings should begin with an engineering environment that is tightly integrated with third-party plant design tools such as Intergraph's SmartPlant Instrumentation software.

Today's project "speed of implementation" often results in having project teams distributed around the globe using secure communication technologies to support 24/7/365 project activities. That certainly equates to CapEx savings, but it also means the ICSS architecture must include a secure common database and engineering environment uniquely designed and sufficiently robust to facilitate parallel and geographically distributed engineering activities.

In such a diverse and fast moving, often changing engineering environment, the ICSS engineering database must include excellent management-of-change and audit trail features. Without these two features in place, resources can be wasted in re-engineering and re-testing. Though it may not seem very important during the early engineering phases, having these two features in place helps ensure that the ICSS solution that is delivered on-site is what was "Approved for Construction."

The characteristics of modern ICSS' share a lot of similarities including open communications between subsystems. Facilitating this communication openness requires a strong, yet flexible common engineering platform and the use of an advanced global database designed to eliminate the need for managing and synchronizing multiple data resources.

The ideal ICSS will also be both flexible - right type of device for the application; and scalable – ability to add controllers and logic solvers with no impact on system performance.

Additional characteristics of a modern ICSS include:

- Advanced alarm management,
- Security management,
- Access control,
- Multiple fieldbus support – AS-i, Profibus, DeviceNet, FOUNDATION, etc.
- Management of change tracking,
- Audit trail / Version control,
- Data historian,
- Sequence of events recording,
- Asset management capabilities,
- Process simulation, and
- Training systems.

Intelligent Field

A strong ICSS foundation begins with intelligent field devices – devices that can anticipate problems, report current device health, control variability, and perform multiple measurements to reduce the number of devices required – even take local control when necessary to maintain plant availability.

Intelligent devices contain a wealth of diagnostic information, available remotely to allow you to streamline your maintenance efforts. No longer is a transmitter or valve a one-way communication device.

However, gaining access to all this data can become a challenge in the traditional DCS architecture. Without open communications architectures found in today's ICSS systems, mapping all of the available data becomes a cost barrier and results in valuable information being stranded at the field device.

Because leading instrument manufacturers build intelligent devices to global industry standards, ICSS systems are designed to allow the freedom to choose the best devices from the best suppliers, without being limited by proprietary technologies.

Examples of the diagnostic data and predictive maintenance alerts that are now available to ICSS system are:

- Plugged impulse line
- Reverse Flow
- Calibration error
- Empty pipe
- Device health
- Sensor probe suspect
- RTD drift
- Travel pressure high
- Travel accumulation
- Cycle counter
- Valve signature
- Drive signal
- Signal Saturation
- PV out of range
- Stem position
- ...and more

The ICSS solution lets you put control anywhere – where it best suits the application and level of performance you require.

> *Reliability can be maximized by taking the PID logic out of the process controller and distributing it to the fieldbus valve controller. If the transmitter is on the same segment, the entire loop is involved in the field – the transmitter's A1 (PV) and signal select blocks, plus the valve controller's PID and AO blocks.* – John Rezabek, BP Chemicals

The ICSS will include function blocks built with fieldbus in mind. These function blocks should be exactly implemented per the IEC61158 specification. The parameters on the function block are those confined by the Fieldbus Foundation and are consistent throughout. Unlike existing DCSs or PLCs where complicated mapping for field function blocks to legacy data structures are required, the ICSS's function blocks are exactly the same as field device function blocks. So the system is easy to engineer and easy to maintain.

Peer-to-peer control among field devices broadly distributes control and provides a higher degree of single-loop integrity. With the ICSS you have the complete freedom to choose where control should operate: in the controller or in the intelligent field device. It's an easy drag-and-drop exercise to select where you want control to take place.

Recently, industrial wireless communications have been made secure and robust enough to deploy intelligent devices where it is too costly to run wires, thereby extending the ability of the system to monitor remote process data and provide a communication pathway so that desired or required data is no longer stranded in the field.

Smart SIS

An equal but physically separate entity of the ICSS, the smart SIS uses a new approach to help reduce risks and uses the intelligence embedded in the total SIS loop: sensors, logic solvers, and final control elements to increase overall system availability.

> *Today's safety systems need an integrated safety approach where transmitters are part of the safety system and perform autocalibration, diagnostics, validation and remote monitoring, connecting with an intelligent fieldbus such as HART or Foundation fieldbus.* – Wayne Labs, CONTROL Magazine, May, 2005.

Based on an interpretation from the Offshore Reliability Database (OREDA), 50% of SIS malfunctions – failure to perform on demand – result from valve malfunctions; 42% from sensor malfunctions; and only 8% from logic solver malfunctions.

The smart SIS uses a holistic approach by diagnosing the sensors, logic solvers, and final elements ability to perform on demand – a smart SIS solution approach.

Areas of improvement that a smart SIS can address include:

Risk Reduction

The smart SIS begins and ends with field devices. Smart field devices:

- monitor the entire SIS loop from sensor through the final control elements
- provide non-disruptive actuator partial-stroke testing and spurious trip prevention
- imbedded HART multiplexers proactively communicate maintenance alerts from intelligent sensors and actuators

- support advanced diagnostic capabilities for sensors, logic solvers and final control elements for both self-test and detection of abnormal situations in the surrounding process.

Easier regulatory compliance

The smart SIS, including sensor, logic solver, and final element, is designed in accordance with IEC61508 and is TUV or FM certified*. To help you address the IEC61511 standard more easily, a smart SIS should have:

- safety logic signature authorization
- change management of safety logic and field device configuration/calibration.
- Security authorization of online trip point or bypass changes.

Increased availability

A smart SIS increases the availability of an operating process. It:

- increases system availability through redundancy as required.
- Minimizes risky manual final element testing through automatic periodic testing
- Reduces operator response time with advanced alarm management
- Manages bypasses during startup sequences.

A high-functioning SIS requires a high-functioning programming environment. Expect the smart SIS to include:

- A full palette of TUV-certified smart function blocks designed specifically for SIS functions. Special blocks like MooN voter blocks with bypass management reduce what used to be pages and pages of ladder to engineer, test, and commission into a simple drag-and-drop specification activity.
- Voter simplifies device upset and diagnostic condition handling to avoid spurious trips while automating bypass management.
- Cause and Effect Matrix (CEM) block greatly simplifies the logic solver configuration.
- Step sequencer saves hours of engineering over conventional ladder logic approaches.
- State Transition Diagram provides simple fill-in of state, transition inputs, and desired outputs saving hours of engineering.

All of the function blocks should be certified by TUV for safety applications.

Other capabilities making the smart SIS software intuitive include:

- Built-in sequence of events handler with automatic first-out trapping.
- Built-in bypass handling
- Built-in override bundling
- Automatic compliance to IEC 61511 standard
- File based off-line simulation and Operator Training Solutions
- Built-in alarm state engine per EEMUA 191 standard
- Optional operator interface.

Entire SIF Loop

Safe operations include many aspects—material handling procedures, process operations and safety instrumented systems (SIS). Yesterday's SIS solutions considered only the logic solver and left it to your maintenance organization to manually test the entire safety loop.

Measurement

Sensors for pressure, temperature, flow and level play an important role in your risk reduction strategy. It's important to consider improvements in measurement technology as well as installation and maintenance practices.

The health of your safety loop is only as reliable as the weakest component so in an ICCS solution, the entire safety loop – sensor, logic solver, and final element.—consists of smart devices.

The use of smart devices throughout the safety loop:

- **Deliver predictive diagnostics** – By replacing switches with transmitters, you take the first step towards reducing undetected failures. Smart transmitters have far fewer dangerous undetected failures than switches. In addition, the latest generation of smart measurement devices extends the embedded diagnostics beyond the device and into the process.
- **Extend health diagnostics –** Today's leading smart transmitters go beyond detecting component failures. They evaluate the performance of the complete measurement system, extending diagnostics to detect formerly undetectable dangerous failures outside the physical bounds of the transmitter—providing both transmitter and process diagnostics.

The end result is greater credit for failure on demand calculations, easier compliance with IEC 61511, higher safe failure fractions, less redundancy, and less proof testing, less often.

Process Measurement Devices – IEC 61511 defines two approaches for selecting the right device for your safety measurements. Both methods have merit and are used extensively.

- **Certified** – These sensors have been designed and analyzed to meet the guidelines that are set in IEC 61508 by a 3rd party certifier, such as TUV or Exida. This takes the burden of proof from the user to get the product certified.
- **Prior use** – The user must prove the device is safe for use in a specific application and environment. This method requires that you have sufficient failure data to be able to investigate and calculate the probability of failure on demand and the safe failure fraction. Ensure that your device manufacturer can provide the reliability date you need for these calculations.

Final Elements – Because final elements have been shown to have the highest rate of failure-on-demand, the next step is to equip the final elements with digital valve controllers that provide the diagnostics to extend the proof test interval, while delivering higher reliability and safety.

Final Element Testing – Process manufacturers have gone to great lengths, adding bypass valves, manual jamming devices, and expensive pneumatic panels to facilitate proof testing of final control elements.

Beyond the increased capital expense, safety valve testing often involves the installation and subsequent removal of mechanical valve interlocks. This can expose maintenance personnel and operators to hazardous locations in the process. And if the interlocks are not removed after the testing, the performance of the safety instrumented system may be severely compromised. The majority of plant incidents are caused by personnel and procedural error, so removing the need for manual proof tests while maintaining the overall SIS integrity is key in SIS applications.

Digital valve controller – Digital valve controller instruments provide automated performance monitoring and testing by enabling remote partial stroke testing while the safety valve is online. This keeps personnel safely away from the valves' locations. In addition, these instruments have extensive diagnostics to monitor travel deviation, pressure deviation, valve packing friction and more. Information is communicated back to the ICSS asset manager software.

Logic solvers – Monolithic PLC and multiplexer technologies can now be replaced with state-of-the-art logic solvers that support digital communications for continuous health monitoring of every complete Safety Instrumented Function (SIF).

Ideally you want the ICSS solution to consider the entire safety loop. When the entire safety loop is considered – sensor to final element diagnostics and testing – your process benefits from a significant decrease in spurious trips, thereby increasing reliability from sensor to final element.

Key capabilities of the ideal ICSS safety logic solver will include:

- Certified for applications up to SIL 3
- 24V DC redundant power

- User definable I/O in any combination of HART AI, HART two-state output, DI, DO
- Line fault detection on all I/O
- Failsafe and Fault Tolerant CPU Configurations Available
- Consistent 50msec application execution time base
- Modifications, Downloads and addition of new modules on-line
- Geographically Flexible architecture
- -40 to 70C temperature rating
- ISA G3 (corrosive environment rating)
- NAMUR NE21 electromagnetic compatibility rating.
- Common Database and HMI Communications Network

Additionally, the flexibility of the ISCC system provides for increasing the availability of the process with a fault tolerant configuration of SLS logic solvers. The two modules should work in an "Active Redundancy" configuration with no conditions of master/slave or hot backup functionality. This ensures bumpless transfers, allows automatic online proof testing, updating and replacement of the logic solvers thus avoiding process downtime.

Integrating BPCS & SIS

Historically design of the DCS system included the engineering of communication subsystems in order to integrate sub components from different manufacturers. These subsystems were limited by the ability of the communication interface to transfer the information back to the operator and maintenance personnel. Several options were available however; this bottleneck often stranded much of the data potentially available for safe and efficient operations of the facility.

<u>Disparate Systems</u>

With the advent of open standards, integration of Basic Process Control System (BPCS) with an SIS's from a different manufacturer became easier.

Introduced in 1996 the OPC standard provided a mechanism for high data transfer rates in real-time from an SIS to a BPCS. For those with smaller data transfer needs, the Modbus protocol based serial link was an alternative.

However, these interface options carried a significant cost for engineering and testing. In addition to the mapping of data points between the two systems, the standalone concept brought with it a need to manage and synchronizing as many as 3 disparate databases.

Finally, the fiscal requirements for long term support of the system communication subsystem had to become part of the system maintenance plan.

<u>ICSS Solution</u>

> *As data-driven systems become larger, making more extensive use of data, the Identification and management of data integrity becomes a significant factor in the demonstration that the required system integrity has been achieved.* – Alastair Faulkner, CSE International Ltd.

In the ideal ICSS solution the BPCS and SIS communicate over a common data network and configured from a common engineering environment: BPCS and SIS data and alarms are displayed on the same operators interface while the power supplies, communication channels, hardware, and real-time operating systems remain completely independent of the BPCS and any of its components and sub-systems, thereby maintaining the separation promoted by mainstream safety philosophies..

All operations, engineering and maintenance functions for the two systems are integrated including:

- alarm handling
- configuration
- time synchronization
- user security
- device health monitoring.

The integrated configuration environment simplifies and streamlines the engineering effort. This integrated approach eliminates the need for expensive data mapping, and handshaking logic that is common in existing solutions. The result of the ICSS's integrated-yet separate architecture allows operators to more effectively manage the process from a single operating environment.

Reducing CapEx

Keep in mind that ICSS solutions aren't simply a product or a specific control and safety system brand name. ICSS has become a proven strategy for building an all digital architecture, a blueprint for building solutions that optimize plant performance by:

- Leveraging digital intelligence – The intelligence (smarts) available in today's field devices represents a new source of previously unavailable information, including diagnostics that enable predicting and preventing problems before they impact process availability.
- Connecting the plant - Open communication standards link devices, systems, and applications in a secure, robust plant-wide network designed to ensure that process and equipment information is available wherever it's needed.
- Controlling your process - The more you know about your process and its operating equipment, the easier it is to improve production and achieve high-integrity process availability.

- Optimizing your assets - Digital technologies permit tapping into the architecture's information flow in order to increase uptime, optimize performance, and reduce maintenance costs.

After having converted several hundred DCS-centric architectures to ICSS solutions, overall plant efficiencies typically improve by at least 2%.

Some of these efficiencies actually begin before the system arrives on site.

Historically, “Experiential Preference” has been the best measurement for reducing the capital cost of an automation project. However, in an ICSS project fiscal savings can come from a variety of sources spread along the entire project execution timeline. Additional CAPEX improvement opportunities may include the following:

Project related savings resulting from the selection of an ICSS solution from a single supplier can include:

- Project management:
  - Minimal number of people to contact
  - Efficient coordination meetings
  - Coordinated procurement and expediting
  - Coordinated cabinet designs, wiring, etc
- Integrated FAT/SAT
- Consolidated packing and shipping
- Coordinated construction support

In addition, advanced ICSS solutions include the following characteristics and features designed to assist with CapEx savings during:

Implementation

- Flexible, modular, and scalable design for applications of any size, anywhere – and for easier system change orders.
- Security Access used to audit user privileges and system access
- Common database and engineering environment.
- Intelligent point configuration, for example, each time a point is initially configured, that point contains all the appropriate attributes (i.e., process variable, setpoint, output, alarms, data historian, etc.).
- Complete palette of pre-engineered function blocks.
- Integrated/intelligent alarm management.
- Integrated Event Historian
- Integrated FDA Compliant Audit Trail for Management of Change

Configuration Management

Your team's engineering efforts are automatically chronicled and saved as part of the ICSS configuration audit trail. This audit trail should provide you with a complete version-to-version comparison – not just who, when, where and why the changes were made, but most importantly, what was actually changed.

The audit trail should automatically record changes to device's configuration and include the following information for each event:

- Date and time of the event.
- User who made the change.
- Exactly what was changed.

Some of the time reductions reported by ICSS users include:

- 90% in configuration.
- 75% in loop-checkout.
- 85% in commissioning.
- 28% in transmitter count when multivariable transmitters are installed.

Further reported reductions realized by the use of the ICSS's all digital fieldbus network include:

- 93% in I/O cards.
- 74% in wiring.
- 84% in terminations.

Regulatory compliance

Nearly every process plant is subject to some form of regulatory compliance. In the U.S. these include FDA's 21 CFR Part 11, OSHA's 29 CFR 1910.119. Additionally most plants have committed to complying with ISO 9000. And where safety system conformance is required the governing safety standards include ANSI/ISA 84 and IEC 61511.

The ideal ICSS solution should provide considerable assistance in complying with mandated and voluntary regulations by completely integrating change management and generating, on demand, detailed documentation from its:

- Device audit trail.
- Device calibration history.
- Control and/or safety configuration audit trail.
- Process history.
- Event history.

Commissioning

Commissioning costs are dramatically reduced as the ICSS auto-recognizes intelligent field devices and control hardware. The ICSS automatically sets up the fieldbus, making deployment of the devices fast and trouble free.

Additional commissioning areas realizing cost savings from an ICSS solution include:

- Proof testing.
- Training costs.
- Documentation costs.
- Automated performance monitoring, testing, and documentation.

Specific to the safety system, the ideal ICSS solution will include:

- Third-party certified function blocks.
- Secure bypass handling, with configurable time expiration.
- Automated partial-stroke testing.
- Sequence of event handler with first-out trapping.
- Off-line simulation.
- Executable cause & effect matrices - eliminate design to configuration interpretation errors.
- Graphical indications of bypasses, overrides, and degraded modes of operation.

Certainly an ICSS solution lowers upfront costs of engineering, installation and commissioning, but it will also reduce ongoing maintenance and management costs to satisfy your safety and regulatory requirements.

Reducing OpEx

After commissioning the process, plant efficiency comes from making more with less – more product on-spec – with less staff, less variability, and more process availability (less unscheduled downtime). The ICSS is designed to deliver results in operations and maintenance to improve plant efficiency and profitability.

Part of those 2% improvement mentioned earlier, is the result of having access to the advanced digital diagnostics embedded in field-based measurement devices, such as analyzers, transmitters, etc.

Predictive maintenance is key, whether in he control room or the maintenance shop, use of asset management software's advanced diagnostic information allows keying in on the field devices that really need attention. Beyond the advanced warning of faltering equipment, asset management software suggests corrective action and streamlines every faucet of maintenance work, including troubleshooting, diagnosing and calibration. This means the with predictive maintenance, unnecessary trips to the field can be reduced by as much as 63%

Surveys conducted by different sources at different times reveal that more than 80% of maintenance is reactive (too late) or preventive (unnecessary).[1] In fact, typical maintenance practices for reactive, preventive, and predictive maintenance have not changed in over 15 years.[2] This is primarily the result of an insufficient tool-set capable of improving maintenance practices – a situation that a modern ICSS can change.

Not only has digital technology improved the accuracy, repeatability, and stability of primary measurement sensors, embedded digital technologies also enable intelligent predictive diagnostics – a critical first step to achieving high-integrity process availability.

Frequently, under-performing assets often go unnoticed until quality and/or rate became affected. When analyzers and transmitters are fitted with intelligent predictive diagnostics under-performing assets and potential problems that may eventually result in off-spec product and/or an unscheduled shutdown are monitored and assessed by asset management software. If a problem is detected, such as a plugged impulse-line, an alert is automatically generated.

Similarly, using embedded sensors and advanced diagnostic algorithms, control valve-signature diagnostics can determine which control valves do and don't need attention during the next scheduled shutdown.

Being able to embed digital technologies in field analyzers and transmitters is analogous to how digital technologies have changed how we live, work, and stay connected.

For example, the digital technologies that allow the service technician at your automobile dealer to pinpoint that your car's fuel injectors need cleaning is similar to the digital technologies that can detect that an in-line pH sensor needs cleaning.

It is hard to imagine but even as recent as 2002 independent survey's revealed that over half of today's maintenance activities are unnecessary. This includes routine equipment checks as well as preventive maintenance on equipment that does not need it. For example, one analysis showed that over 60% of all instrument work orders did not result in corrective action because there was nothing wrong with the instrument. Or consider that a diagnostic evaluation of 230 valves that were already scheduled to be rebuilt during an upcoming shutdown revealed that only 69 of those valves actually needed to be rebuilt.

---

[1] Reliability Magazine, 2002 (No longer in existence.)

[2] "The Business Case for Reliability," available at www.mrginc.net

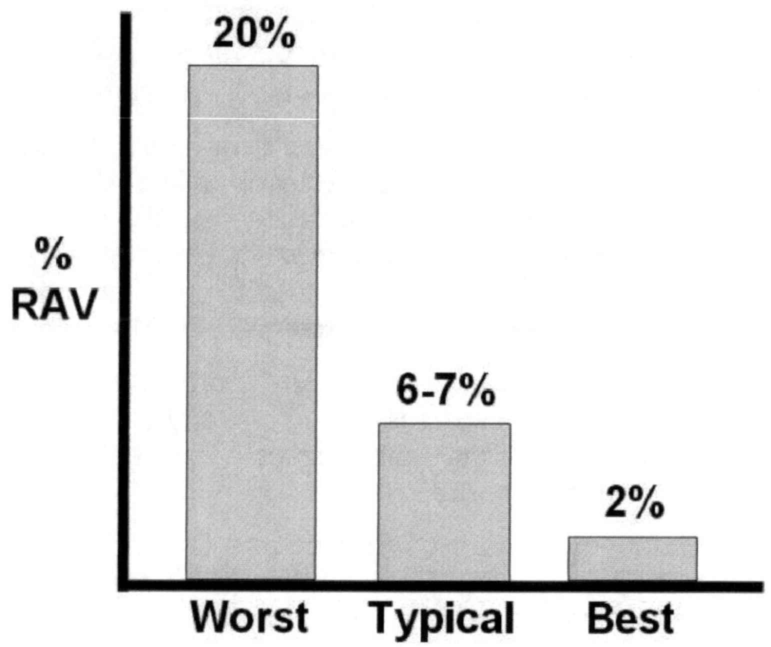

Figure 2

What we know is that the use of robust ICSS solutions can go a long way toward ensuring that instrument technicians are spending their time where they are needed and when that is happening, process availability will also improve.

However, survey results aren't enough. In order for us to sell the benefits of investing in an ICSS solution to management we need hard numbers.

Maintenance cost as a percentage of Replacement Asset Value (RAV) is a frequent benchmark of maintenance productivity. For example, a plant spending $5,000,000 annually to maintain assets that could be replaced for $100,000,000 has a 5% RAV.

The graph "Benchmarking Maintenance Costs" shows typical as well as worst- and best-in-class %RAV [2], figure 2.

For a plant with $250,000,000 in maintainable assets, moving from typical to best-in-class could mean more than $10,000,000 in annual savings. The goal is to use the maintenance-budget and -personnel more efficiently in order to reduce overall OpEx while advancing toward high-integrity process availability and minimizing process variability.

The key to minimizing process variability in order to produce on-spec product is the health and precision of the field devices and equipment.

Performance of control is best measured as close to the process as possible. With ICSS intelligent field devices its performance can be calculated in the device. Using all digital fieldbus communications, the system collects the variability data to provide process engineers with a snapshot of the sources of variability in your plant – by current measurement, by sift or by month. And it does this instantaneously with no configuration.

With this precise data available you can tackle those problems that have the greatest economic impact.

The ideal ICSS solution permits assigning alarms on variability index trip values for the loops that can influence variability most. For the plant manager, this should all be automatically rolled up into a summary display that includes key performance metrics.

Operations have achieved a reduction in process variability due to tighter controls on the process. This has resulted in a more consistent product with respect to quality, yields, and cycle-times. – Dean Hale, DSM Catalytica Pharmaceuticals

Identifying and predicting problems in the entire safety loop – sensors, logic solvers, final elements, and surrounding process – is also critical.

The ICSS helps monitor the health of the entire safety loop by:

1. Detection - Continuously monitoring the health of each safety loop device and the associated process.
2. Notification – When a problem is detected the ICSS generates an alert. Within the system alert types are pre-configured to be sent via narrowcasting to the appropriate personnel. Some alerts may be sent directly to "on-duty" plant personnel. Other alerts may be sent to an expanded list such as the on-duty plant personnel plus the unit and maintenance managers. Each alert includes information identifying the root cause of the detected problem along with context sensitive guidelines on possible actions and corrective measures.
3. Correction – Once the correct personnel arrive, additional information from the intelligent devices will be stored in the ICSS' asset management software database thereby facilitating quick access to detailed device diagnostics and problem correction.

The bottom-line is the ICSS architecture provides a platform for more reliable safety loop operations, from early detection through notification and correction.

It's not only possible to check valves, but we can also do so more safely, at less cost, and with greater efficiency. – Patrick Flanders, Saudi Aramco

Because its been shown that more than 50% of SIS malfunctions can be contributed to valve malfunctions, the use of partial-stroke testing has become a valuable means of detecting related malfunctions.

During an automated partial-stroke-test, expect the ICSS to verify that the valve physically moves by measuring pneumatic supply, actuator pressure, and valve position.

Benefits of automated partial-stroke-testing include:

- Less human error.
- Better maintenance practices.
- Better documentation.
- Less risk.

Automation removes the guess work associated with when a safety valve needs to be maintained. The ICSS provides the ways and means to better understand the overall electro-mechanical condition of valves.

Valve signatures generated during each partial-stroke-test provides maintenance personnel with insight into:

- Valve friction.
- Air-path leakage.
- Valve sticking.
- Actuator spring-rate
- Inherent diaphragm pressure range.

This information provides maintenance personnel the ability to obtain the correct repair parts and schedule repairs.

Because of the actual measurements and actins taken during partial-stroke testing the ICSS can automatically develop detailed documentation thereby making regulatory compliance much easier.

Interlock Testing

Checking process and safety interlocks is one of the most resource intensive tasks of commissioning and in some cases represents a recurring requirement. Typically technicians in the I/O room and/or in the field connect signal generators to simulate an interlock condition. In the control room additional technicians man the operator interface and verify that each interlock performs its pre-defined actions. Given the hectic environment of commissioning, efficiency is often degraded by additional communication traffic, misunderstood tag identifications, etc. The asset manager portion of the ICSS and the "wonders" of intelligent digital technology permit quickly and efficiently facilitate interlock testing and verification from a single location.

Ron Harlow, Principal Automation Engineer at Syngenta in St. Gabriel, Louisiana shares that before they can change-over to a new herbicide they must re-validate 20 sets of safety instrumented functions involving approximately 60 inputs. He says that the use of the ICSS's asset manager to verify interlocks has reduced the time to verify these safety interlocks from 2-3 days to 1 day.

When you are trying to make the most of an ever-dwindling staff and budget, or looking for even more ways to trim operations and maintenance costs, an all digital ICSS solution with built-in predictive intelligence can provide significant assistance.

Culture, Organization & Competency

Repeatedly we've mentioned how an ICSS solution can reap 2% plant efficiency improvements but those benefits can only be achieved if the technology is harmonized with the correct culture, organization and competencies.

There may be some among us that remember when automobile mechanics worked in the service bay of the local gas station or in a garage in back of their house. As cars began to become more sophisticated independent mechanics were replaced by service departments at the automobile dealers and mechanics began to receive special training for specific automotive systems. When cars became fitted with a variety of advanced technologies and associated sensors service departments changed once again.

Today no service technician gets anywhere close to your car until the diagnostic technician has run a series of tests, analyzed the results, and printed out the details of why the "Check Engine" warning light on your instrument panel is 'ON'. With the detailed diagnostic results in hand, the service manager is able to assign the right service technician to the job.

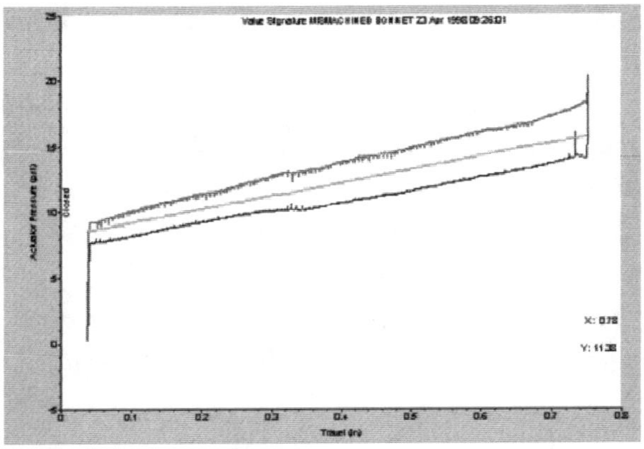

Valve signature diagnostics reveal conditions that if left undetected will increase process variability.

Modern automobile service departments provide a very good model for organizing plant maintenance departments desiring to take advantage of the digital ICSS solution.

As we discussed in the "Intelligent Field" section of this article, digital technologies enable intelligent predictive diagnostics of analyzers, transmitters, and control valves. For example, control valve stiction – a combination of <u>stic</u>king and fric<u>tion</u> – can be a major cause of process variability. Modern control valves, fitted with digital positioners are able to detect the presence of control valve stiction under actual operating conditions. Depending on the amount of stiction present, there are several things that can be done to improve the situation without shutting down the process.

We wouldn't expect the operator to recognize there's a control valve stiction problem, but we would expect them to recognize there's a problem maintaining process stability. That recognition alone would get them to the correct group of graphic displays where they would see a system generated warning indicator next to a particular control loop – the equivalent of the "Check Engine" light.

The operator notifies the diagnostic technician, describes the process symptoms, and gives the technician the loop identifier. The diagnostics technician can login to the control valve, gather data, analyze the data and turn over the findings to the service manager. After a brief consult with the unit manager and the shift supervisor the service manager dispatches a service technician trained in re-tuning control loops. Because the service technician knows the cause of the misbehaving control loop, she knows just what to do to minimize the stiction problem.

Certainly there will need to be follow up but this illustrates the importance of harmonizing the architecture of the organization using the technology with the architecture of the technology itself.

As we have learned, ever analyzer, transmitter, control valve, controller, logic solver, fieldbus, etc. of the ICSS solution is conducting its own diagnostics and reporting the results to the database of an asset management application – the application where predictive intelligence begins to take place.

Asset management solutions are a collection (suite) of applications with each application targeted at specific asset types such as mechanical equipment, electrical systems, process equipment, instruments, and valves.

Working with data automatically populated into the asset management database, each application runs appropriate performance checks. Any performance check that exceeds pre-defined limits is flagged in a report that is sent to the diagnostic technician, thus providing technicians a list of devices that require further diagnostics and analysis. Once the advanced diagnostics is complete, the technician forwards it to the service manager who determines the appropriate next course of action.

For example, there may be a backup device available that could be placed into service; it may be necessary to expedite repair parts; or it may be necessary to schedule third-party services.

What's important to recognize is that because the ICSS technology has been harmonized with both the using (operations) and supporting (maintenance) organizations high-integrity process availability is achieved.

Simply stated, the features and capabilities of the ICSS technology can not be fully maximized unless the culture, the organization and the competencies of the entire organization – engineering, operations and maintenance – are thoroughly and completely developed.

Summing Up

Your company is committed to installing (or replacing or upgrading) the control- and safety-systems at one of your plants. You've seen the very aggressive project schedule and recognize that speed of implementation is critical. You've heard and read about these new highly-integrated, all digital system solutions. You recall hearing that these "smart" systems are a lot easier to implement than their predecessors, but you've heard things like that before. Beside's, even if these new systems are easier to implement, isn't it likely that all that sophistication will end up costing more to operate and maintain?

The fact of the matter is, choosing an all digital ICSS architecture with its "smart" devices and embedded predictive diagnostics is more likely to meet an aggressive project schedule and end up costing less to purchase, engineer, install, commission, and maintain.

Is it Time for a New Approach to PID Tuners

Mark L. Darby, CMiD Solutions

Michael Harmse, IPCOSAptitude Ltd.

Abstract: As typically practiced today, PID tuning is performed on a single-loop basis, via trial and error, or based on the results of an open-loop step test. For interacting loops, this is a time-consuming task that often results in compromised performance, or at least leaves open the questions: "How much performance is being left behind?" and "How robust is the tuning?" This contribution presents a multivariable approach to PID tuning, which takes advantage of modern identification methods and incorporates a control-relevant optimization to determine optimal PID tuning parameters. A key feature of the approach is that robustness margins to dead time and gain errors are explicitly addressed.

1 Introduction

Most applications of PID controllers today involve an interactive, multivariable process. The interaction is due to the inherent nature of typical chemical processes as well as complexities due to heat integration, recycle streams and intermediate buffers. As a result, tuning multiple interacting PID controllers for such situations can be a challenging task, further complicated by the sheer number of PID loops in the typical plant.

PID controller tuning is typically performed on a single-loop basis using a trial and error approach, or based on the identification of a low-order model from an open-loop step test and corresponding tuning rules linking process model parameters to PID tuning parameters. With the first approach, decisions are typically made as to which loops are the most important and therefore tuned tighter and the others tuned slower. If significant interactions are observed, then heuristic detuning is performed, and the least important loops are made even slower. A challenge with the second approach is that a low order model may not accurately represent the true process dynamics, due to the afore-mentioned complexities; in addition, such tuning rules are typically limited to the single input single output case.

A key concern once tuning is performed is how well will the loops perform in the future. Process plants are, to at least some degree, nonlinear. Tuning that performed adequately for a given production rate or a particular feed type may perform poorly later when operation changes. It is obviously desirable to ensure stable and adequate performance of the overall system when plant responses change.

This paper presents a new approach to tuning interacting PID loops, taking advantage of modern identification methods, advances in optimization theory, and the availability of fast multi-core processors. The procedure for developing PID tuning for multivariable systems consists of the following steps:

- Plant test of the multivariable process.
- Model identification.
- Model reduction to a state-space model with explicit dead time.
- Constrained optimization to identify optimal PID tuning parameters.
- Simulation of the optimized process.

The last three steps have been incorporated into a software package called AptiTune™. It allows the user to interactively determine the best tuning for his/her application based on *process safety* and *operability* considerations and *robustness* specifications, tailored to the *specific DCS*. The goal is to shift iterations from the process to the tool, and to achieve plant performance that closely matches the optimized results. This is referred to as "one shot tuning".

In the following, the workflow and optimal tuning approach are further described. Tuning results are presented for an application of AptiTune™ involving selected loops in a Desulphurizer Unit.

2 Motivating Example

Here an example, shown in Figure 1, is considered to motivate the proposed approach. From the control configuration shown in the figure, process interactions can be expected to exist in the following:

- Tray temperature (TC1) and overhead pressure control (PC1).
- Feed preheat via LC1.OP and its effect on tray temperature (TC1.PV) and overhead pressure (PC1.PV).
- Interaction from reflux drum LC2, if drum level is controlled by reflux flow (FC2).

Here, PV refers to the process value (feedback variable) and OP refers to the output of the controller (the manipulated variable). As a result of these interactions, we see that we have at least a 2x2 interacting problem, possibly as high as a 5x5 problem. For illustrative purposes, assume that the reflux drum level is controlled with distillate flow. When interaction is small between certain subsystems, the overall problem can be split up. For this example, we could treat the distillation tuning problem as a 4x4 system or as multiple subsystems: a 2x2 (TC1 and PC1), a 1x1 (bottoms level, LC1) and a 1x1 (reflux drum level). For a given subsystem, we would need to develop a model between all the controller outputs (or manipulated setpoints if the PID output is connected to a cascaded controller) and all the controlled variables in the subsystem.

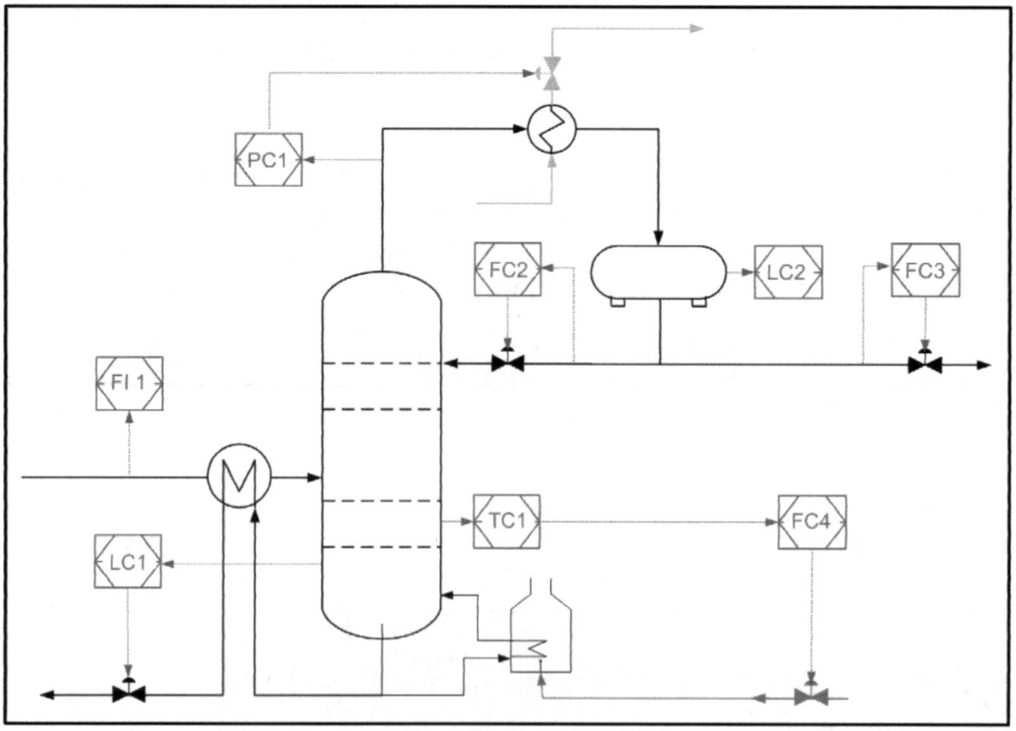

Figure1. Example Distillation Column.

To tune this overall system, we would need to account for interactions. In addition, we would want to ensure the following:

- Avoid excessive kick from tray temperature output (TC1.OP) to reboiler fuel gas (FC4) to prevent flame out on the low side or tube damage on the high side
- Adequate performance for disturbance rejection and setpoint changes
- Limit the amount of PV overshoot in TC1 and PC1.
- Take advantage of buffering capacity in the bottoms sump to smooth out bottoms flow fluctuations and minimize interactions between feed preheat and the other loops.
- Take advantage of buffering capacity in the reflux drum.

3 Work Flow

The work flow parallels that used for model predictive control (MPC): Test » Model » Tune.

A plant test is performed on the various subsystems where PID tuning is desired. The goal is to identify a dynamic model that accurately captures the input-output behavior of the plant. With an accurate model, we have high confidence that performance of the optimal tuning on the actual plant will match the simulated responses. The plant test can be performed open-loop or closed-loop.

<u>Open-Loop:</u> Controllers are placed in manual mode and steps made to the controller outputs. Four to six steps of varying duration is typically sufficient, as long as the steps are of sufficient amplitude.

<u>Closed-Loop:</u> Controllers are kept in automatic and steps are made to the controller setpoints.

In both cases, it is important to vary the step durations in order to capture the relevant dynamics of the process. It is recommended that steps be made of various durations, ranging from 10% to 100% of the estimated open-loop response time Methods based on PRBS (pseudo-random binary sequence) or GBN (generalized binary noise) can be used to automate the testing. Testing time is often on the order of 1/2 day to 1 day for problems of size 5x5 or smaller, with settling times in the order of 1 hour or less. It is important to make moves larger than the amount of valve stiction. Note that some DCS systems have standard features for enforcing a minimum move size in the controller outputs for the closed-loop case.

The next step is to analyze the raw data to select the data ranges for model identification. Modern techniques based on FIR (finite impulse response), ARX (auto regressive moving average) and Subspace methods are used. AptiTune™ supports the import of models created by identification tools from MPC packages. It also allows the user to specify a transfer function matrix.

Once a model has been imported into AptiTune™, the next step is to smooth the model (if necessary) and covert the model to a state-space model using a singular value decomposition (SVD) reduction technique (Maciejowski, 1989). State space models exhibit much more process realism than FIR models identified from short noisy step test data sequences. State space models are also much more compact (less parameters), and therefore improves simulation speed. Dead time can also be set explicitly during this step.

The smooth state space model is then used to determine optimal PID tuning parameters for the closed-loop control system. The user is required to enter instrument ranges and to select the desired PID equation associated with a particular DCS. The user can select from popular DCS systems, such as

Honeywell TDC and Experion, Emerson DeltaV, Foxboro I/A, ABB Symphony, Yokogawa and others. Optimal controller parameters can also can be calculated for P-only, I only, and PI options.

Specification and constraints are specified by the user and an optimization is performed. Design cases are logged and the user can analyze the different scenarios in terms of setpoint tracking, disturbance rejection, and noise attenuation. The integrity of the system to one or more controllers being placed in manual can also be analyzed. Finally the robustness against plant-model mismatch is evaluated and displayed in an easy to understand graphical way.

4 Optimization Formulation

The PID controller parameters, controller gains, $k_{c,i}$, integral times, $\tau_{I,i}$ and derivative times $\tau_{I,i}$, are determined by solving the nonlinear, constrained, optimization problem

$$\begin{aligned} &\min_{k_{c,i},\tau_{I,i},\tau_{D,i}} \quad J \\ &st \qquad\qquad\qquad\qquad , \\ &c_j(k_{c,i},\tau_{I,i},\tau_{D,i}) \le 0, \qquad i = 1,\ldots,n,\ j = 1,\ldots,m \end{aligned} \tag{1}$$

where J denotes the objective function and the c_j denote the constraints. The objective function J is a weighted sum of three terms $J = J_1 + \alpha J_2 + \beta J_3$, each of which contributes different aspect of closed-loop performance. The first term $J_1 = \int_0^{t_f} \left| y^r(t) - y(t) \right| dt$ is based on an integrated absolute error (IAE) criterion. Here y^r is an optional user-defined first order trajectory. The second term $J_2 = \int_0^{t_f} \left| y^{sp}(t) - y(t) \right| dt$ denotes the IAE for a load disturbance occurring at the input of the process. The third term $J_3 = \int_0^{t_f} \left| \Delta u(t) \right| dt$ is a criterion used to affect control effort. The weighting coefficients α and β are user specified to reflect the relative importance of the three objectives. For each control loop, the user can specify one or more of the following inequality constraints:

- Maximum OP change following a setpoint change.
- Maximum PV overshoot following a setpoint change.
- Minimum damping ratio.
- Maximum noise amplification.
- Maximum process-model gain mismatch.
- Maximum process-model dead time mismatch.

For buffering applications such as with liquid level, the following may be specified:

- Maximum setpoint deviation and minimum return time following a preset level disturbance.

By specifying appropriate constraints, the user can tailor the tuning to process-specific requirements.

Initialization values for the optimization may be selected as the actual DCS values or calculated based on the Cohen-Coon SISO tuning rule. The degree of difficulty of the optimization depends on the number of controllers, the order of the state-space model and the number of constraints. The problem is non-convex and local minima can occur. Therefore, several optimization methods have been implemented, including a brute force global search, a genetic algorithm and a generalized gradient algorithm (Vlachos et al., 2000).

Robustness plots are provided for each PID control loop. The robustness plots are similar to those proposed and developed and by Garry and Hansen (1987). A robustness plot is shown in Figure 2. The plot shows the stability line (in purple) as a function of gain ratio and dead time deviation, relative to the nominal model. The minimum stability region is shown in red. The robustness spec is shown in light blue. A gap between the robustness spec (the blue circle) and the stability line indicates that the robustness specification is not an active constraint in the optimization. The user can mouse click on any point in the plot to observe both the open loop step response relative to the nominal model (shown in the same frame as the robustness plot) and closed-loop PV and OP setpoint responses in the plots to the right.

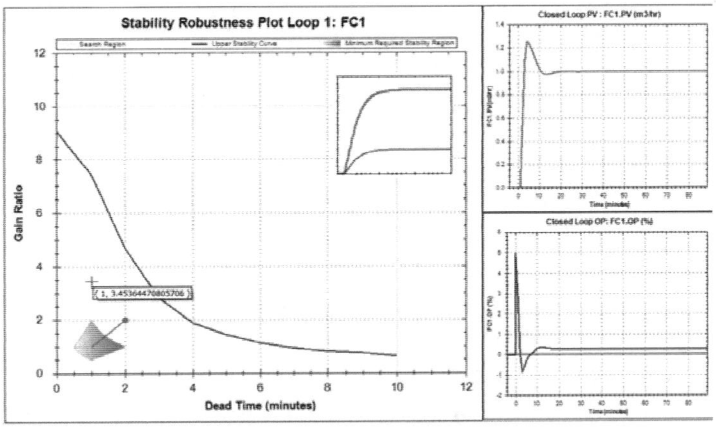

Figure 2. Robustness Plot

5 Industrial Example

The method and software tool have been applied to loops in a Desulphurizer Unit. The client had deemed the performance of the existing loops as adequate, but was interested in determining if performance could be improved. Note that no use was made of the existing tuning parameters.

A diagram of the process is shown in Figure 3. PC001 sets the backpressure on the unit, as the other PCs are kept in manual with the valves 100% open. The PC manipulates fuel gas import, which affects the heating value of the gas, and thus interacts with loops TC001 (reactor temperature) and TC002 (column temperature). In addition, changes to fuel gas flow via TC001 and TC002 affects the pressure loop PC001. As a result, the 3x3 system is interactive. The normal cascade for the reflux level controller is the distillate flow, and the step test was conducted in this mode. An interesting aspect of this problem is that the three loops are associated with different equipment in the unit. As a result of this distribution and the interaction, the settling time is fairly slow, on the order of 90 minutes.

A closed loop multivariable test was performed with the following loops.

Loop	Controller variable	Manipulated Variable
TC001	Reactor inlet temperature	Fuel gas flow setpoint
TC002	Reboiler outlet temperature	Fuel gas flow setpoint
PC001	Unit back pressure	Fuel gas make up valve

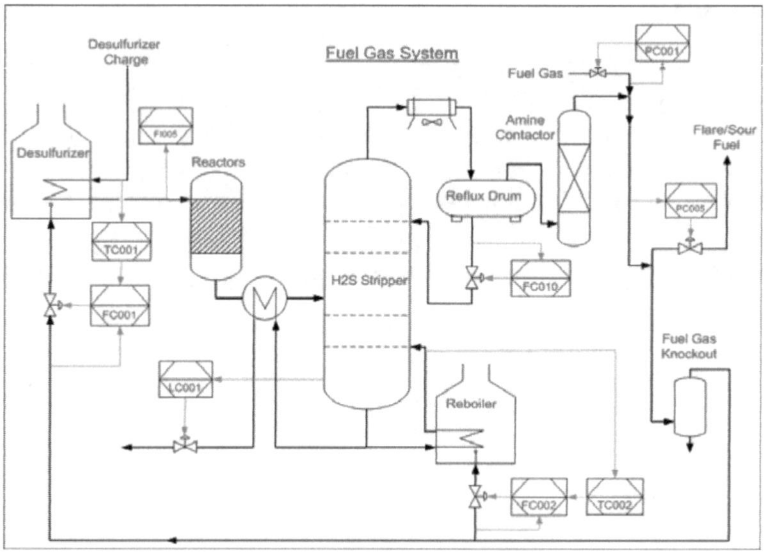

Figure 3. Desulphurizer Unit

Standard TDC3000 capabilities were used to conduct the step test with operator supervision only. OPMNCHG parameters were used to ensure valve changes greater than stiction limits. Setpoint changes of varying step were made to all loops. Since the Time to Steady State was fairly slow (90 minutes), a total of 23 hours of test data was required for this particular application. The test signals are shown in Figure 4.

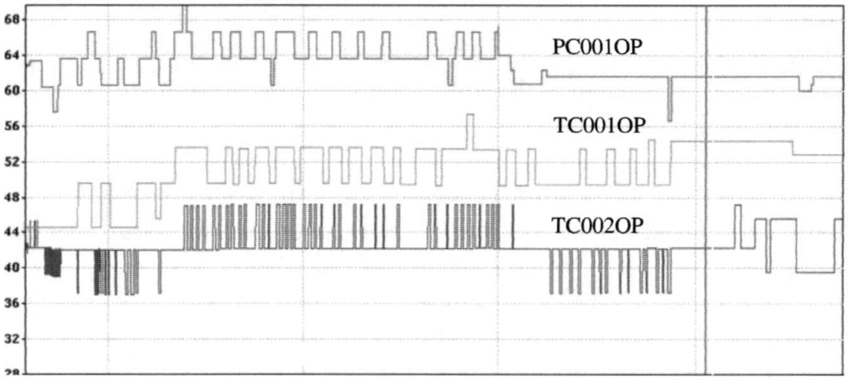

Figure 4. Step Test Trends

Intermediate identification results were made to check progress of the test in terms of model accuracy. These results were used to modify the step width durations (frequency content) of the step signals.

The identification modeling and PID tuning was completed in 2 hours. A 3x3 state space model was identified using a subspace method. The model matrix is shown in Figure 5. Note that the model matrix is fully dense and that several of the response exhibit high-order characteristics: The loop pairing (OP and PV of the same control loop) is determined by the diagonal curves so these model curves need to be strong on typical move scaling. The strength of the off-diagonal model curves provides a good indication of the degree of interaction between the three PID loops. This is the reason why all three loops should be optimized simultaneously, rather than one at a time.

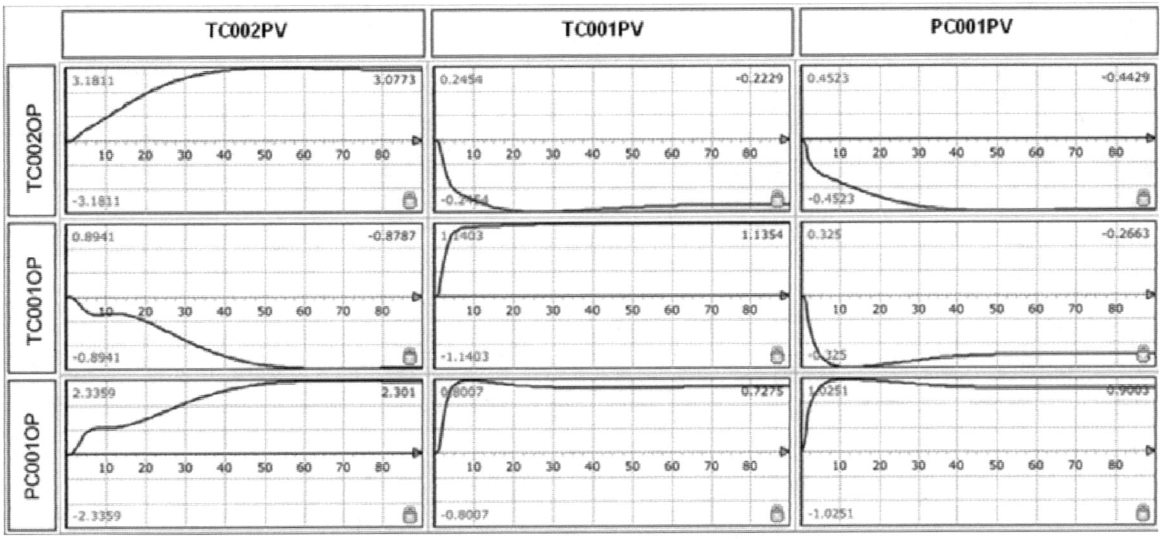

Figure 5. Step Response Model

The following criterion was used to tune the loops in AptiTuneTM

Column TC002	Reactor TC001	Backpressure PC001
•Remove cycle tendency	•Tighter tuning	•Remove cycle tendency
•Good dampening	Good dampening	•PV to SP in 3 min
•Minimal overshoot	•Low PV overshoot	•10% max PV overshoot
•	•Hard constraint on OP	•Moderate kick in OP

The existing tuning showed that both TC002 and PC001 had a tendency to cycle. TC001 was tuned very sluggishly.

Robustness margins for gain and dead time were specified for all loops and were active for TC001 (gain ratio margin 2 and dead time margin of 1) and PC001 (gain ratio margin 1.25 and dead time margin of 1). Closed loop simulated setpoint responses are shown in Figure 6. Each row in the matrix corresponds to a setpoint change in the corresponding controller (diagonal entries). Interaction responses are shown in the non-diagonal entries. Stable responses, without excessive manipulated variable movement, are observed in the simulated responses of all loops.

The tuning parameters for the three DCS loops were updated with the AptiTune™ results. The loops were then tested by making setpoint change and observing responses to normal plant disturbances. The results on the process were found to closely match those in simulation. For example, PV rise times, amount of overshoot, time to reach the peak OP value, time at which maximum PV overshoot occurs, etc, were all within 20% of actual.

Before and after trends, of length 48 hours, are shown in Figure 7. The data for the trends were selected to obtain similar values of feed rate and number of setpoint changes. In analyzing the before and after results, it is necessary to analyze both PV and OP variances, since tuning affects both. As indicated earlier, both column TC002 and backpressure PC001 had a tendency to cycle. This tendency was removed with the new tuning. In both cases, we see that the OP variance is decreased. We see that the cycle in column TC002 is removed. The decrease in the OP variance for pressure PC001 is accompanied with an increase in PV variance. This is consistent with the stated desire to increase robustness of the PC001. The tighter tuning in reactor TC001 is clearly seen.

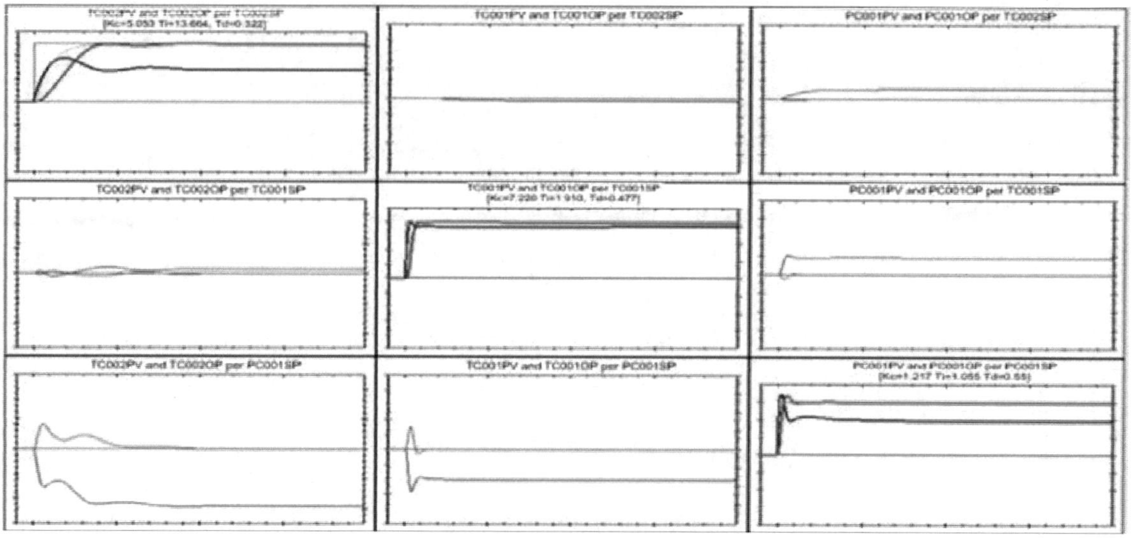

Figure 6. Closed-loop Step Response (Red: PV; Blue: OP).

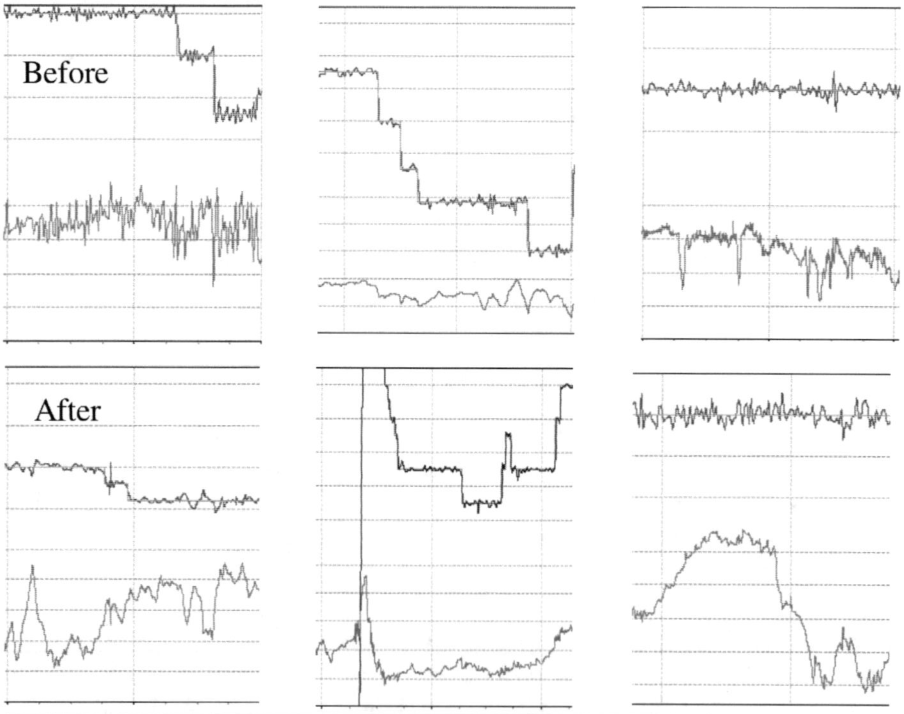

Column TC002	Reactor TC001	Backpressure PC001
•Reduced sp rise time •Tuning less aggressive •Cycle tendency removed	•Tuning more aggressive •Robustness till good	•Tuning more robust •Tuning less aggressive •Cycle tendency removed

Figure 7. Before and After Data, Duration: 48 Hours (Blue: PV; Red: OP).

The tuning determined by Aptitune^{TM} was placed in service in October 2008 and was kept unchanged until a review of the loops was help in January. Since stable control was observed, it was decided to tighten the tuning of backpressure PC001. This was accomplished in Aptitune^{TM} by reducing the gain ratio margin. The resulting tuning, which was observed to perform well, remains in place today.

6 Conclusions

A new approach to tuning multiple interacting PID controllers was presented. The methodology consists of a dedicated plant test, followed the use of modern identification technique to identify an accurate dynamic model. The resulting model is used in AptiTune^{TM} to develop optimal PID tuning parameters based on process safety/operability considerations and plant-model mismatch margins. The advantage of the approach is that if an accurate model is identified, the resulting tuning will yield process responses that closely match the optimized responses, thereby avoiding trial and error methods.

The approach was successfully applied to an interacting 3x3 Desulphurizer problem. The application was completed in 25 hours. Although not mentioned in the main body of this paper, the sump level was also tuned with Aptitune^{TM}. The SISO test and tuning for the level was performed in 6 hours

These results demonstrate that a model based approach to multivariable PID loop tuning is viable. The majority of the effort involved step testing, which was automated in this application and did not burden the APC engineer. The responses calculated in Aptitune^{TM} performed similarly to the actual plant responses. There has been no need to manually modify the tuning that was calculated by the software. The goal of "one-shot tuning" was demonstrated, as well as the ability to take process operability considerations and safety constraints into account.

References

Gerry, J.P. and Hansen, P.D. (1987). Choosing the right controller. *Chemical Engineering*, May 25, 1987, 65-68.

Maciejowski, J.M. (1989). *Multivariable Feedback Design*. Addison-Wesley.

Vlachos, C. Williams, D. Gomm, J.B. (2000). Genetic approach to decentralized PI controller tuning for multivariable processes. *IEE Proceedings. Control Theory and Applications* 146(1), 58-64.

Growth of Wireless Field Networks in Refining

Greg Martin, Senior Process Control Consultant
Advanced Applied Technologies Group
Emerson Process Management
Austin, Texas

Abstract

Wireless technology is currently used in refining for remote applications like in the tank farm where wired solutions are costly because of distances. What will be the role of wireless in five years for the refining industry? The refinery of the future will be an open, smart, combined bus-based and wireless digital plant. Wireless will be used for all kinds of applications, including advanced process control. State-of-the art security is enforced, so that unwanted influences cannot penetrate the wireless network. For applications powered by batteries, it is important to note that battery life continues to improve, but the other side of the signal transmission power requirement equation is just as important. Methodologies have been developed that use exception reporting to conserve power. In other words, signals are transmitted only after a significant difference in the signal is detected. Control algorithms are modified to account for intermittent signal transmission. Sophisticated scheduling is used to ensure bandwidth. Communications use the new WirelessHART™ standard.

This presentation will focus on the enablers of advanced process control in refining using wireless communications.

WirelessHART Network

The Highway Addressable Remote Transducer (HART) protocol is an industry standard. The WirelessHART 7 network uses a flexible, mesh network organization where every device sources and sinks packets. Every device can also act as a router. This allows adaptation to application and environment, and, unlike approaches to in-plant wireless which require direct line-of-sight between the instrument and the communications gateway, the smart wireless network routes communications around obstacles and interferers. Large geographic coverage is possible.

Dynamic bandwidth allocation is provided. Devices request fixed bandwidth for publishing process and control data, but they can request additional bandwidth temporarily (e.g., to transfer a vibration spectra).

The network is self organizing and self healing. All devices provide statistics to the network manager, including signal levels, communication reliability with neighbors, and detection of new neighbors. The network manager grooms connections and routes, and changes connections and routes for reliability. It can add interconnections to make the network more resilient, and flatten the network for increased speed and lower power consumption.

Reliability

WirelessHART self-organizing mesh networks (Figure 1) have demonstrated high reliability in the field. WirelessHART is a secure networking technology operating in the 2.4GHz ISM radio band. WirelessHART utilizes IEEE 802.15.4 compatible DSSS radios with channel hopping on a packet by packet basis. There are five key components of WirelessHART that contribute to end-to-end network reliability:

- Time synchronized communication
- Frequency hopping
- Automatic node joining and network formation
- Fully-redundant mesh routing
- Secure message transfer

Figure 1. Self-Organizing Field Network.

WirelessHART networks are robust/tolerant to almost all interference and coexist with other wireless networks. Robust security is designed in. Demonstrated reliability is greater than 99%.

Self-organizing wireless field networks can be easily installed and deliver significant value without the need for investing in a plant-wide wireless infrastructure.

Wireless Network Security

At the wireless field network level, robust security is provided through advanced, standards-based encryption as well as authentication, verification, key management, and anti-jamming techniques.

Smart wireless solutions employ end-to-end 128-bit encryption using the Advanced Encryption Standard (NIST standard FIPS-197). For authentication purposes, each gateway maintains a "whitelist" – sometimes referred to as an Access Control List or ACS in the IT world - of devices allowed to communicate with it, and individual devices accept messages only from a previously identified gateway or from other gateway-validated devices. Unique “join” keys per device are an important enabler of this security.

Implementation of the WirelessHART standard also adds “session” keys for communication between any two network devices so that other devices can't "listen in". These can be rotated as well.

Message Integrity Codes (MICs) are used to verify messages, both per-hop and end to end. Anti-jamming techniques such as Direct Sequence Spread Spectrum (DSSS) with channel hopping plus multi-path routing help sidestep noise sources, whether malicious or not. And gateway-to-host security leverages well-known standards such as SSL as well as complete encryption/authentication.

Wireless Field Network

A variety of wireless field devices exist, including: pressure, flow, level, temperature, vibration, valve position monitor, multi-input temperature, discrete switch, HART upgrade module, and wireless device router. Some of these are illustrated in Figure 2.

Wireless instruments can be widely and remotely distributed throughout a plant, across roads and ponds, or on mobile platforms like railcars, barges, or trucks where traditional wired data collection is not feasible. In some cases noise from grounding problems can also be eliminated with wireless devices.

Figure 2. Wireless Devices.

Flexible Communications

Multiple, simultaneous communication modes are supported (Figure 3).

The traditional HART communication using an acyclic request-response mode is used for maintenance, commissioning, and diagnostics.

A cyclic publishing process and control data mode, or enhanced burst mode, provides cyclic updates from 1 per second to per minutes. It is used for user driven applications and allows the user to configure based on process requirements, and, when configured, the devices ask for bandwidth. Time stamping and process data status are included in this mode.

Block data transfer mode is used for radar waveforms, vibration spectra, and other large data sets, which are automatically broken into pieces, transferred, and then reassembled. Bandwidth is requested by the network manager, used, and then released.

Event notification, which is new, and based on Command 48, is used for device disconnection, if the device needs maintenance, and change in configuration.

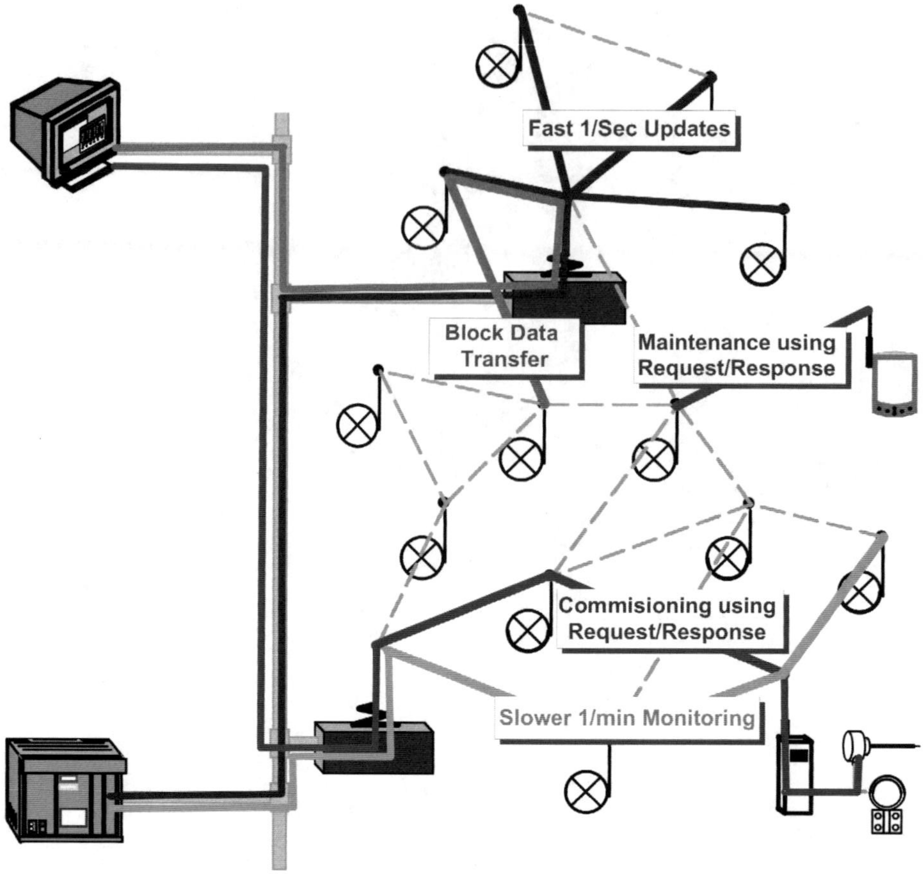

Figure 3. Multiple, Simultaneous Communication Modes Supported.

Gateway - Network Manager

The network manager is responsible for the creation of routes that can be used by plant automation hosts, gateways, other devices, and the network manager itself to perform communications with the application layer in network devices. The network manager continually collects data from devices on the health of connections and traffic patterns and uses this information to adjust routing and scheduling.

The network manager also establishes the communication schedule that the application layer network devices use to transfer process data, alerts, diagnostics and other traffic to the gateway for access by the plant automation host.

For network devices that are actuators, interlocks, or any device that affects the process, the network manager is responsible for the establishment of the scheduled communications that the plant automation host may use to send setpoints and outputs to the user layer application in field devices.

An example link schedule is shown in Figure 4.

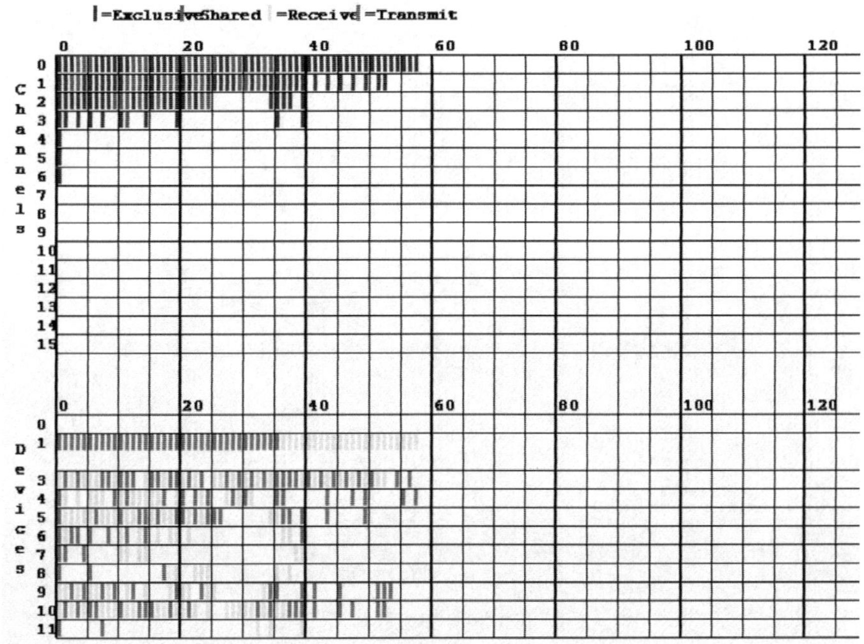

- The top portion shows the overall slot allocations subdivided into channels.
- The bottom portion shows the transmit and receive slots for each device

Figure 4. Link Schedule from the Network Manager.

Control Over Wireless

For some applications it is desirable to utilize wireless networks for control, to reduce cost and because the new wireless measurements meet installation requirements. WirelessHART is designed to support monitoring and control applications. In some cases the use of wireless measurements in closed-loop control may improve process operations.

To insure control performance over wireless, feedback should be executed 4 to 10 times faster than the process response time (process time constant plus dead-time). See Figure 5.

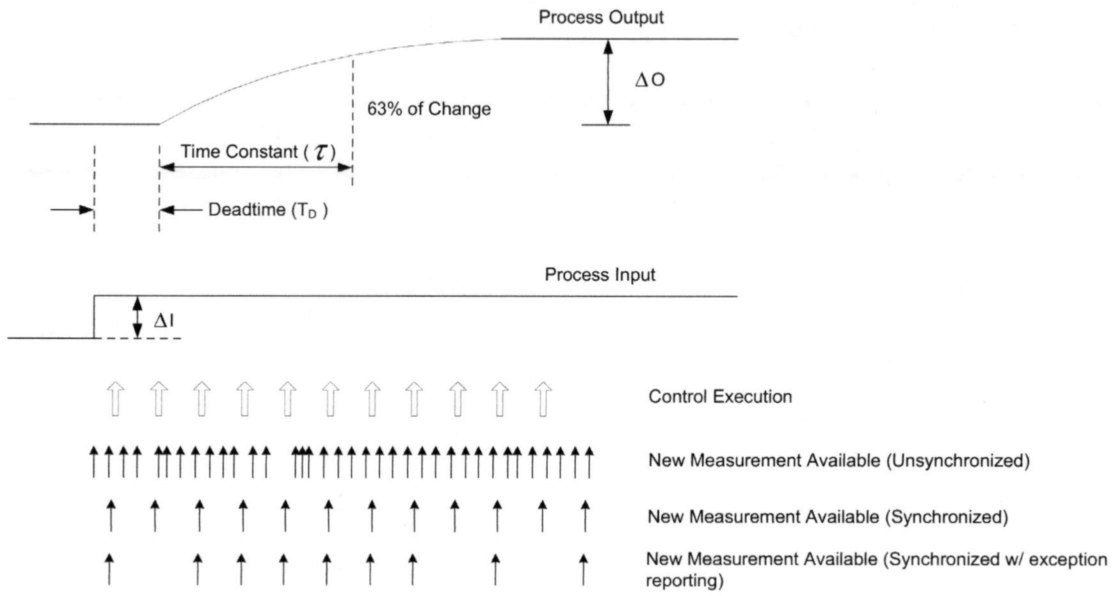

Figure 5. Meeting Control Performance Over Wireless.

Since the measurement system is often unsynchronized with control, measurement values are sampled much faster than the process responds - in many cases multi-loop controllers over-sample measurements by a factor of 2 to 10 times. To reduce transmitter power consumption, it is desirable to minimize how often a measurement value is communicated.

Burst Mode Communications

Process data needs vary as the process conditions change (Figure 6).

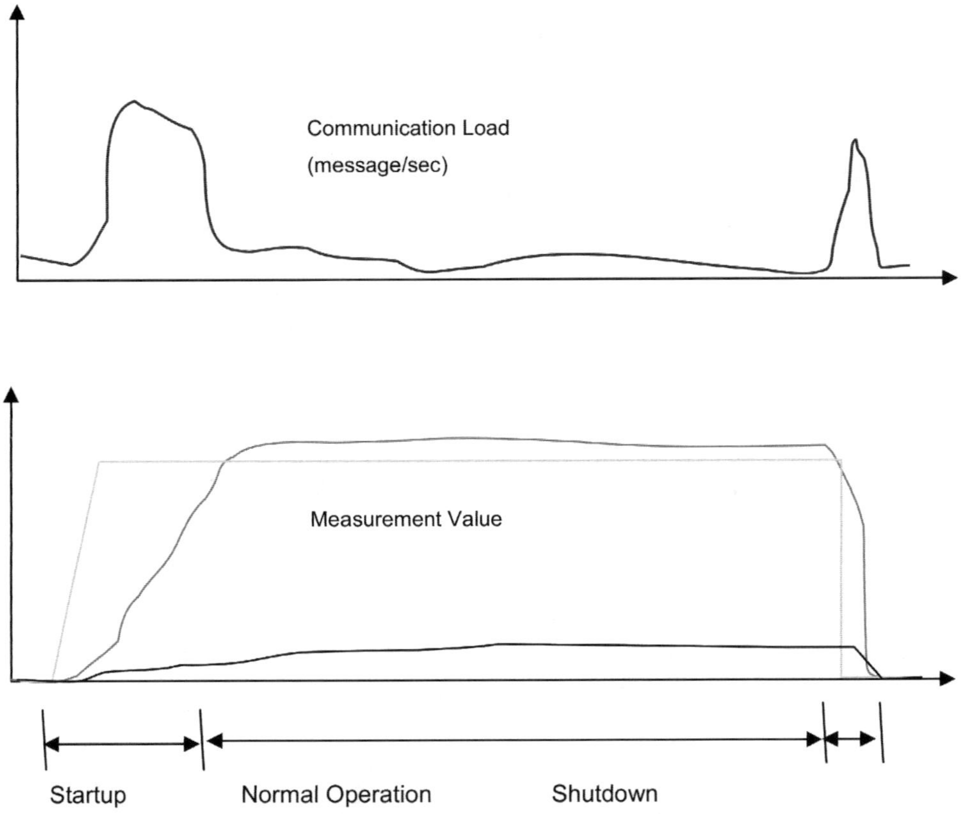

Figure 6. Process Automation - Optimized Communication.

Process values should be communicated based on change in value (dead-band exceeded) or after a timeout period (dead-band not exceeded). Communication is scheduled based on measurement scan rate, so communications on average will be much slower, e.g., 4X slower. The periods of maximum communication load must be accommodated, during startup/shutdown or when a significant process disturbance occurs.

If the magnitude of the difference between the new measurement value and the last communicated measurement value is greater than a specified resolution or if the time since the last communication exceeds a refresh time then the new value will be communicated.

Smart Updates: Window Triggered

Enhanced burst mode allows intelligent publishing of process information, allowing time and value criteria to trigger data publication, without polling by the host application. This enables efficient operation in wired or wireless networks, extending battery life in wireless applications.

Windowed triggered operation sets a dead-band around the last process value. Data is published when the dead-band is exceeded (i.e. process change) or when the maximum time between updates is exceeded. This is illustrated in Figure 7.

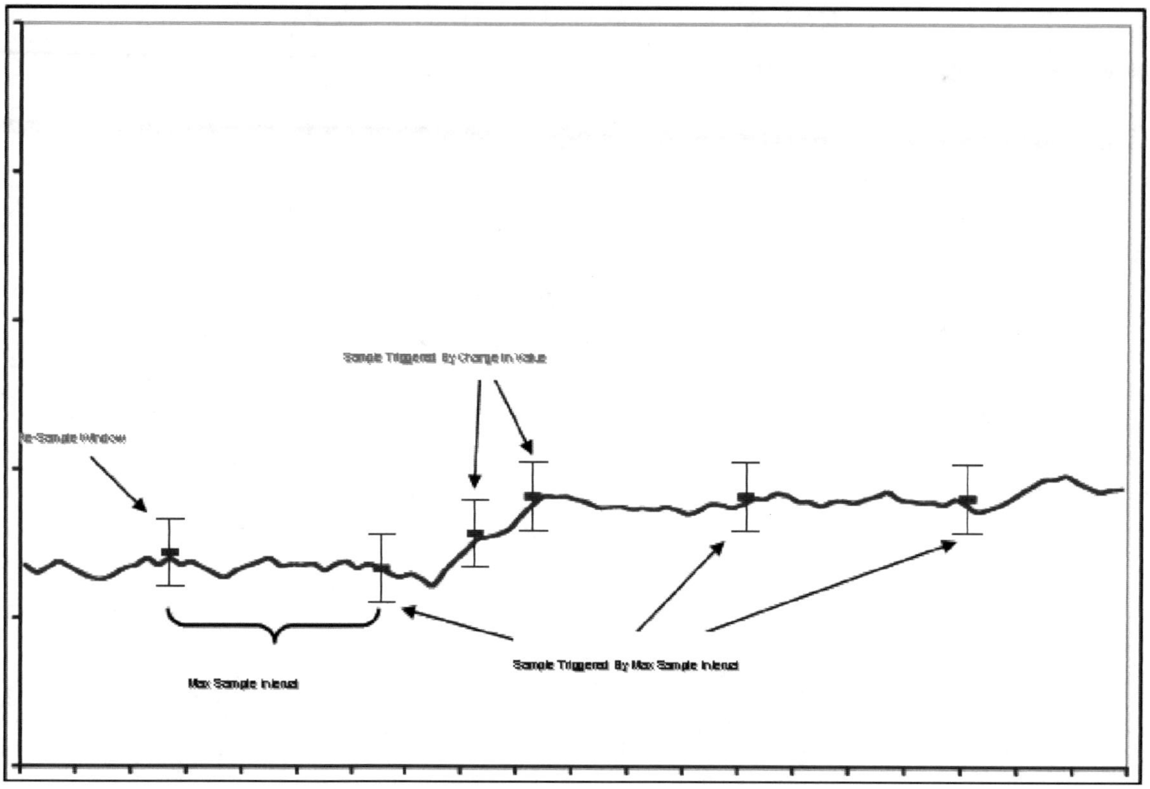

Figure 7. Window Triggered Updates.

Smart Updates: Threshold Triggered

This uses a threshold trigger based on the absolute process value. Given the minimum and maximum sample rates, data is published using the minimum sample rate until threshold crossing where the maximum sample rate is used. This is illustrated in Figure 8.

Publishing can trigger on a low-high transition, as shown in Figure 8. Conversely, it can also trigger on a high-low transition.

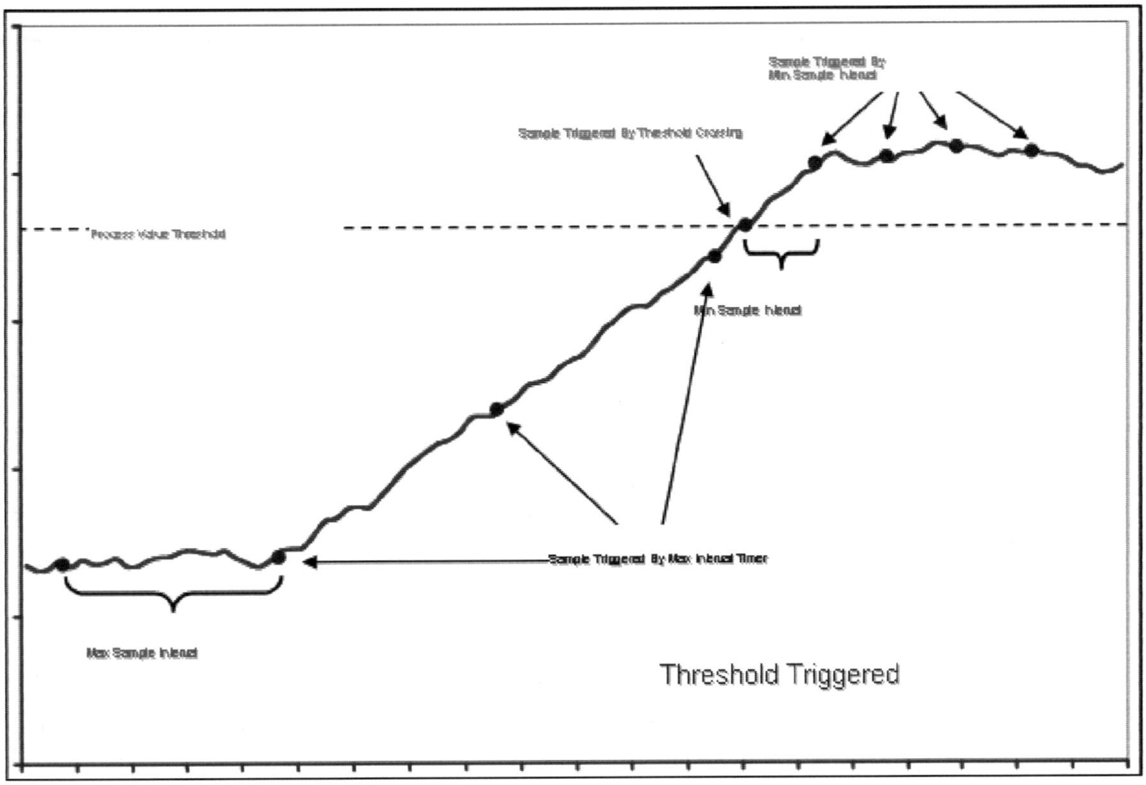

Figure 8. Threshold Triggered Updates.

Communication Results From the Field

In practice both transmitter and receiver need to be on for a communication, but if there is nothing to transmit the transmitter won't be scheduled for a transmission. The receiver only has to stay on for 1ms, then it can then go back to sleep.

Test results using these triggering mechanisms have indicated that 70% of measurements can be handled in 1 hop, with the remaining 30% being split evenly between 2 and 3 hops. The communication load is reduced by 90%.

But What About Control Performance?

Needless to say, the PID algorithm must be modified to accurately account for the intermittent feedback from the triggering mechanisms. Call this new algorithm "PIDPLUS". A performance comparison with a normal PID with synchronous update based on an unchanging control interval is presented in Table 1.

Scenarios / IAE / PIDs	Unreliable Inputs: Setpoint Change	Unreliable Inputs: Process Disturbance	Unreliable Outputs: Setpoint Change	Unreliable Outputs: Process Disturbance
PID	372	366	196	388
PIDPLUS	169	333	190	267

Table 1. Comparison of Control Performance for PID vs. PIDPLUS.

In Table 1 the numbers are integrated absolute error (IAE) values. The larger the number, the poorer the performance.

The performance of the PIDPLUS is as good as or better than the comparable PID algorithm for all cases studied.

The PIDPLUS algorithm and methodology in whole are the subject of a patent pending.

Start Anywhere and Grow the Wireless System

Figure 9 illustrates wireless flexibility and scalability. The small image on the left shows a tank level application, a common starting point for a wireless system because of remote locations and associated costs of implementing a wired solution. Subsequent applications can be added as shown, all being automatically linked via the self-organizing network.

Figure 9. Grow the Wireless Network As Needed.

Summary

Wireless technology is the game changer in many industrial automation scenarios, including refining. The reliability, security, and cost-effectiveness of this technology is well-utilized in our ever-changing economic environment. Business drivers for wireless are lower installed cost, incremental measurement for process improvements, and new field diagnostics.

References for Additional Reading

Martin, G. D., "Smarted PdM", PlantServices, October, 2008, pp. 49-51.

Martin, G. D., "Wireless Networks Improve Refinery Operation", Hydrocarbon Processing, March, 2009.

37g Smart Instruments Provide Advanced Diagnostics and Early Event Detection to Improve Reliability and Availability of Refineries and Petrochemical Units

Ravi Kant, Emerson Process Management, 12001 Technology Dr., Eden Prairie, MN 55344

The focus of this paper will be to present advanced industry oriented solutions in refining and petrochemicals that utilize the unique capabilities of smart instruments. Predictive intelligence, advanced diagnostics, wireless solutions and other latest functionalities embedded in the smart instruments can be utilized for "Early Event Detection" of developing operational problems and provide solutions that improve reliability and availability of operating units. Some examples include applications for early detection of flooding in distillation columns, catalyst transfer problems in FCC units, flame instability detection in direct fired heaters, detection of high tube skin temperatures in ethylene crackers, and condensation detection in compressor inlet lines. The paper will also present results of some field trials that have been recently conducted. The preliminary results are quite encouraging and are expected to provide significant benefits that include operational enhancements, prevention of unit shutdowns, and improved profitability.

How to Maximize Return on Simulation Investment*?

Laurie Wang[1], Sanjoy Saha[1] and Rajat Basu[2]

Key words: Process Design, Simulation, UniSim Design, Lifecycle Solution

Introduction

Simulation software is used widely for process design to ensure safety, reliability and efficiency of plant operations, as well as for troubleshooting and debottlenecking of real operation challenges to reduce operational risk and improve productivity.

The purpose of the presentation is to demonstrate how an investment in simulation software that can provide comprehensive coverage throughout the entire lifecycle of a plant, from Front End Engineering Design (FEED), to operator training, to operation optimization can be utilized for maximizing the returns. The returns of the simulation investment are typically achieved through reduced downtime, improved product yield and quality.

A case study will be shared in this presentation to illustrate how a simulation tool helped to turnaround a challenging situation in a production facility.

Challenges

Have you used a process model for troubleshooting a real operation problem? Have you ever built steady state and dynamic models separately for a same process? Have you ever wondered whether the outcome of these models is consistent? Has your company ever had the need to train plant operational staff for start-up and shutdown procedures and other upset conditions for a new or existing facility and spent millions for the training service? Have you seen somebody design the control scheme for a new process without a process model? Does it sound familiar to you that somebody starts a Real Time Optimization project by looking at the thermodynamic and physical properties of the system in order to build a base process model from scratch? If you answer YES to any of these questions, you are not alone. Companies invest millions these days in process design, simulation, operator training, process optimization, and advanced process control, all aimed at excellence in design and operation. Quite often, however, the groups in charge of these individual tasks don't get chance to share their work, forming unfortunate silos within companies. It causes unnecessary repetition of the modeling effort and increased demand on resource. By taking advantage of the commonality of process models between design and operation excellence, and thus minimizing such inefficiency in modeling effort across the lifecycle of a plant, companies can achieve a much better return in their simulation investment. That is exactly where many business leaders are turning their attention to these days, especially in the economic downtime.

UniSim Lifecycle Solution Vision

A key enabler for sharing process modeling information across disciplines is a set of well integrated tools which allow users to seamlessly expand models at each stage of the plant lifecycle to meet their specific needs. That's where UniSim Lifecycle Solution comes in. Supplemented by service availability, this solution offers

---

* NOTICE: All statements, information and data given herein are believed to be accurate and reliable, but are presented without guarantee, warranty or responsibility of any kind, express or implied. Statements or suggestions concerning possible use of our products are made without representation or warranty that any such use is free of patent infringement, and are not recommendations to infringe any patent. The user should not assume that all safety measures are indicated herein, or that other measures may not be required. The values presented herein are typical values and are not to be interpreted as product specifications. User assumes all liability for use of the information and results obtained.

[1] Honeywell Process Solutions
[2] Honeywell Specialty Materials

users three well integrated suites of products for process design, simulation, operator training and operation optimization.

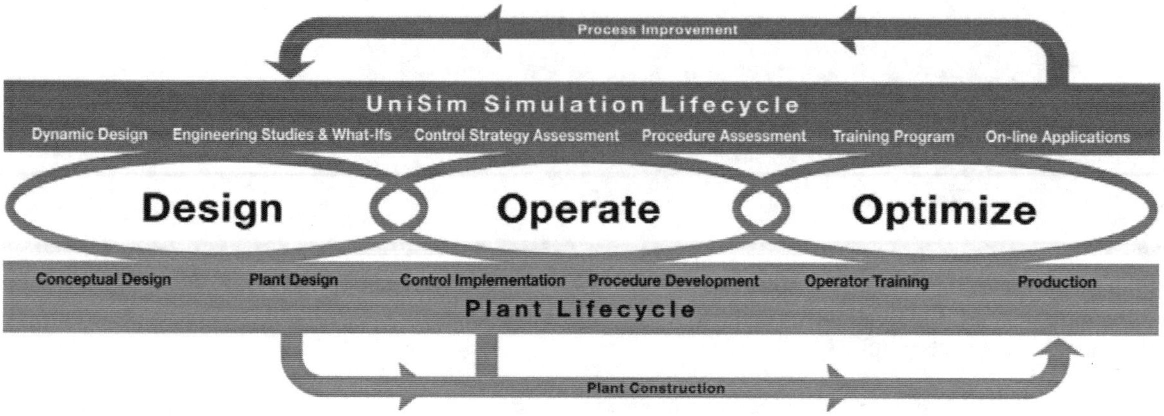

The **UniSim® Design Suite** enables steady-state & dynamic modeling for plant design, troubleshooting, process improvement, and what-if studies. The **UniSim® Operations suite** is used for operations improvement, capturing knowledge and sharing it through operator training. The **UniSim® Optimization Suite** ensures responsiveness to changes in business demands and process operations with linkages to real-time control/optimization, planning and scheduling. The models established in the design stage can be leveraged into the Operations Suite and Optimization Suite for operator training and on-line and off-line optimization. So once created, the models can be used repeatedly and extended for multi-purpose applications.

In addition to reduced cost for model building effort, the lifecycle solution also brings the benefits of model consistency across the applications, with reduced amount of repeated data entry which saves time and minimizes transcription errors. The consistent thermodynamic models across applications also ensure consistent, meaningful and realistic simulation results.

Let's take a closer look at how a simulation model grows and expands through the lifecycle of a plant.

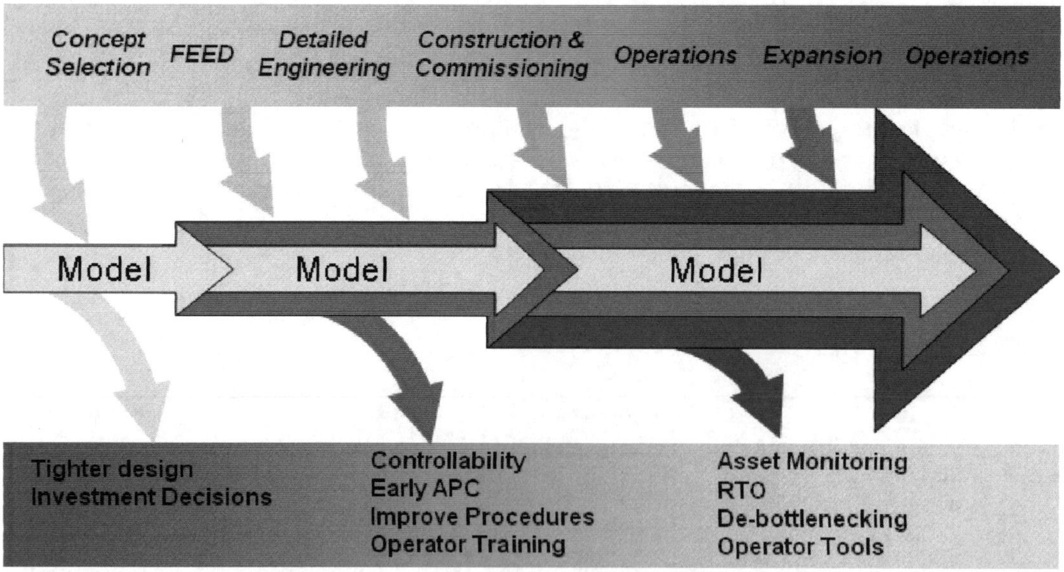

In the early stage of the design phase, there are a lot of unknowns to process engineers. However, they do know what will go into the process and what products need to be produced. So the model at this stage will include properly established thermodynamic groundwork. It sets the foundation for a model which accurately represents the expected behavior of the designed process. Combined with the chosen production technologies/process, the model will help for tighter and optimized process design, and will impact investment decisions.

As the process moves to the Front End Engineering Design phase, the model will be enhanced to include all the process information in terms of feed specs, product requirements, major equipment required (such as separators, tanks, heat exchangers, pipelines and columns), as well as their operating conditions. Equipment related information is added in during the detailed engineering phase. With sufficient information, the model can be expanded from steady-state to dynamics mode to allow for controllability study (control scheme), procedure improvement, as well as early adoption of APC application

As the model being further expanded through the construction phase, it can be used in parallel to build operator training tools for the designed process at the specific facility. As a result, operators learn what to do during start-up, shut-down, normal and abnormal conditions before commissioning. What this means is that operations staff will be prepared before your plant comes on-line, avoiding training on a live plant. The training models can also be used to validate the design in abnormal conditions.

It gets better. All that work that went into your design models and preparing your operations can be brought forward for use in process optimization. Through simulation tool's tighter integration with Advanced Process Control (APC) solutions, one can help assure that APC comes online faster and more efficiently and with less risk. Through the entire lifecycle of a plant, the model grows and expands going through different phases, while retains its consistency (data, thermo, etc.) throughout the whole process. Thus a simulation tool that supports lifecycle solution helps to pave the way for your plant to achieve operational excellence on day 1.

Process model for operational excellence (a case study)

Not all operating companies have readily available simulation models for all their processes. But for those do, they have a distinct advantage. The truth is that it is not always easy to troubleshoot a real operational problem on site. Sometimes, it is not even practical and that is where a process simulation model can make unique contributions.

Here is a case study from a Honeywell production site. The facility is used to produce multiple high value specialty chemicals. It involves complicated processes and several distillation trains, one of which is shown below.

An urgent market need came up in spring 2008 for a high value product. To satisfy the pressing market demand, high productivity of this product needed to be maintained. In other words, the plant has all the incentive to maximize its production for this valuable product.

In the distillation train, there were three columns (#1, #2, #3) particularly involved in separating this product from impurities.

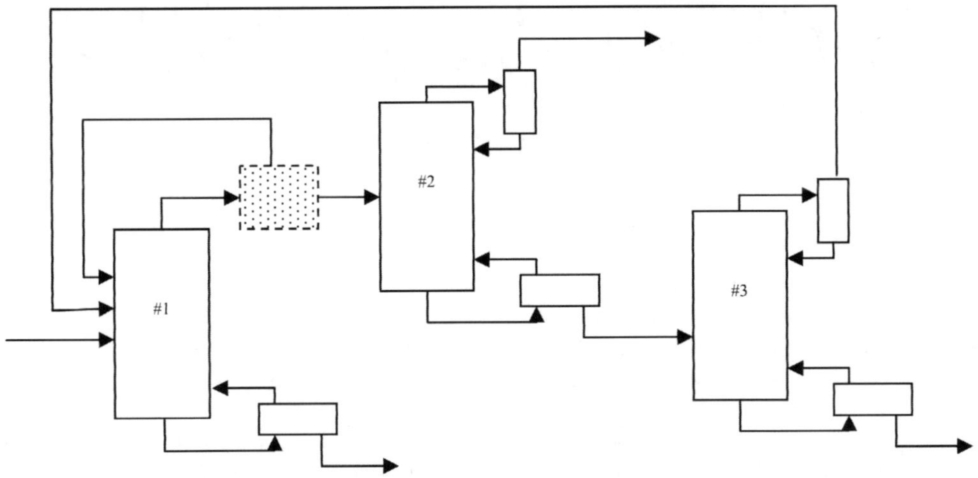

The problem production engineers were facing was that they could not meet required product purity (99.8%). To make the matter worse, this production needed to stop soon due to scheduled equipment maintenance. This means that time was money, literally. It was critical for them to resolve this problem most urgently in order to maximize the production with desired product purity, in order to meet the market demand and maximize the profit.

So the production engineers tried different things on site: such as increasing feed rates till the system reaches its limit; venting the mixture of the product and impurity from column #2, and purging the mixture of the product and impurity in column #3. The consequence of such practice was that the required production volume and product purity were met at the cost of the yield loss of this high value product. Such trial and error operation was carried out for quite some time without success. The reality was that with an operation of this nature, there were so many factors that could have impact on the productivity and product purity. But it was impractical to try them all on the product line.

Meanwhile, the problem was reported to the technical service team. The experts on the team used their process model for trouble-shooting. This model built in UniSim Design realistically represented the real behavior of the production line. What-if study was conducted for different scenarios within a day. Everything pointed to having higher reflux rate in column # 1 for in-spec product.

With this knowledge, plant engineers opened up a by-pass valve to get reflux in column #1 to the desired level. The results made everybody smile. The plant was able to produce in-spec product at a high production rate.

A picture is worth a thousand words...For a two week period, the level of impurity in product had remained high. On April 3, as the reflux rate was increased, the level of impurity dropped immediately as shown in the figure.

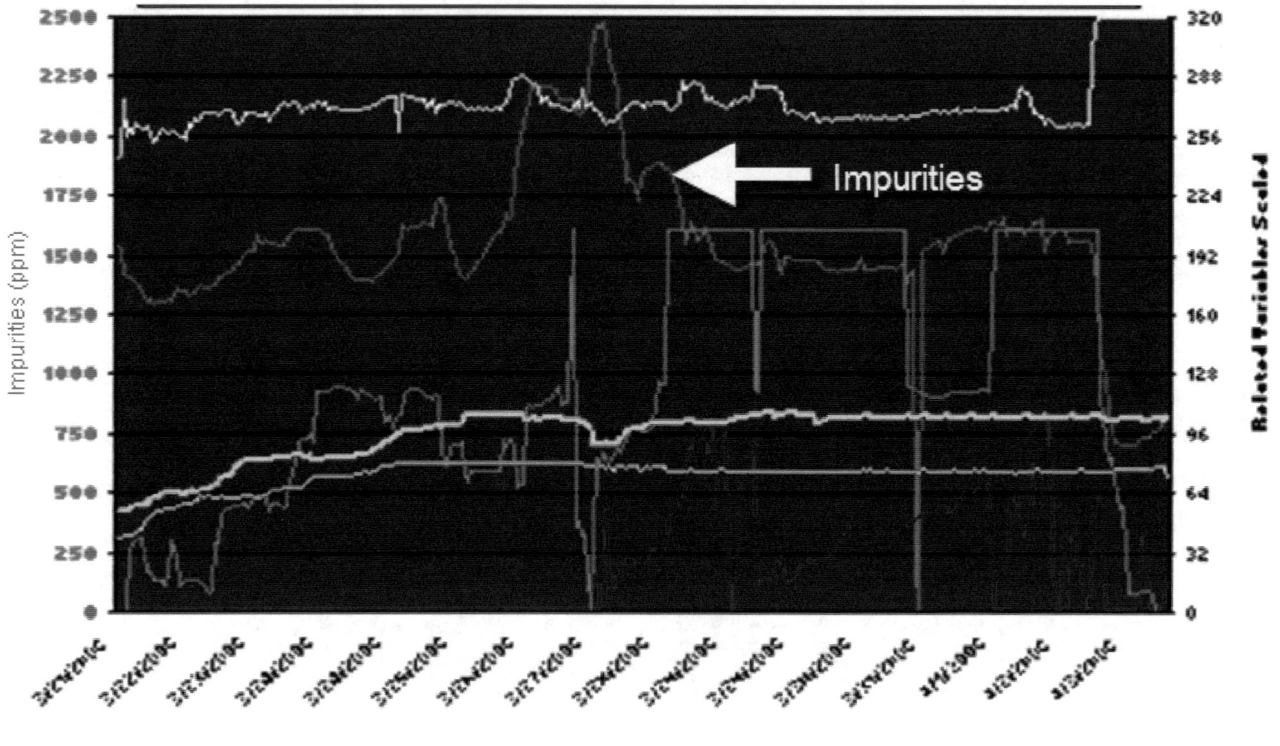

Simulation Lifecycle Solution Showcase

Here is a story where the customer has taken full advantage of the lifecycle solution of a simulation tool, from process simulation, to operator training system, to advanced process control. To protect customer IP, the screenshots used here are taken from UniSim Design sample files, the concept illustrated would be the same.

To start with, a steady-state simulation case was created in UniSim Design. Its results were validated against the plant data available to the customer, to ensure that the model accurately represents the true behavior of the plant operation. In the following step of expanding this model for dynamic applications, valves were added to the flowsheet to ensure proper pressure-flow relationship across the flowsheet. Regulatory controllers were also added to the model, so that process variables can be manipulated to achieve the desired operating conditions. The unique and consistent platform of the simulation tool enabled users to make easy transition from steady-state models to dynamic applications. Again, once completed, the dynamic model was validated with the plant disturbances and was found to represent the disturbances and their mitigation back to steady state condition very well. A key to utilizing models to your benefit is to have validated models available to engineers at all times and model development should be started right from the design phases.

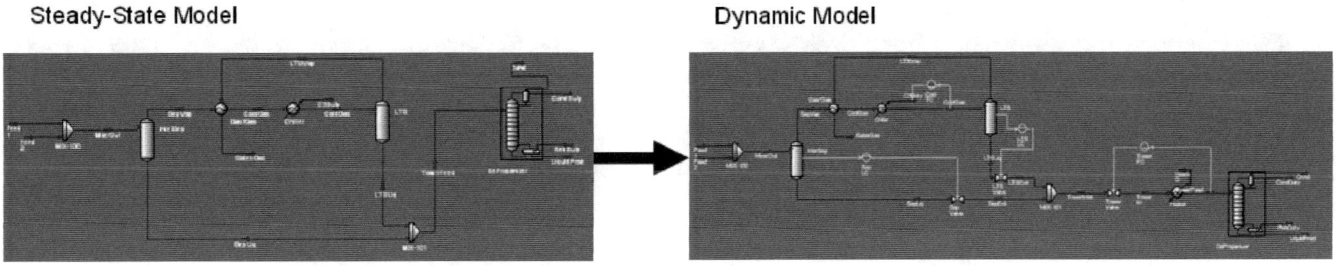

Turn a steady-state model to a dynamic model

- Add valves to ensure proper pressure-flow relationship across the flowsheet
- Add controllers to control
    - Flow rate
    - Temperature
    - Liquid level
    - ....

The dynamic model was then used to create an Operator Training System (OTS) for the same facility. The Operator Training System built in UniSim Operations contains three elements:

- UniSim Engineering Station, where the rigorous UniSim Design dynamic model is located. The model provides an accurate calculated response to control changes or disturbances.
- UniSim Instructor Station, where instructor can effectively design and manage the training required. It allows the trainer to invoke a training sequence, such as start-up, shut down, planned scenarios, or any number of upsets, including instrument failure, equipment malfunction, etc. The instructor station also holds the operator performance as comparative trends.
- UniSim/Experion Operator Station, where operators respond to the scenarios set by instructor through UniSim-Experion DCS Interface. With the same interface and identical look/feel as they see in a plant control room, the operators will have a real hands-on experience.

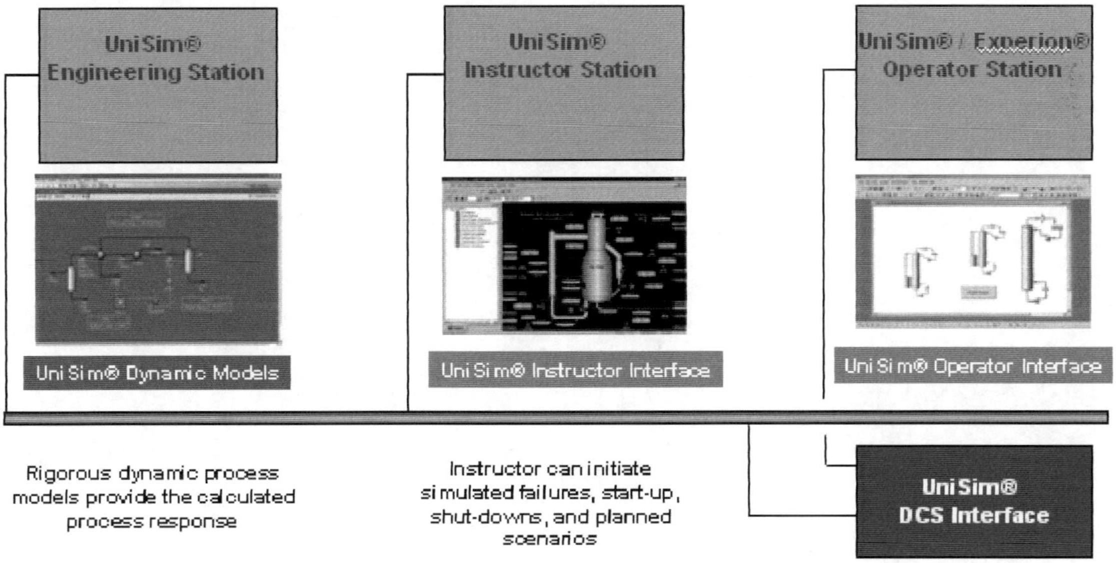

Once the OTS model was in place, the company used the OTS to

1. Provide unstructured training in which operator becomes familiar with how the display looks, how the plant behaves, and how the controls perform. Again the beauty of this system is that there is no risk to failure, and it is endlessly repeatable, fun, and builds confidence.
2. Provide structured training: This addresses specific situations, such as start-up and shut-down, known recurring issues, and the impact of instrument or equipment failures.

3. Help the engineers and control room operators to develop better procedures.
4. Help design and test out new control strategies. In fact, it has motivated the control engineers from the company to step up to a new level.

This customer has been using Honeywell's UniSim technologies for less than three years and has reported an estimated financial benefit of $700K/y.

In Summary, a systematic approach towards simulation investment can result in significant financial benefit. The key is proper selection and utilization of the tool with emphasis on the value chain of the plant lifecycle.

38b Combined Use of High and Low Fidelity Models for Refinery-Wide Simulation

Hiren Shethna, Simulation Product Development, Aspen Technology, 200 Wheeler Rd, Burlington, MA 01803

Use of models to make refinery-wide decisions, such as optimize crude purchase is sought after. In the last few years, process simulators have made significant advances to enable process engineers extend scope of process simulations beyond individual process units. It is however true that process simulators have usually been thought in terms of rigorous first-principle models to model process units – which in turn is expected to bring more accurate results for design. Scaling up these first principle models across the refinery to build refinery-wide models to model refinery operations requires significant amount of investment in terms of personal as well as computing infrastructure. This therefore results in very limited use of such models to make refinery-wide decisions by process engineers. There have been developed approaches which combine use of simplified as well as rigorous first principle models to make refinery-wide decisions. This approach includes using first principles models where necessary and filling in gaps with lower fidelity models. Key challenges in such an approach involves maintaining consistency at boundaries where once transitions from a high fidelity to a lower fidelity model and vice-versa. Some examples will be demonstrated within this paper.

AIChemE Spring National Meeting 2009

27th April 2009 – Tampa

DRIVING EXCELLENCE

Stream Routing Optimisation Using Rigorous Refinery-Wide Simulation

Simon Rogers

Contents

- Introduction
- Methodology
- Case study
- Summary

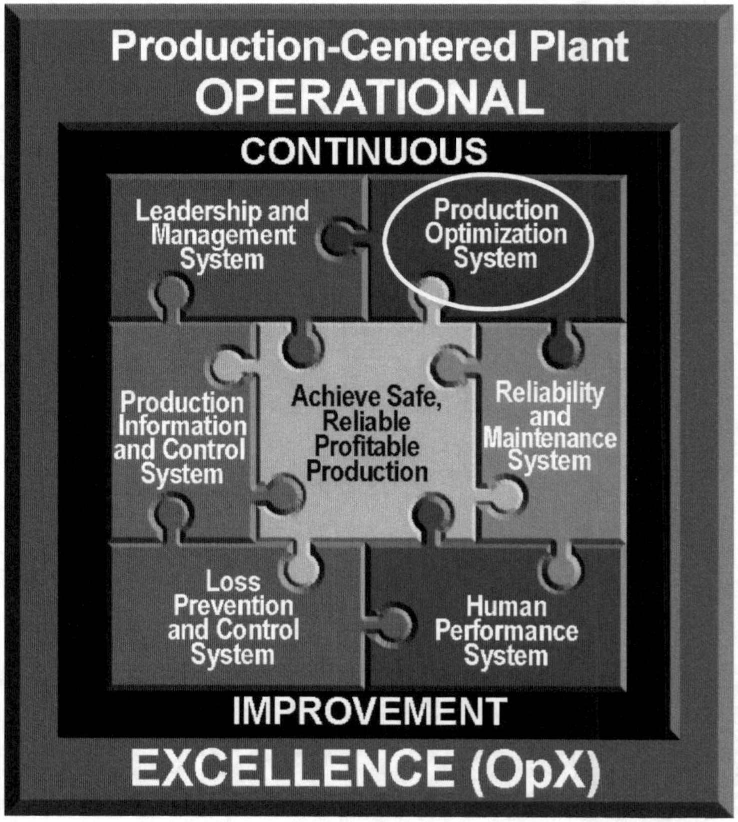

DRIVING EXCELLENCE

Challenging Environment

- Heavier Feeds
- More Diesel
- Less Emissions
- Tighter Specs

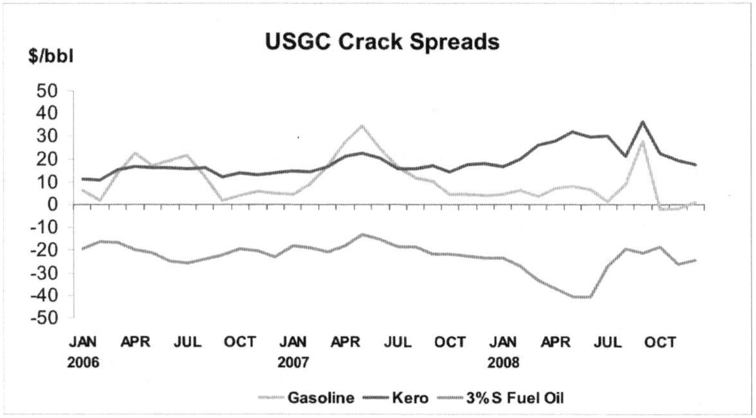

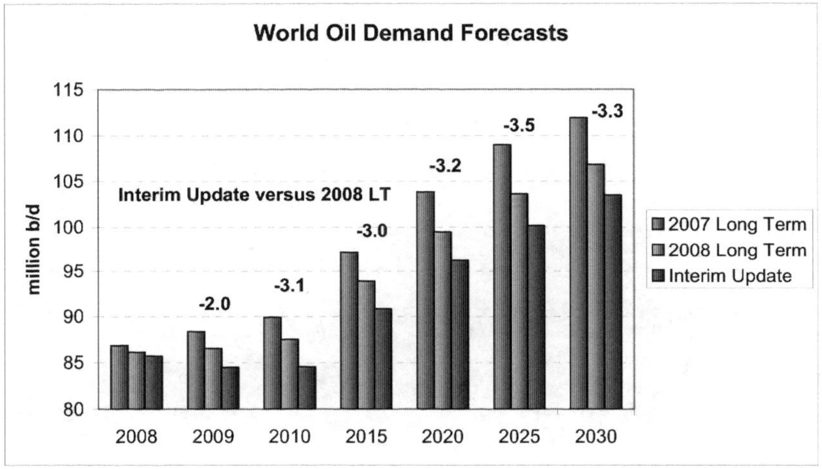

KBC
DRIVING EXCELLENCE

Molecular Management

- Hundreds of decision variables
  - Feed quality
  - Product rates
  - Key operating parameters
  - Blending recipe
- Thousands of constraints
  - Safety / Equipment
  - Environmental
  - Specifications
  - Demand
- Millions of species
- Highly non-linear

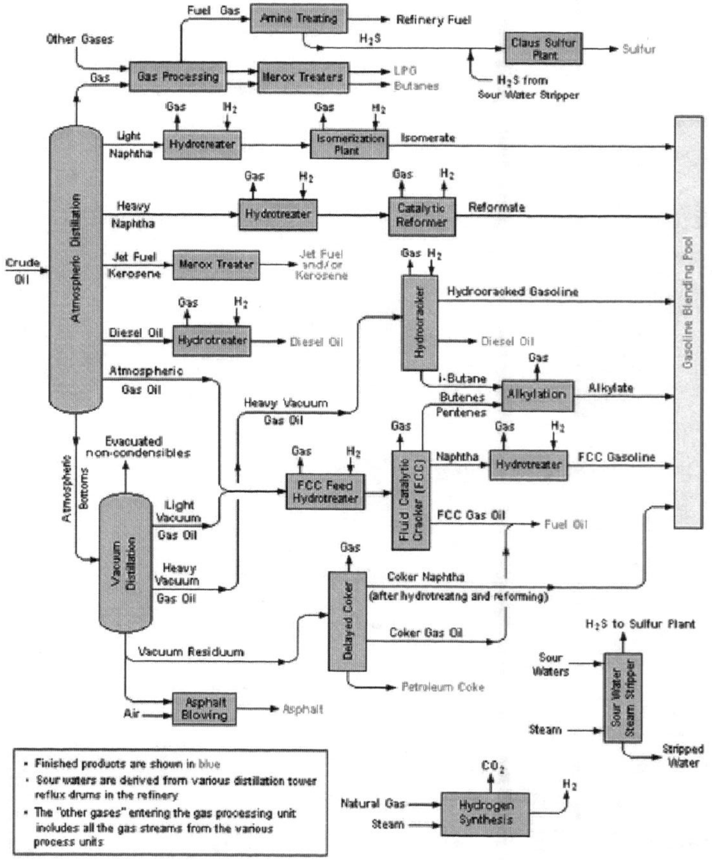

© Milton Beychok

KBC
DRIVING EXCELLENCE

Current Practice

- Monthly plan using LP model
  - Empirical (risky to extrapolate)
  - Mainly linear
  - Poor fractionation
  - Few operating parameters
  - Developed using simulation
  - Limited number of properties
- **Limited cross-function communication**

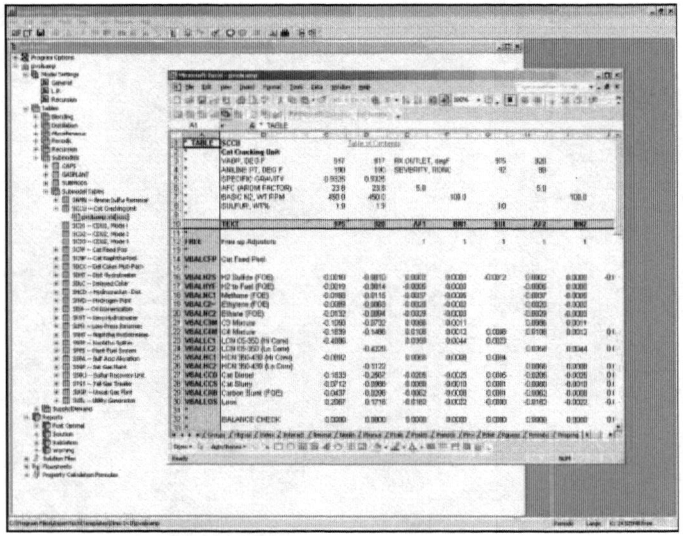

© Beijing Zhongyou AOTO Science and Technology Limited

"The World is more complex than it seems, ... our knowledge less certain than we think" – Wm. Theodore de Bary

KBC
DRIVING EXCELLENCE

People

- Vision
- Sponsorship
- Optimisation Team
  - Dedicated staff
    - Broad refinery knowledge
    - Simulation expertise
  - Part-time staff
    - Operations
    - Planning
    - Engineering
    - Maintenance
    - Control and instrumentation
    - Lab

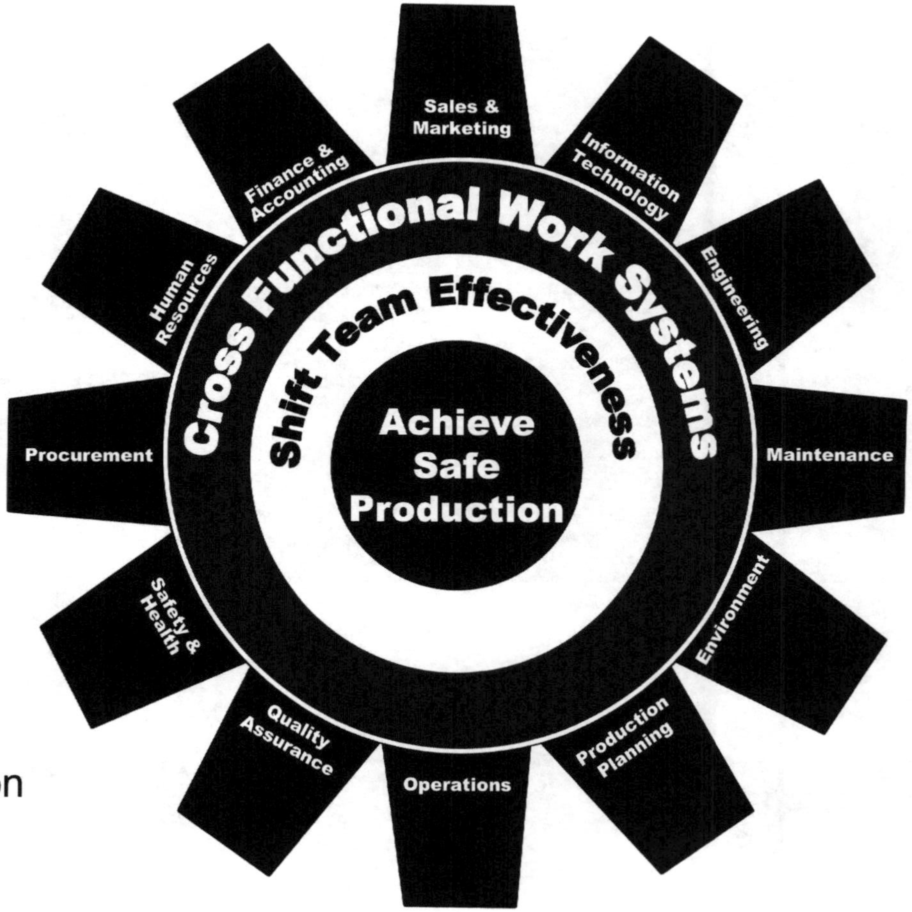

KBC DRIVING EXCELLENCE

Tool – Petro-SIM

- First principle thermodynamics
- Proprietary hydrocarbon characterisation
- Kinetic reactor models
- Intuitive GUI
- Calibrated

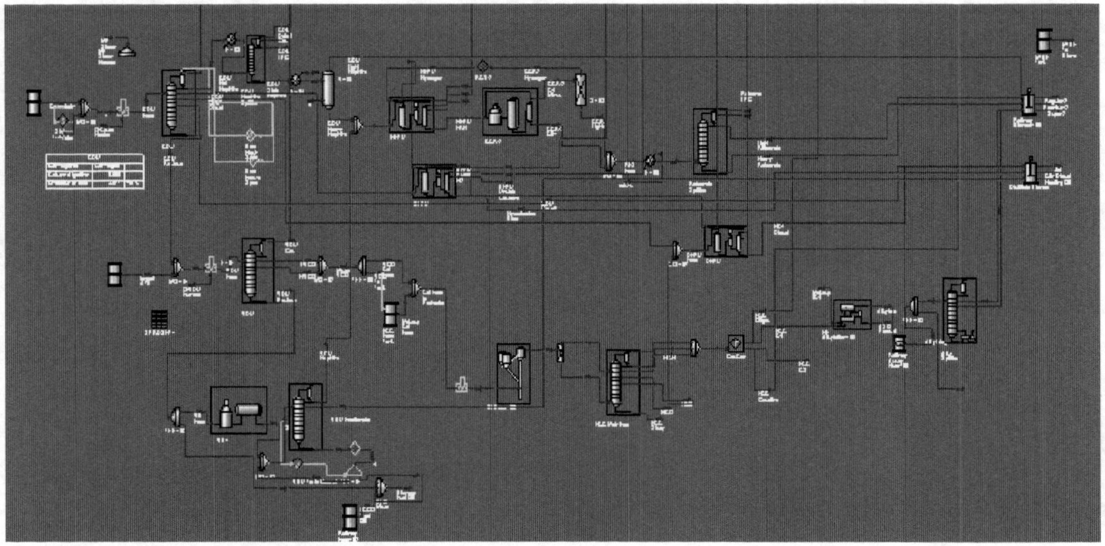

KBC
DRIVING EXCELLENCE

Achieving Operational Excellence

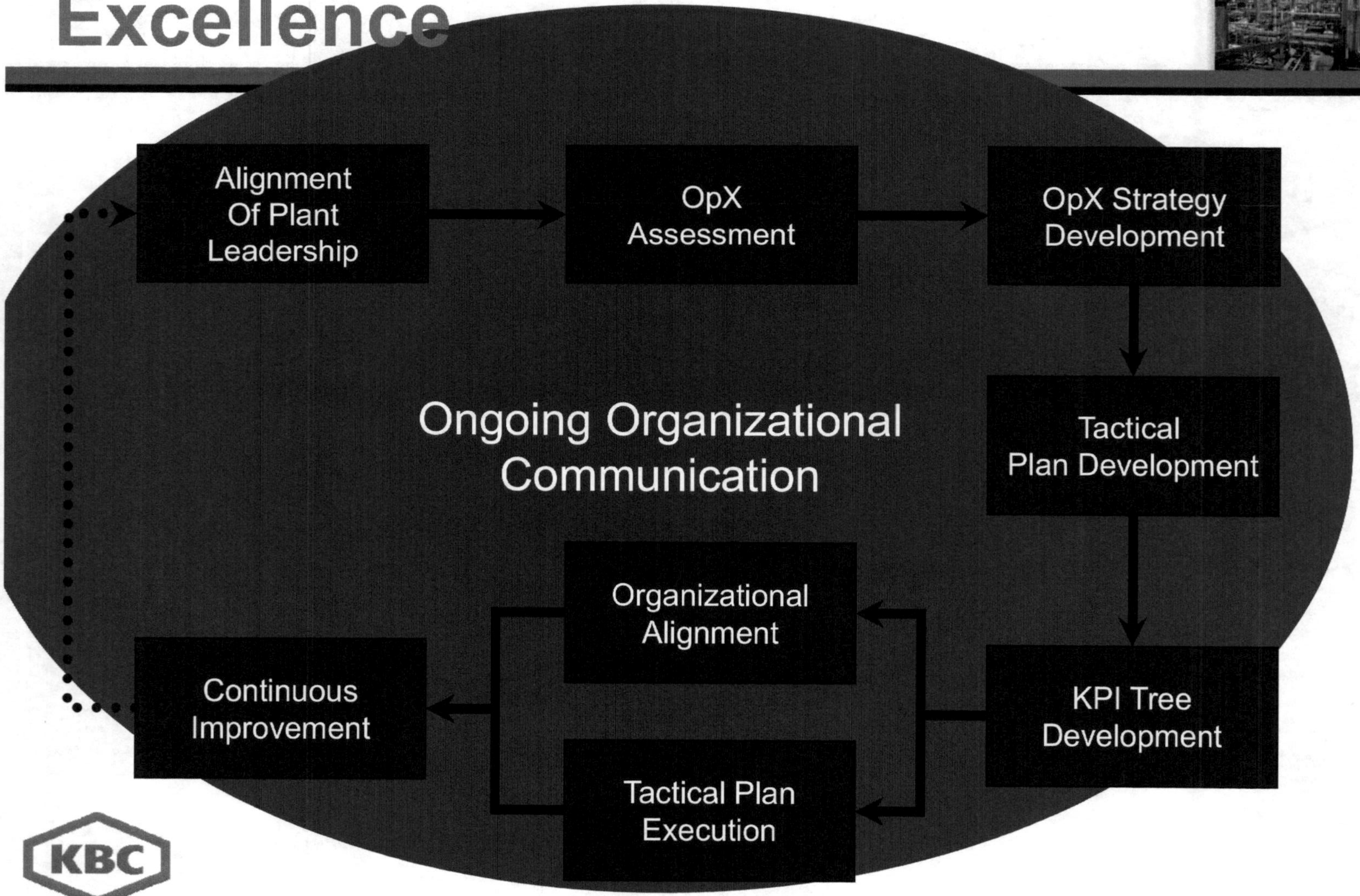

Production Optimisation System

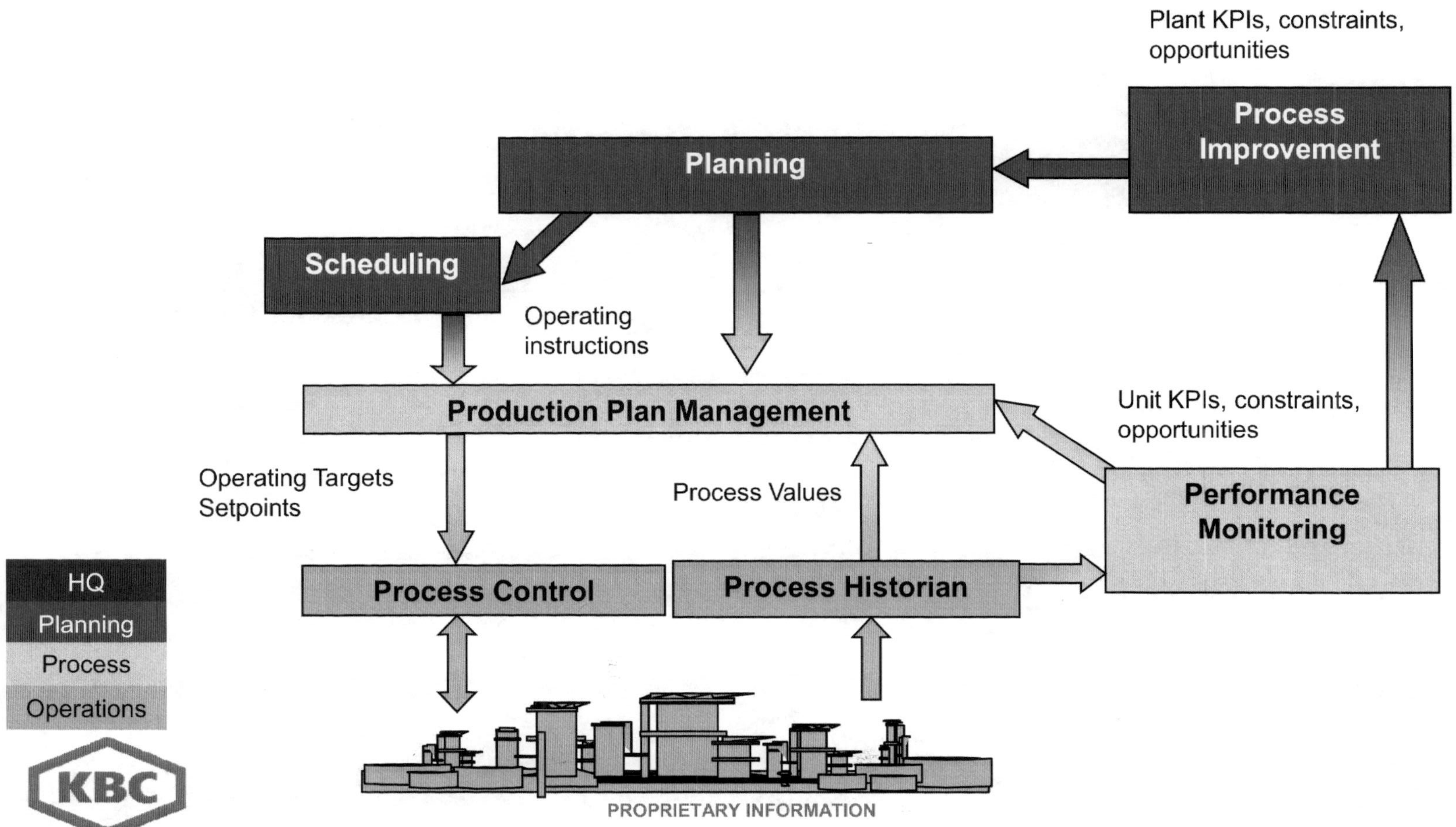

KBC
DRIVING EXCELLENCE

Process Improvement

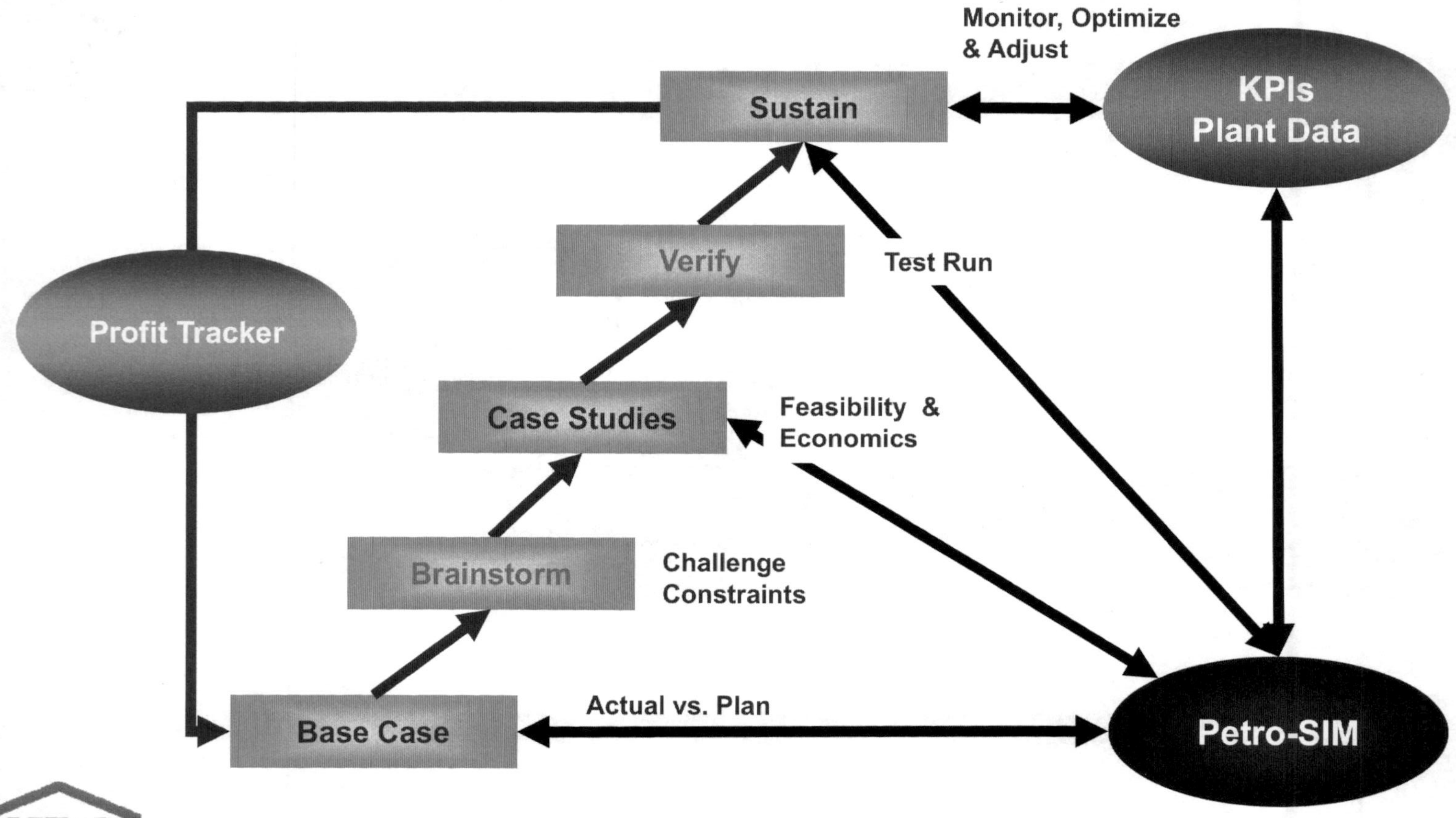

DRIVING EXCELLENCE

PROPRIETARY INFORMATION

Contents

- Introduction
- Methodology
- Case study
- Key success factors
- Summary

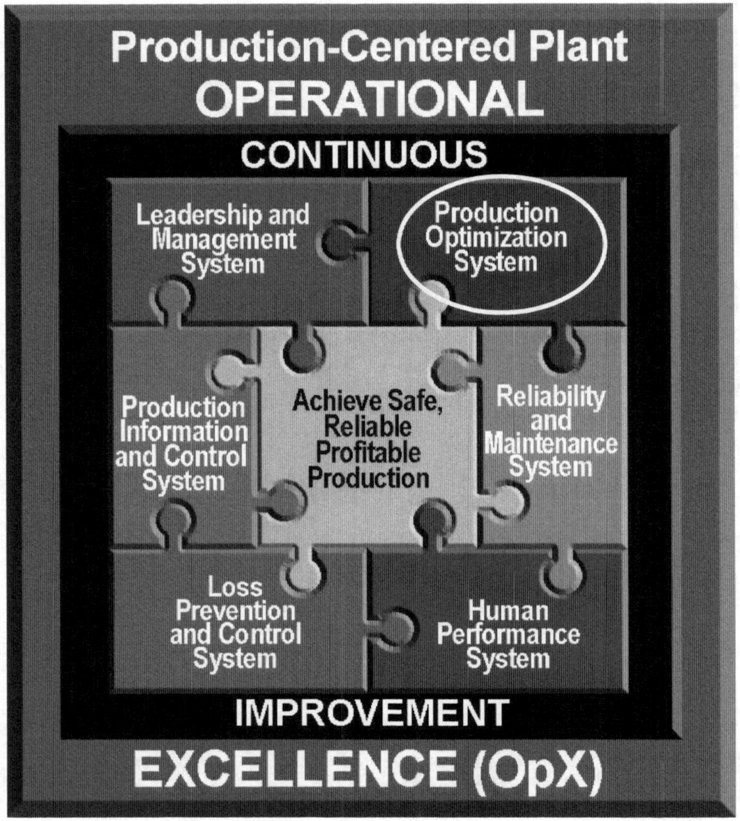

DRIVING EXCELLENCE

Asian Refinery

- Complex (reforming, RFCC, HCR, MHCR, Coker, Olefins)
- CDU limited by Coker (Hydrocracking and Hydrotreating)
- <u>Sour crude</u> most profitable
- 10 MTPA
- Optimisation program

	Ident	Impl
Total, MM$/yr	179.34	72.97
CDU	14.88	4.66
CCR	32.73	16.59
FCC	17.09	10.45
DCU	26.08	17.93
Reroute	81.45	19.47
Others	7.11	3.89
Total, c/bbl	240.38	97.81

KBC DRIVING EXCELLENCE

VDU Deepcut

- Maximize FCC / HCR feed
- HVGO to FCC from HCR
- Sour crude to refill Coker
- Monitor for cracking
  - DP, TMT, gas make, bromine
- Diesel and naphtha increase
- Gasoline decreases

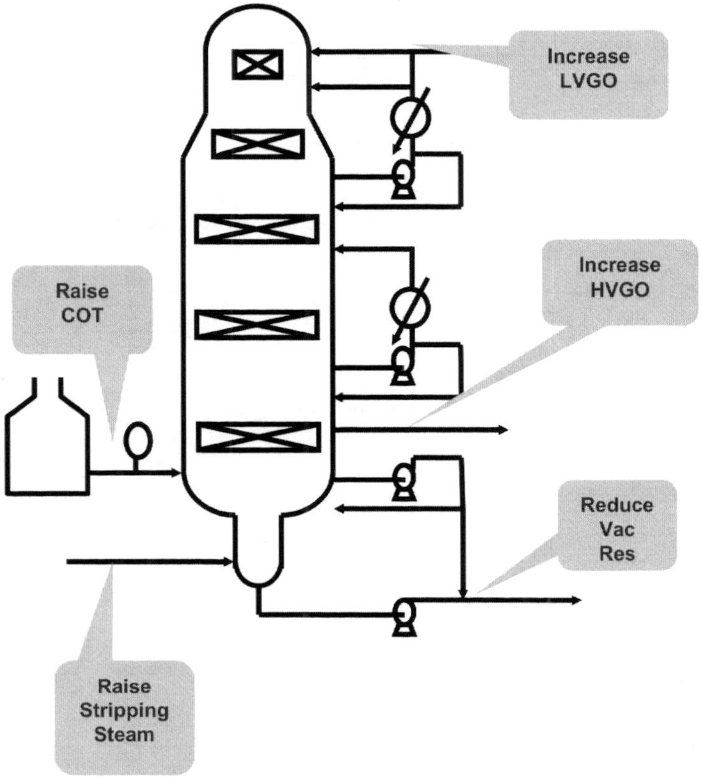

$12.7M p.a.

DRIVING EXCELLENCE

CGO from FCC to HCR

- FCC feed less olefinic
  - Conversion increases, less coke
- HCR feed heavier
  - Severity increase to maintain Tail Oil yield/quality
  - Improve filter operation
  - Improve coker fractionator washing
- Diesel & Naphtha increase
- Gasoline & Fuel decrease

$12.1M p.a.

KBC DRIVING EXCELLENCE

Sweet VR from PDA to FCC

- Lower PDA feed rate
- FCC
  - Conversion decreases
  - Higher feed rate up to sulphur limit
  - Higher catalyst makeup
- Less fuel oil
- More diesel

$7.4M p.a.

DRIVING EXCELLENCE

LVGO from HCR to MHC

- MHC in cracking mode (part-time)
- Sour crude to consume additional HCR capacity
- Additional sour VR to PDA
- Switching procedure
- Diesel and Naphtha increase
- Fuel oil decreases

$22M p.a.

KBC DRIVING EXCELLENCE

Lube Extract from FCC to HCR

- Uses spare HCR capacity to remove aromatics from FCC feed
- Upgrades FCC coke to diesel
- Monitor HCR diesel (BMCI / Cetane)
- Piping change

$7.6M p.a.

KBC DRIVING EXCELLENCE

FCC Slurry to Coker from Fuel

- Upgrade fuel oil
- Dahlman filters to remove catalyst from slurry
- On-line spall of coker to increase throughput
- Piping change

$10.3M p.a.

KBC
DRIVING EXCELLENCE

Rerouting Summary

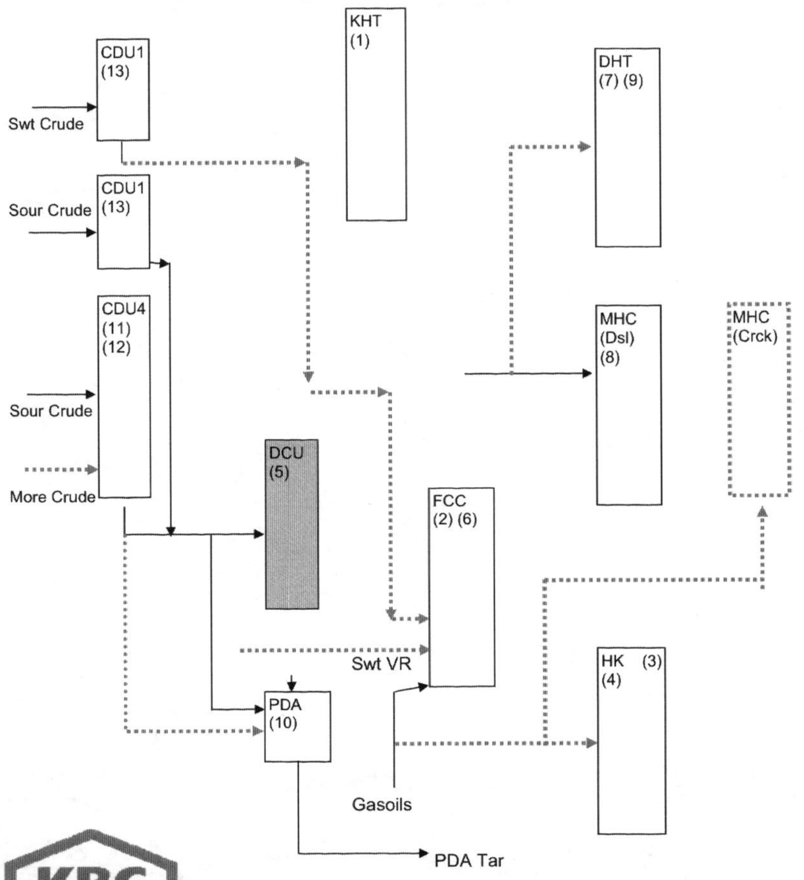

1. Shutdown KHT in winter
2. ALL sweet VR to FCCs
3. Maximize Coker GO to HK
4. Extr 2 and 3 to HK
5. Extr 4 to Coker
6. Minimize gasoil to FCCs (+ MHC cracking mode)
7. Use DHT/MHC for Diesel
8. LCO/Coker Diesel to MHC from DHT
9. Remove coker naphtha from DHT feed
10. Balance sour VR between PDA and Coker
11. Increase VDU4 Cut-2
12. Deepcut VDU4 to FCC
13. Block Operate CDU1

KBC
DRIVING EXCELLENCE

Overall Refinery Effects

Product	Delta t/h
Naphtha	58.8
Gasoline	-8.4
Jet	0
Diesel	51.3
Fuel Oil	-23.3

Margin Improvement: $81.5M p.a.

DRIVING EXCELLENCE

Additional Benefits

- Faster and better response to change:
  - Crude supply
  - Regulations
  - Specifications
  - Demand
- LP accuracy
- Process design / engineering
- Target setting / optimisation

KBC DRIVING EXCELLENCE

Key Success Factors

- OpX Methodology
  - Change management
  - Optimisation work process
- Tools
  - Rigorous simulation (Petro-SIM)
- People
  - Dedicated staff
  - Team work
  - Refinery and simulation knowledge

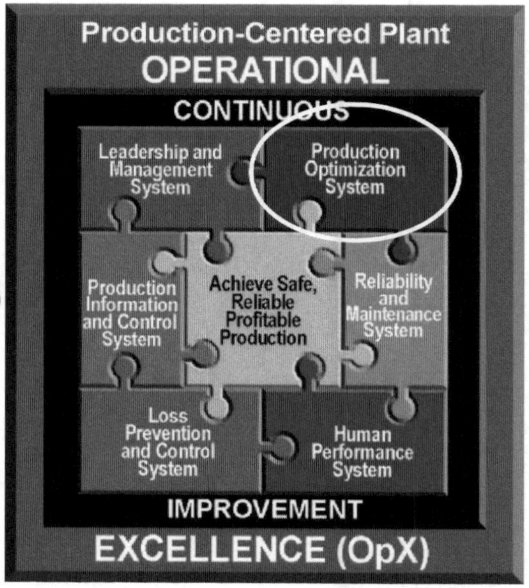

DRIVING EXCELLENCE

Summary

- Challenging environment & problem
- Molecular management requires
  - Skilled team
  - Rigorous refinery-wide model
  - Operational Excellence
- Benefits > $1 / bbl

"He who would learn to fly one day must first learn to stand and walk and run and climb and dance; one cannot fly into flying." - Nietzsche

KBC DRIVING EXCELLENCE

MODEL-BASED SCALE-UP IN PRACTICE

Alejandro Cano, Process Systems Enterprise, Inc., Cedar Knolls, NJ
Sean Bermingham and Costas Pantelides, Process Systems Enterprise, Ltd., London, UK

Introduction

This paper describes a recently developed methodology for Model-Based Scale-up and its application to industrial practice, based on Advanced Process Modeling technology.

Advanced Process Modeling (APM) is a technology that uses high-accuracy mathematical models of process equipment and phenomena to provide predictions of the performance of the equipment over a wide range of conditions, including different equipment sizes.

The predictive power of an advanced process model results from the combination of first-principles descriptions of the chemistry and physics taking place in a system, with the validation of such models by fitting adjustable parameters to observed data such as laboratory, pilot plant, or operational measurements.

Once validated, advanced process models can be used to optimize the design and operation of medium-scale and large-scale equipment, limiting the conditions to be tested in a pilot plant only to the most promising alternatives. This greatly reduces the cost and risk of scale-up.

Model-Based Scale-Up Methodology

The model-based scale-up methodology practiced at PSE consists of two major components. First, high-fidelity first-principles models of equipment at the laboratory and pilot-plant scale are used to provide a quantitative characterization of the key physical phenomena (e.g. chemical reaction, heat and mass transfer etc.) taking place in the process. A model-based model-targeted approach to experimentation is employed which combines advanced parameter estimation techniques and model-based experiment design to arrive at accurate values of the model parameters with the minimum number of experiments. Both the models and the approach are implemented in PSE's gPROMS® modeling environment (www.psenterprise.com).

The second element of our model-based scale-up methodology is the use of hybrid models which combine gPROMS models of the key phenomena identified at the first stage with computational fluid dynamics (CFD) models of mixing and fluid flow. This allows the accurate description of full industrial-scale equipment, allowing its design and operation to be optimized.

Model validation through model-targeted experimentation

Advances process models are constructed from equations describing the physical and chemical phenomena that take place in the system. These equations usually involve parameters that can be adjusted to make the model predictions match observed reality. One example of such model parameters are heterogeneous reaction kinetic constants, which need to be adjusted for each different catalyst formulation.

Model-based scale-up requires that the process of fitting adjustable model parameters to laboratory or plant data provide two main results: (a) the values of the parameters for which the best fit is obtained between model predictions and observations, and (b) a statistical evaluation of the parameters obtained, including confidence intervals and a measure of cross-correlation among parameters.

Parameters with tight confidence intervals are necessary for successful model predictions at scales different from those for which the data were collected. This requirement dictates the types of experiments that need to be performed. Most experimentation work conducted for the purpose of obtaining a predictive model is *model-targeted experimentation*, where the purpose of the experiments is not to optimize any aspect of the equipment of process, but to derive the most accurate possible model. The predictive model can then be used for optimization of various aspects of the design.

Model-targeted experimentation is thus very different from traditional pilot plant experimentation. In traditional pilot plant experimentation, conditions are usually selected that mimic the conditions in which the target commercial plant will operate. For pilot plant reactors, this means selecting conditions that will yield high raw material conversion and high selectivity to the desired product. These conditions are not necessarily the most informative conditions from the perspective of generating a truly predictive model. For instance, kinetic constants derived from low conversion experiments are much more reliable than kinetic constants derived from high conversion experiments, in which the final conversion may be limited by thermodynamic equilibrium. For complex reaction networks, experiments in which intermediates or even the main product itself are fed to the reactor can greatly reduce kinetic parameter uncertainties for secondary reactions.

Another feature of model-targeted experimentation is that the most informative data may come from small scale experiments designed to be free of mixing and heat transfer effects normally found at larger scale experiments. A well-designed model-targeted experiment will involve well-defined conditions, such as perfect mixing and constant temperature. This helps to isolate the influence of the parameters of interest and means that values obtained for parameters will be valid for scale-up. Once validated models of microscale phenonena (such as intrinsic kinetics) have been obtained, these models can be embedded in equipment models in which heat and mass transfer limitations are added.

Hybrid Computational Fluid Dynamics / Advanced Process Models

There are many examples in the chemical process industries where non-ideal mixing or complex geometry are significant factors affecting the performance of industrial scale equipment. Equally, there are many processes (such as crystallization and polymerization) that can only be described accurately using very complex, detailed models of the underlying physical and chemical phenomena.

Modeling mixing effects is a core capability of computational fluid dynamics (CFD) software such as ANSYS Inc.'s Fluent® package. While Fluent is capable of handling a range of reacting flows on its own, there is a practical limit on the number of reactions and species that can be considered. PSE's gPROMS modeling system, on the other hand, can model processes of arbitrary complexity, but is generally limited to geometries that can be described using bounding planes with simple descriptions in Cartesian, cylindrical, or spherical coordinates.

By combining APM and CFD technologies it is possible to obtain a powerful predictive capability within practical use of computer resources in situations where highly complex chemical phenomena, sometimes requiring the solution of large numbers of implicit equations, and fluid dynamics both play an important part. Typical examples are the scale-up of crystallizers and reactors in which complex chemistry takes place.

Several approaches for combining APM and CFD models are described in Reference (1). The common characteristics of all the approaches is that they combine CFD descriptions of fluid dynamics and mixing with APM descriptions of all other phenomena, such as crystallization and heterogeneous reactions. Two of these approaches are particularly relevant to model-based scale-up: (a) multizonal approach for weakly coupled systems and (b) surface linked models.

<u>Multizonal (multicompartment) approach for weakly coupled systems</u>

There are systems such as crystallization or fluidized beds reactors in which the fluid dynamics have a strong effect on the rates of conversion, but the conversion rates have relatively weak effects on the fluid dynamics.

In this approach, the hundreds of thousands or millions of cells in a CFD domain are mapped into several zones. Each zone is represented on the APM side as a single well mixed compartment. This allows highly complex APM models to be used to describe the physics and chemistry; for example, a typical crystallization well mixed compartment might use thousands of equations to represent multidimensional population balances including terms for nucleation, growth, and attrition for each population. The multizonal approach makes it possible to include a high degree of process detail, and hence accuracy, while still obtaining reasonable computational times (2,3).

PSE's gPROMS Multizonal software automates the creation, linking and execution of the combined APM and CFD models in a way that is transparent to the user. The following steps are involved (Figure 1):

- The CFD domain is divided into one or more zones, taking into accound physical boundaries. The domain can also be further subdivided using CFD results. In Fluent, this can be done by marking cells based on important flow field characteristics, such as temperature
- The user decides how many well-mixed zones to use on the APM side to represent each of the zones of cells in the CFD model.
- The Multizonal tool automatically computes the volume of each well-mixed zones, as well as the flowrates between all combinations of well-mixed zones, and sends the information to gPROMS.
- gPROMS solves the zone-network model, using detailed chemistry and physics

- Optionally, gPROMS can send back to the CFD model updated values of the density and viscosity of the fluid in each zone, which are then used to revise the solution to the fluid dynamics problem. New interzonal flowrates are computed and sent back to gPROMS for a new solution of the zone network model, iterating until convergence.

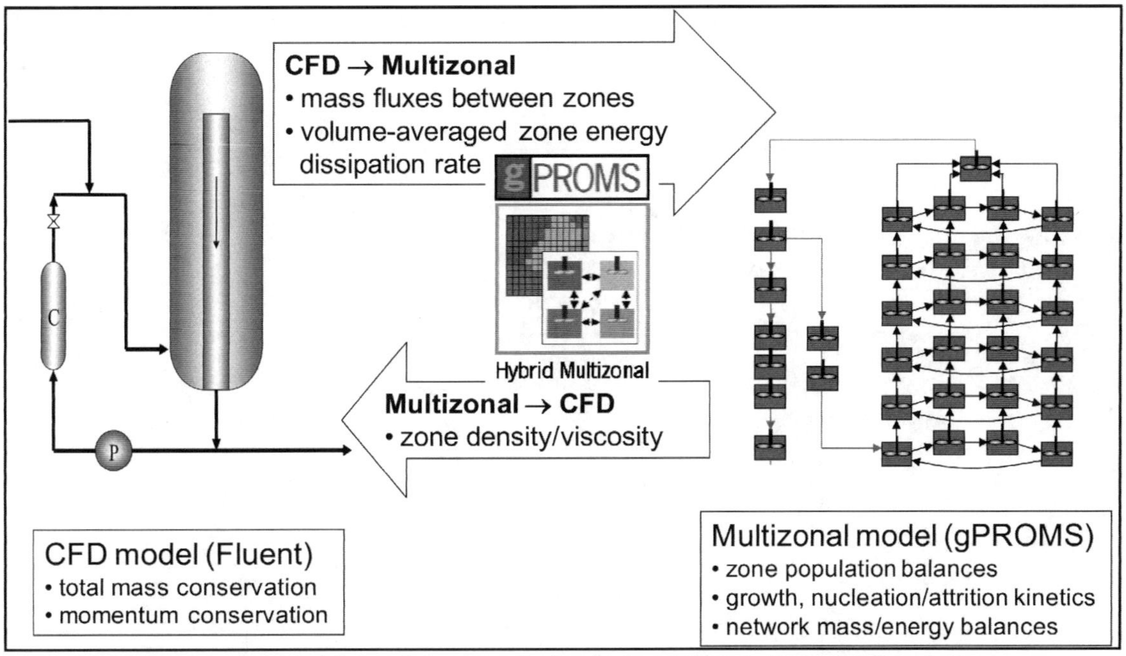

Figure 1. Multizonal approach for hybrid CFD/gPROMS simulation

<u>Surface linked models</u>

In this case, the APM and CFD models are linked at a heat exchange surface. The processes on one side of the surface are represented by APM models, while a CFD model is used to represent the fluid dynamics on the other side of the surface.

In a typical application, the hybrid approach is used to construct a high fidelity model of a multitubular reactor. These reactors typically consist of thousands of tubes filled with catalyst particles embedded within a shell in which a cooling fluid is used to remove the heat of reaction. Each of the tubes may be subject to a different cooling environment, due to local variations of coolant temperature and velocity. These differences in local cooling can lead to large differences in tube performance. Small changes in design, such as the dimensions and location of baffles, can have a dramatic effect in tube performance variability.

Rather than including each of the thousands of individual tubes in the CFD model of the reactor shell, the tube bank is represented as a porous medium. Several locations within the tube bank (50-200 are typical) are chosen for detailed whole-tube simulations. For each representative location selected by the user in the tube bank, the CFD package sends to gPROMS the values of the coolant temperature and the heat transfer coefficient along the entire surface of the tube (Figure 2). gPROMS uses this information as boundary conditions for the solution of a detailed model of a single tube packed with catalyst, which may include complex physics and chemistry, such as a large reaction network and intraparticle diffusion resistances. gPROMS calculates the heat fluxes at the tube surface, and this information is sent back to the CFD package. PSE's multitubular interface uses the heat fluxes calculated by the solution of the tube model at the various representative locations to interpolate a heat source term throughout the CFD domain. This updated heat source information is used by the CFD package to generate a new solution to the flow problem, which allows new values of fluid temperature and heat transfer coefficients at tube surfaces to be calculated and sent back to gPROMS. The calculations are repeated until convergence is achieved.

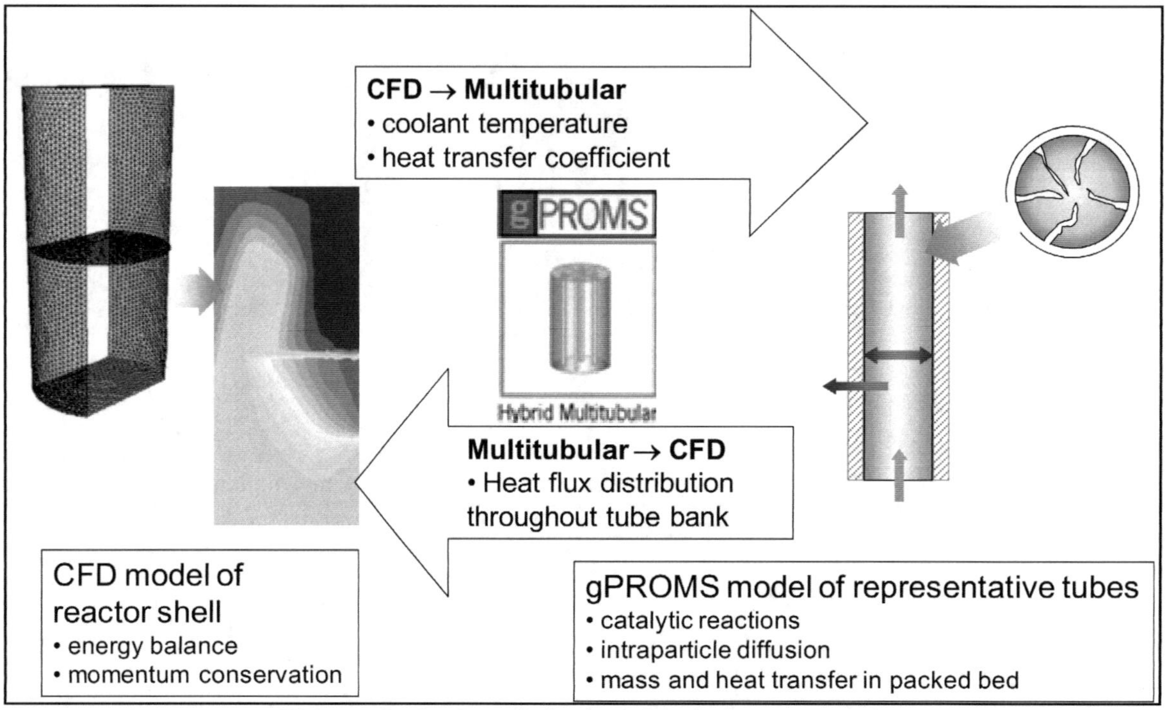

Figure 2. Surface linked multitubular approach

Industrial Practice Examples

We describe the above methodology as practised in several industrial reactor and crystallizer design projects. Two case-studies drawn from PSE's experience are be used to illustrate specific aspects of the tools and methods.

Scale-up of multitubular reactors

The starting point for multitubular reactor scale-up projects at PSE is our Advanced Model Library for Fixed Bed Catalytic Reactors (AML:FBCR). This model library contains first-principles models of several elements of a multitubular reactor:

- Two dimensional fixed bed models, including:

- psedohomogeneous catalytic bed (in which concentrations at catalytic surfaces are assumed to be the same as in the fluid outside catalyst particles)
  - Bed of catalyst particles (including the solution of intraparticle concentration and temperature profiles at each radial and axial location)
  - Bed of intert particles
- Coolant models (e.g. cooling jacket, fixed temperature coolant, inter-baffle compartments)

Partial differential equations representing continuity, momentum, species, and energy balances are solved in the models. Phenomena represented include:

- Axial/radial convection in tube-side fluid
- Radial dispersion of species
- Multicomponent diffusion across the solid particle boundary layer
- Intra-pore convection and multicomponent diffusion
- Catalytic reaction on pore surfaces
- Heat conduction within catalyst particles
- Heat transfer across particle boundary layers
- Radial conduction and dispersion of heat
- Heat transfer between packed bed and wall
- Radial heat conduction in tube walls
- Heat transfer between tube and coolant at the tube outer surface
- Convective heat transfer in coolant

Configuration of the library for specific projects requires the definition of the list of chemical components, adding reaction rate expressions, and adjusting two sets of parameters to obtain a validated model: (a) constants in the kinetic rate expressions, and (b) adjustment factors for heat transfer correlations. The preferred method for adjusting these parameters is the sequence shown in Figure 3:

1. Fit kinetic parameters by performing small-scale experiments. These experiments are designed to provide isothermal conditions. Experiments are performed at various inlet gas compositions and reactor temperatures. After the first few experiments are performed, the parameter estimation feature of gPROMS is used to obtain parameter values, together with a statistical analysis of the results. If the statistical analysis indicates that the parameter uncertainties are too large, more experiments are performed. The experiment design feature of gPROMS can be used to determine the combination of experimental conditions that are expected to provide the most information. This feature takes into account: (i) the model, (ii) the experiments already performed, (iii) the accuracy of the analytical instruments, and (iv) bounds on experiment decision variables provided by the user to solve a constrained optimization problem whose objective function is the minimization of the size of the joint confidence region on the parameters, and allows a reduction in the number of conditions that need to be tested with respect to traditional factorial design approaches. The experiment design feature can recognize, for example, when there is high uncertainty in the kinetic

parameters of a secondary reaction, and would then recommend feeding directly the component participating in this reaction to reduce the uncertainty.

2. Once a validated kinetic model has been obtained, the next step is to adjust heat transfer parameters. This is done with tubes of similar diameter to those under consideration for the commercial reactor design, although their length is typically shorter than the length of commercial reactors. Thermocouples are used to obtain bed temperature measurements are various locations along the tube. Experiments are performed at various reacting fluid velocities (even if they result in low conversion), as the objective of these experiments is to obtain accurate values for adjustable parameters in flow-dependent heat transfer correlations. Tubes of various diameters should be tested to discriminate between effective bed radial conductivity and the bed to wall heat transfer coefficient.
3. The result after successful completion of steps 1 and 2 is a predictive model of a single tube, which can now be combined with models of the cooling environment at the tube outer surface. Preliminary designs are obtained using the optimization feature of gPROMS on gPROMS-only models of the tube and the shell, which assume that all tubes in the shell see the same cooling conditions. The effects of tube-to-tube variations in cooling environment are examined by performing hybrid CFD-gPROMS simulations, using PSE's gPROMS Multitubular interface to CFD packages.

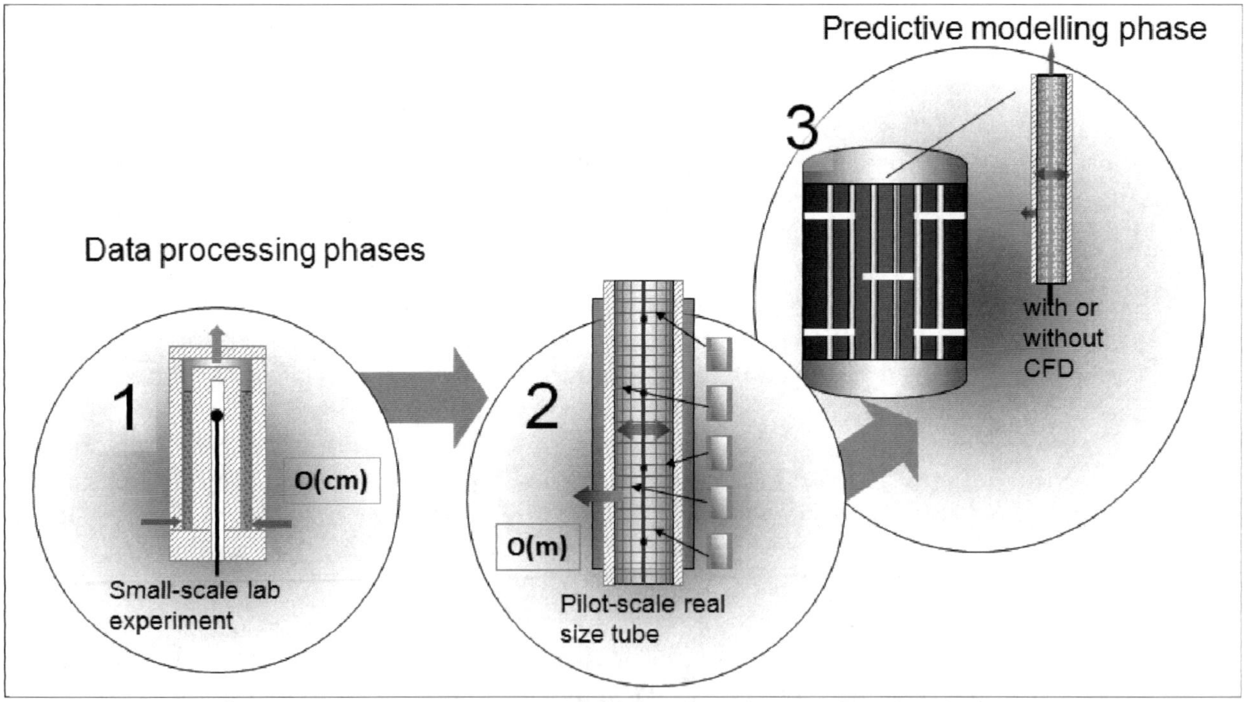

Figure 3. Sequence of steps in model-based scale-up of multitubular reactors

Case Study: Scale-up of multitubular reactor for highly exothermic reaction

A U.S. refinery needed to design a multitubular reactor for the production of a commodity chemical using one of its byproducts as raw material. For a number of reasons, including contractual timelines and delays in the procurement of pilot plant equipment, only three weeks of pilot plant operation were available before the reactor design had to be submitted to the reactor manufacturer. The project was complicated by the availability of several alternative catalysts.

PSE's approach to solve the problem consisted of the following steps:

(a) Review of the literature, to identify plausible chemical reaction mechanisms

(b) Evaluation of small scale experiments already performed on behalf of the client. Unfortunately, most of these experiments had been performed at high conversion, and only a handful of experiments could be used for kinetic parameter fitting.

(c) PSE designed a small set of low-conversion experiments whose purpose was to derive accurate kinetic parameters. Some of these experiments including co-feeding one of the reaction byproducts, which the literature had identified as a component that absorbed strongly on the catalyst surface.

(d) Parameter estimation was carried out throughout the four weeks available for pilot plant testing.

(e) Model predictions were used to design a dual bed structure: A moderately active catalyst was used in the first half of the reactor, while the most active available catalyst was used in the second half of the reactor. This set-up was tested in the last few days of pilot plant testing available, yielding higher than 98% conversion and higher than 99% selectivity, with moderate hot-spot formation, just as predicted by the model.

(f) The validated model was used to optimize the commercial reactor design. The reactor is now being built and is expected to start operation in early 2010.

Crystallizer design

PSE's Advanced Model Library for Solution Crystallization (AML:SC) is used as the starting point for building predictive models of crystallization equipment. This library contains models of crystallizers and crystallizer zones that can be combined to represent complex equipment. The models solve the partial differential equations that govern crystal population balances for an arbitrary number of crystal populations, as well as the continuity, energy and species conservation equations for species in the liquid phase. Crystal population balances include primary and secondary nucleation, growth and dissolution, attrition, and agglomeration, as well as exchange with other zones. Use of experimental data to fit parameters is facilitated by sensor models, which convert crystal number densities to quantiles, volume fractions, or chord length distributions.

Configuration of the library for specific design projects requires:

- Choosing the configuration of the crystallizer
- Providing physical property data, such as solubility, enthalpies, densities, and viscosities. Most of these data is readily available
- Fitting parameters for the nucleation, growth/dissolution, and attrition models.

Parameter fitting is usually done by collecting data in small scale (bench top) crystallizers under well mixed batch crystallization conditions. The parameter estimation of gPROMS can be used to process simultaneously data collected from different experiments, such as:

- Liquid phase concentrations measured during the batch
- Particle size distributions measured at the end of the batch
- Chord length distributions measured by on-line probes.

Our experience has shown that a key requirement of a predictive model is the formulation of the growth/dissolution step: separate consideration of mass transport across the boundary layer surrounding a crystal and the kinetics of surface growth provides a much better fit to experimental data than models using a single power law expression to represent crystal growth. This is an illustration of an important principle of model-based scale-up: the models must be based on first principles, and must include the most important physical and chemical mechanisms present in the system of interest.

After obtaining validated parameters for the various mechanisms (with tight confidence intervals) the models can be used to predict the performance of large scale equipment. We have used two approaches: (1) Representing large scale equipment within the gPROMS environment as several interconnected well mixed zones (with or without propellers), and (2) building a CFD model of the large scale vessel, and using the Multizonal technique to build automatically a network of well mixed zones, with zone volumes, inter-zone mass fluxes, and energy dissipation rates within each zone provided automatically by the CFD model (Figure 1).

Case Study: Optimization of crystallizer for Bisphenol A production

In this project done at a Japanese Petrochemicals Producer plant, there was an existing process using two large scale (200 m^3) forced cooling crystallizers in series. Consistent operation of the equipment was made difficult by the production of crystals that were too small in size. Advanced Process Models were built to troubleshoot and optimize the process:

(a) A first-principles model was developed that included crystal nucleation, attrition, and crystal growth, with rate expressions that included dependency on the turbulent dissipation rate

(b) Small scale experiments (2-Liter vessel) were used to fit adjustable parameters

(c) A CFD model of the existing equipment was built and linked to a network of well mixed zones built automatically by the gPROMS Multizonal interface.

(d) The hybrid model gave excellent predictions of the crystal size distributions observed in the commercial process, without the need to adjust any parameters beyond those adjusted through the fitting of small scale experiments.

(e) The predictive model was then used to optimize various aspects of the design and operation, including discovery of the fact that the second crystallizer was not needed after optimization of the operating point. Enhanced throughput resulted in benefits of approximately 10 million dollars per year.

Conclusion

Advanced Process Models and their hybrid CFD extensions are becoming well established as technologies with predictive power in the chemical industries. The predictive power of APMs makes their use possible for equipment and process design applications, including scale-up. Model-targeted experimentation reduces the uncertainty in model predictions at a fraction of the cost and effort associated with traditional pilot plant experimental programs, in which experiments are too frequently designed to mimic the operation of the commercial process. High-quality parameters obtained from small-scale experiments can be used to validate advanced models, thus enabling the testing of a very large number of alternatives on the computer. Only the most promising alternatives are selected for final pilot plant testing.

References

1. Matzopoulos, M. (2007), "APM Advantages," *Hydrocarbon Engineering* **12**(3), March, pp. 123-128.
2. Urban, Z. and L. Liberis (1999), Hybrid gPROMS-CFD Modeling of an Industrial Scale Crystallizer with Rigorous Crystal Nucleation and Growth Kinetics and a Full Population Balance," *Chemputers 1999 Conference*, Düsseldorf, Germany.
3. Bezzo, F., S. Macchietto and C.C. Pantelides (2003), "General Hybrid Multizonal/CFD Approach for Bioreactor Modeling," *AIChE Journal* **49**, p. 2113, Düsseldorf, Germany.

38e Dynamic Simulation: An Efficient Approach to Hydro-Processing Unit Design

Dipanjan Bhattacharya[1], **Jaleel V. Valappil**[2], David Chaplin[1], and Ram Tekumalla[3]. (1) Bechtel Oil and Gas, Houston, TX 77054, (2) Bechtel, 3000 Post Oak Boulevard, Houston, TX 77056, (3) Advanced Simulation Group, Bechtel Oil, Gas & Chemicals Inc., 3000 Post Oak Boulevard, Houston, TX 77056

One of the critical aspects in the design of a hydro-processing unit is the analysis of the reactor and separator loop depressurization. Depressurization helps in lowering the reaction and separator loop pressure by venting the loop contents to the flare. This results in safe shutdown and prevents possibilities of thermal runaway. Depressurization can often be the deciding factor for equipment design downstream of the reactor. Conventional relief load sizing calculations do not account for the transient changes in temperature and pressure during a de-pressuring event. A dynamic process model for the unit is an accurate and powerful tool to predict temperature and pressure profiles for all exchangers and equipment that fall in the flow path between the reactor and the separator.

This paper discusses a dynamic simulation study to determine the relief loads for several hydro processing units of a major grass-root refinery. From the simulations, increasing temperature and decreasing pressure for each item of equipment was estimated as a function of time. These data were then analyzed to determine the peak mechanical stresses on the piping and exchangers during the entire event. Mechanical design conditions of the equipment were increased where necessary.

The second objective was to size the depressurization system which was designed to reduce the unit pressure to 50% of design pressure (or 100psig, whichever is lower) over a period of 15 minutes. The sizing took account of several cases such as power failure, reactor runaway etc.

Finally, the study estimated the stripper feed condition at different time points during the depressurization scenario. This was conducted to evaluate the column relief load in a situation where temperature is increasing, but flowrate is decreasing. The possibility of vapor breakthrough to the stripper from the separator during depressurization was also analyzed. In some cases this can be the controlling relief load for the downstream column, as the feed stream gets significantly more pre-heat when the upstream heat-sinks, such as the unit feed liquid, are lost.

38f Emerging Technologies in Mainstream Process Simulator Use

David Hill and Aaron Herrick. CHEMCAD Technical Support, Chemstations, inc., 2901 Wilcrest DR STE 305, Houston, TX 77042

Use of steady state process simulators has become routine in engineering practice. Process simulators allow for fast, efficient, and accurate development of new designs and evaluation of existing operations.

The leading process simulators have advanced features, beyond steady state modeling, which are beginning to be used in mainstream engineering. Some of these features (and commercial case studies) will be discussed.

Best Practice in Facilitating Engineering Workflow through the Integration of Honeywell's UniSim Design and SmartPlant Enterprise from Intergraph

Laurie Wang[1]*, Sanjoy Saha*[1] *and Jeff Johnston*[1]
David Lott[2] *and Frank Joop*[2]

Key words: Workflow, Integration, UniSim Design (Process Design and Simulation), SmartPlant Enterprise (Information Management)

Introduction

Process engineering focuses on the design, operation and maintenance of chemical and material processes in a vast range of industries. The design solution cannot be explicitly determined and therefore is an iterative process that converges into a consistent design that must operate against a range of conditions (cases). This creates a need for a flexible engineering solution to create various design alternatives, revisions and cases. The information with different level of details gets added to the engineering design through the Front End Engineering Design (FEED) process. Therefore, it brings increased productivity and adds values to your business for the engineering solution to have the capacity to update the designs with the new information throughout the FEED process with consistent models.

SmartPlant Enterprise (SPE) is a collection of products from Intergraph used to support elements of process engineering. However, it does not include any process engineering calculation programs. The UniSim Design simulator from Honeywell has been integrated with SPE to publish the heat and mass balance data for multiple design alternatives, revisions and cases. The process engineer will be able to use the simulation data in any product integrated with SPE to perform downstream tasks like P&ID and Instrumentation. The UniSim Design graphic is also published in a native SPE format which can be used as a starting point for creating the process flow diagrams. This presentation describes how you can benefit from the best practice in engineering workflow through the integration of UniSim Design and SmartPlant Enterprise.

Process Engineering and Simulation

Process engineering has two parts: Process Design before the plant is built and Process Improvements after it is in operation. Process simulation is a vital part of both. The outcome of process design decides the behavior of the resulting plant. Meanwhile, a simulation model which represents the true behavior of the plant can be used to identify optimized design for similar processes. Most commercial process simulators handle both process design and simulation. That's a part of the reason why these two things are often mentioned together.

Process simulation is mathematical modeling which has always been a part of the engineering workflow. People used to do it on paper and with a slide rule. During the last three decades, the tremendous value and potential that process simulator provides have been gradually recognized and embraced by many. It has lead to a transformation of the real impact of process simulation on the industry, which was unimaginable in less than a generation ago. People from engineering companies and consulting firms use process simulation to achieve capital projects excellence through optimized designs leading to lower costs. Operating companies benefit from process simulation through faster startups resulting in earlier profits, improved operational safety and efficiency with managed and eliminated risks, as well as optimized process conditions and equipment performance. Technology providers use process simulation to design and develop new processes and offer process simulation models to their licensees as a part of their engineering/operational knowledge transfer. Universities use process simulation tools to teach students basic engineering concepts and demonstrate the impact of changes in key process variables on operation safety and efficiency. Nowadays, many/most engineering workflows include the user of process simulators.

Engineering Workflow

[1] Honeywell Process Solutions
[2] Intergraph

From FEED to the operation optimization, process simulation is intimately involved across the entire lifecycle of a plant. Different tasks are accomplished by different people, often from different departments or companies. Updating and sharing information can become an overwhelming task that costs tremendous resource and unavoidable headache. Along this process, duplicated work is inevitable, and inconsistency is a common threat to the fidelity of outcome of the engineering workflow.

Please take a moment and ask yourself a few questions. Have you ever head of a process engineer making a change to the design? Why does he do it? What does he change? How often does he change it? What is the volume of changes? What is the impact of the changes to the rest of the project? This should help you draw a clear mental picture of how complicated and chaotic the process can get. That's where the best practices in facilitating engineering workflow come in.

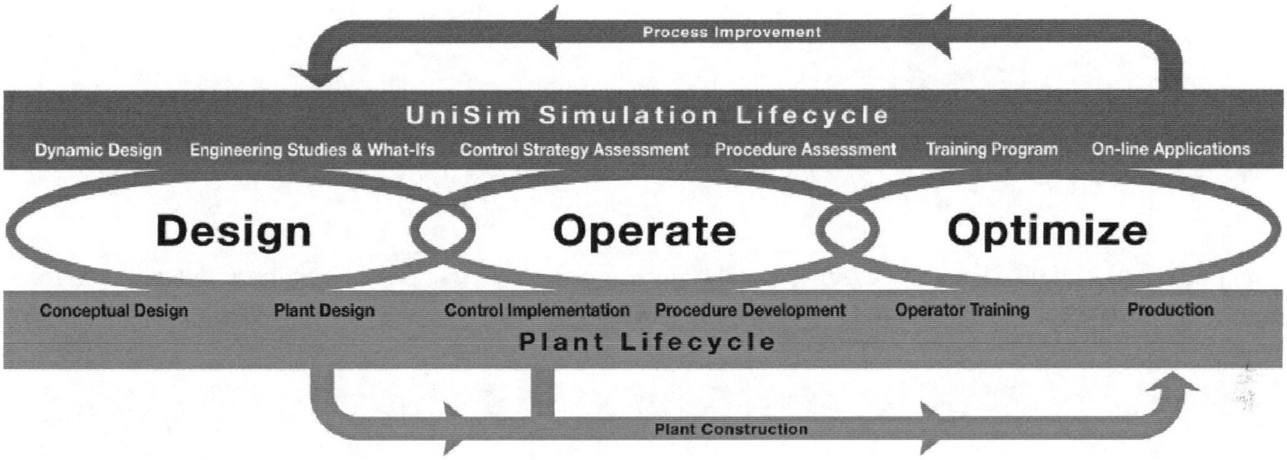

Best Practices in Facilitating Engineering Workflow

The best thing you can do to minimize the needs for changing a design is to make it right in the first place. When changes are necessary, keep an accurate track record of the changes made, in terms of when, what, who, and why. What you need to know is that there are built-in tools available in process simulators, such as UniSim Design from Honeywell, to help you discover hidden errors, identify sources of problems, keep an accurate record of changes and model revisions, consolidate multiple input to ensure consistency across the design project, and manage the level of model control.

<u>Simulation Balance Tool</u>

The Simulation Balance Tool in UniSim Design was developed, with customers' direct input, to help you easily identify the obvious and not so obvious problems in a flowsheet. If you work on a project for a long time, you may not remember what you did a few weeks or even a few months ago, or the assumptions you have applied at the time. The new input you have introduced could cause conflict with what you did before, leading to wrong results which you many not even notice. In an even worse situation where you are asked to take over the work somebody else did. You may have a hard time understanding what is going on with the models/flowsheets you are given, let alone knowing whether the results are right or not. To make sure your models are converged to the right and meaningful results, you should take advantage of the availability of the Simulation Balance Tool. It can be set for selected balance checking across the flowsheet, including Mass/Mole/Energy/Component balance, as well as transition temperature and transition phase fraction balance.

Take the following natural gas dehydration model as an example.

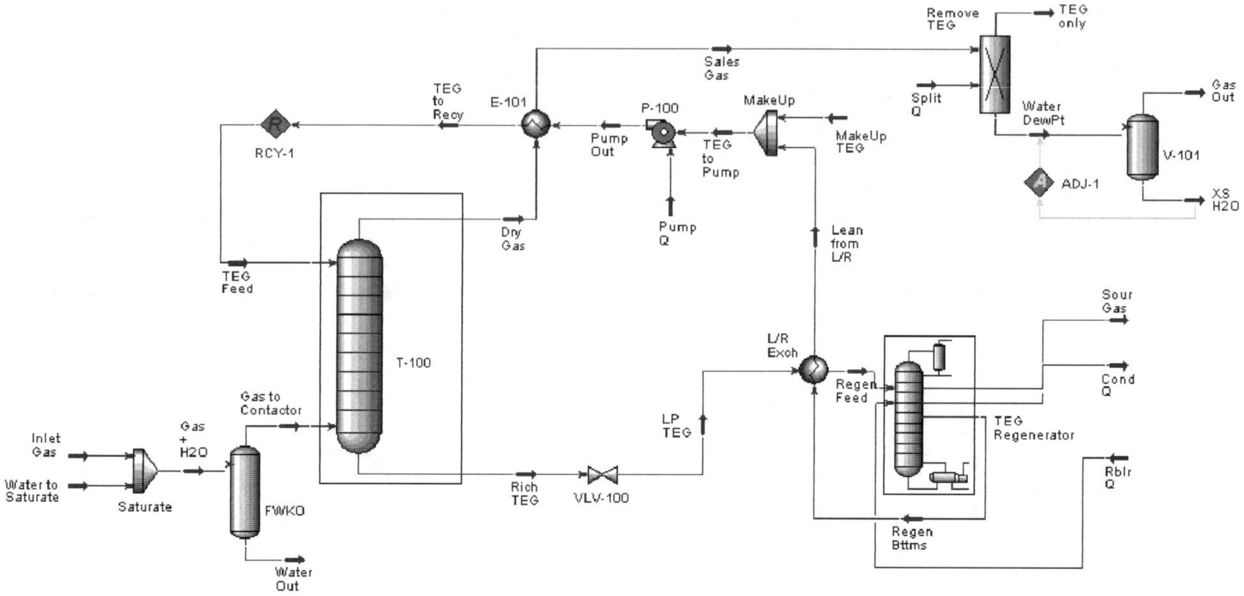

From the first glance, this converged model doesn't give any clue for suspicion. But if the Simulation Balance Tool feature is turned on, it will prompt warnings to the potential areas for mass and energy imbalance which have exceeded the user defined tolerances:

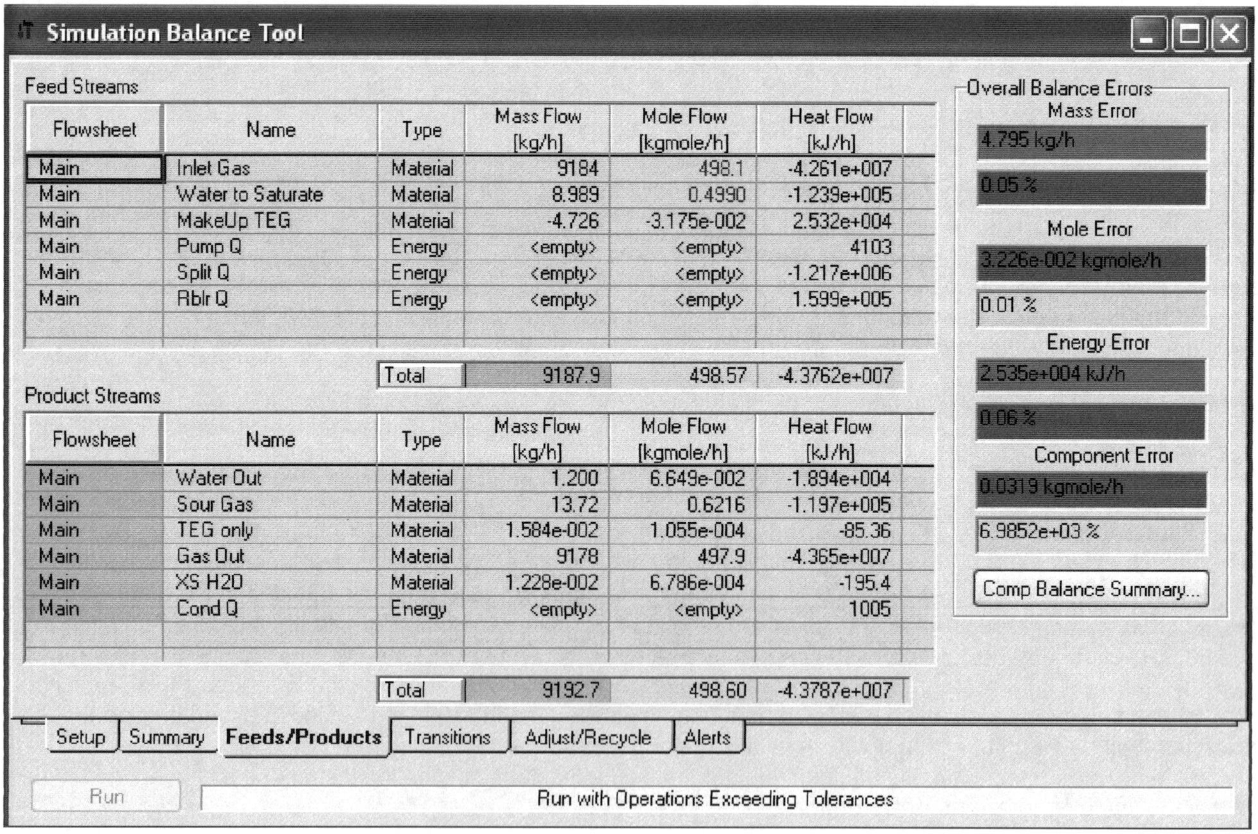
Simulation Balance Tool

Feed Streams

Flowsheet	Name	Type	Mass Flow [kg/h]	Mole Flow [kgmole/h]	Heat Flow [kJ/h]
Main	Inlet Gas	Material	9184	498.1	-4.261e+007
Main	Water to Saturate	Material	8.989	0.4990	-1.239e+005
Main	MakeUp TEG	Material	-4.726	-3.175e-002	2.532e+004
Main	Pump Q	Energy	<empty>	<empty>	4103
Main	Split Q	Energy	<empty>	<empty>	-1.217e+006
Main	Rblr Q	Energy	<empty>	<empty>	1.599e+005
		Total	9187.9	498.57	-4.3762e+007

Product Streams

Flowsheet	Name	Type	Mass Flow [kg/h]	Mole Flow [kgmole/h]	Heat Flow [kJ/h]
Main	Water Out	Material	1.200	6.649e-002	-1.894e+004
Main	Sour Gas	Material	13.72	0.6216	-1.197e+005
Main	TEG only	Material	1.584e-002	1.055e-004	-85.36
Main	Gas Out	Material	9178	497.9	-4.365e+007
Main	XS H20	Material	1.228e-002	6.786e-004	-195.4
Main	Cond Q	Energy	<empty>	<empty>	1005
		Total	9192.7	498.60	-4.3787e+007

Overall Balance Errors

Mass Error	4.795 kg/h
	0.05 %
Mole Error	3.226e-002 kgmole/h
	0.01 %
Energy Error	2.535e+004 kJ/h
	0.06 %
Component Error	0.0319 kgmole/h
	6.9852e+03 %

Comp Balance Summary...

Setup | Summary | Feeds/Products | Transitions | Adjust/Recycle | Alerts

Run | Run with Operations Exceeding Tolerances

It will also display a Balance Error Summary to highlight the sources of the problems. In this specific case, the problems are closely associated with the Recycle block and the column flowsheets.

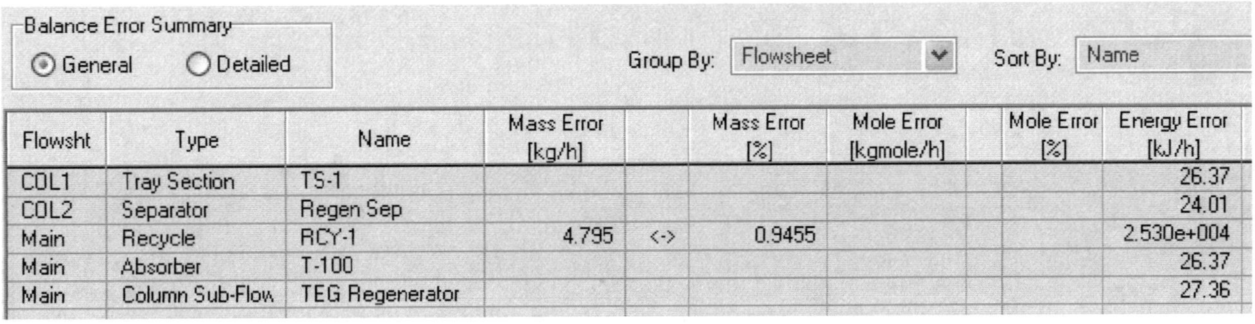

Balance Error Summary: (•) General ( ) Detailed — Group By: Flowsheet — Sort By: Name

Flowsht	Type	Name	Mass Error [kg/h]		Mass Error [%]	Mole Error [kgmole/h]		Mole Error [%]	Energy Error [kJ/h]
COL1	Tray Section	TS-1							26.37
COL2	Separator	Regen Sep							24.01
Main	Recycle	RCY-1	4.795	<->	0.9455				2.530e+004
Main	Absorber	T-100							26.37
Main	Column Sub-Flow	TEG Regenerator							27.36

So by simply tightening the tolerance for mass and energy transfer at the Recycle block and adjusting the Heat/Spec Error Tolerance setting at the Absorber, you will be able to fix these problems in a matter of seconds.

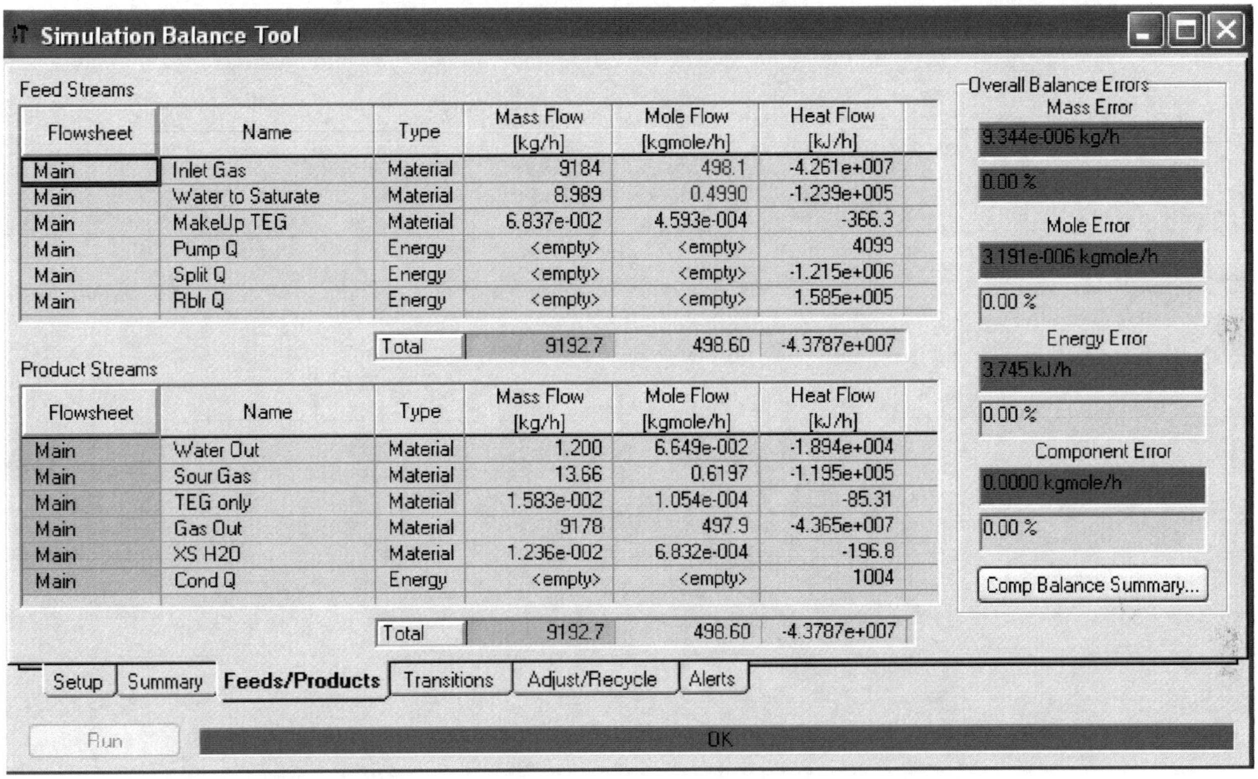

Simulation Balance Tool

Feed Streams

Flowsheet	Name	Type	Mass Flow [kg/h]	Mole Flow [kgmole/h]	Heat Flow [kJ/h]
Main	Inlet Gas	Material	9184	498.1	-4.261e+007
Main	Water to Saturate	Material	8.989	0.4990	-1.239e+005
Main	MakeUp TEG	Material	6.837e-002	4.593e-004	-366.3
Main	Pump Q	Energy	<empty>	<empty>	4099
Main	Split Q	Energy	<empty>	<empty>	-1.215e+006
Main	Rblr Q	Energy	<empty>	<empty>	1.585e+005
		Total	9192.7	498.60	-4.3787e+007

Product Streams

Flowsheet	Name	Type	Mass Flow [kg/h]	Mole Flow [kgmole/h]	Heat Flow [kJ/h]
Main	Water Out	Material	1.200	6.649e-002	-1.894e+004
Main	Sour Gas	Material	13.66	0.6197	-1.195e+005
Main	TEG only	Material	1.583e-002	1.054e-004	-85.31
Main	Gas Out	Material	9178	497.9	-4.365e+007
Main	XS H2O	Material	1.236e-002	6.832e-004	-196.8
Main	Cond Q	Energy	<empty>	<empty>	1004
		Total	9192.7	498.60	-4.3787e+007

Case Scenario Management tool

Similarly, the **Case Scenario Management tool** in UniSim Design was designed and developed with the direct involvement of customers, who have the pressing needs for managing unavoidable changes from different sources and ensuring model consistency across the design process. This tool helps users to track model revisions, compare selected process variables in different cases, set up scenario cases for comparison against a defined master model including the values of compared variables and flowsheet topology. In the future this tool will offer users the choice of updating their models based on their own updating criteria.

The following screenshots illustrate how easily users can use the **Case Scenario Management tool** to

- Select file attributes to display on the Scenario Setup page;
- Select process variables for comparison;
- Set up scenario cases for comparison and
- Display results of topological comparison for all the scenario cases.

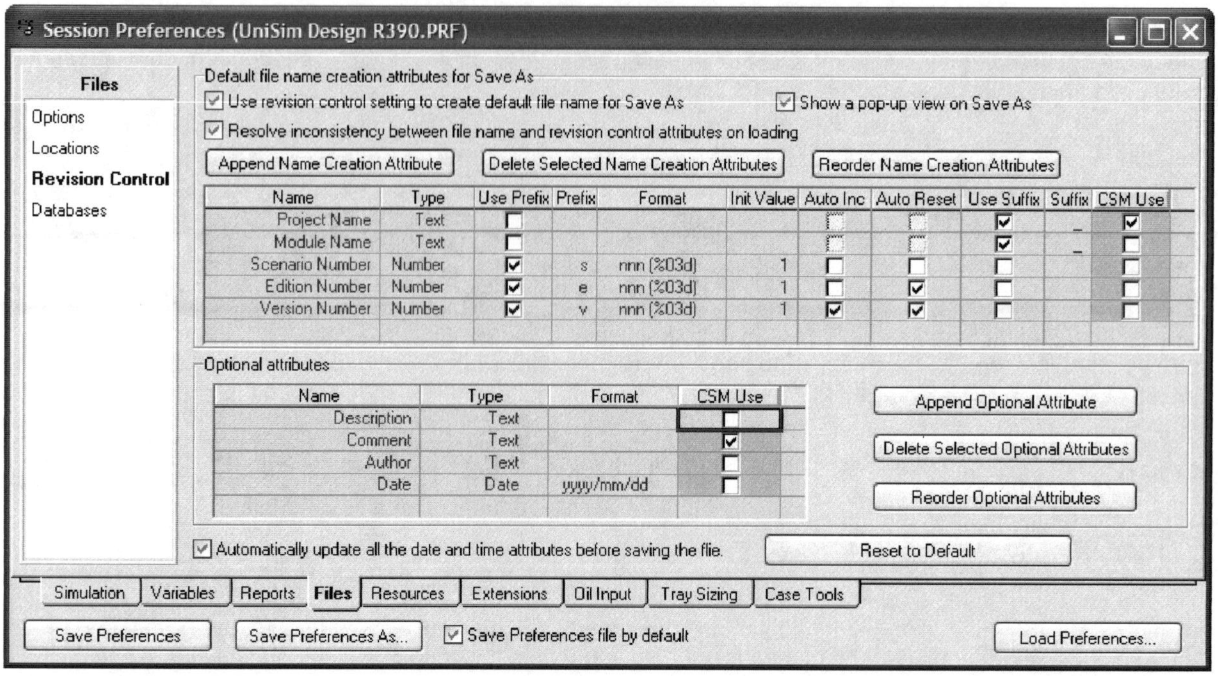
Session Preferences (UniSim Design R390.PRF)
Files
Options
Locations
Revision Control
Databases
Default file name creation attributes for Save As
Use revision control setting to create default file name for Save As
Show a pop-up view on Save As
Resolve inconsistency between file name and revision control attributes on loading
Append Name Creation Attribute
Delete Selected Name Creation Attributes
Reorder Name Creation Attributes
Name | Type | Use Prefix | Prefix | Format | Init Value | Auto Inc | Auto Reset | Use Suffix | Suffix | CSM Use
Project Name | Text
Module Name | Text
Scenario Number | Number | s | nnn (%03d) | 1
Edition Number | Number | e | nnn (%03d) | 1
Version Number | Number | v | nnn (%03d) | 1
Optional attributes
Name | Type | Format | CSM Use
Description | Text
Comment | Text
Author | Text
Date | Date | yyyy/mm/dd
Append Optional Attribute
Delete Selected Optional Attributes
Reorder Optional Attributes
Automatically update all the date and time attributes before saving the file.
Reset to Default
Simulation
Variables
Reports
Files
Resources
Extensions
Oil Input
Tray Sizing
Case Tools
Save Preferences
Save Preferences As...
Save Preferences file by default
Load Preferences...

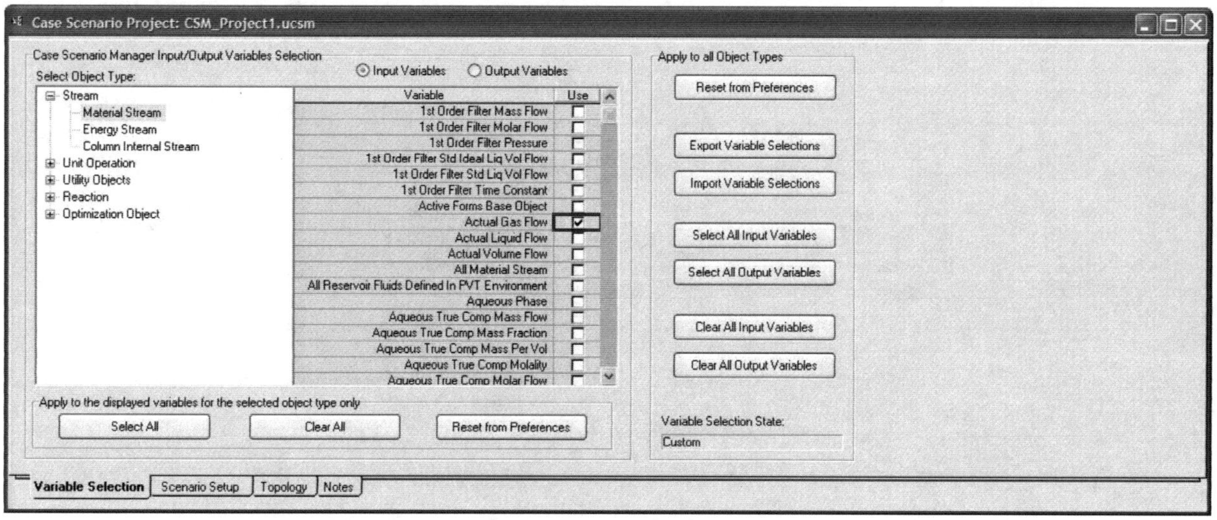
Case Scenario Project: CSM_Project1.ucsm
Case Scenario Manager Input/Output Variables Selection
Select Object Type:
Input Variables
Output Variables
Stream
Material Stream
Energy Stream
Column Internal Stream
Unit Operation
Utility Objects
Reaction
Optimization Object
Variable
Use
1st Order Filter Mass Flow
1st Order Filter Molar Flow
1st Order Filter Pressure
1st Order Filter Std Ideal Liq Vol Flow
1st Order Filter Std Liq Vol Flow
1st Order Filter Time Constant
Active Forms Base Object
Actual Gas Flow
Actual Liquid Flow
Actual Volume Flow
All Material Stream
All Reservoir Fluids Defined In PVT Environment
Aqueous Phase
Aqueous True Comp Mass Flow
Aqueous True Comp Mass Fraction
Aqueous True Comp Mass Per Vol
Aqueous True Comp Molality
Aqueous True Comp Molar Flow
Apply to the displayed variables for the selected object type only
Select All
Clear All
Reset from Preferences
Apply to all Object Types
Reset from Preferences
Export Variable Selections
Import Variable Selections
Select All Input Variables
Select All Output Variables
Clear All Input Variables
Clear All Output Variables
Variable Selection State:
Custom
Variable Selection
Scenario Setup
Topology
Notes

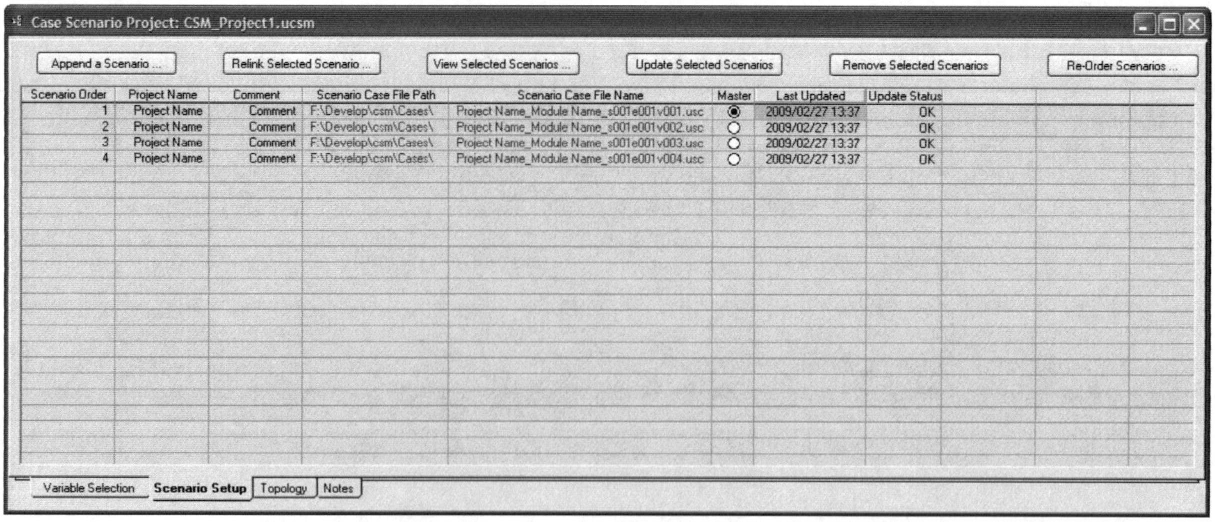
Case Scenario Project: CSM_Project1.ucsm
Append a Scenario ...
Relink Selected Scenario ...
View Selected Scenarios ...
Update Selected Scenarios
Remove Selected Scenarios
Re-Order Scenarios ...
Scenario Order | Project Name | Comment | Scenario Case File Path | Scenario Case File Name | Master | Last Updated | Update Status
1 | Project Name | Comment | F:\Develop\csm\Cases\ | Project Name_Module Name_s001e001v001.usc | 2009/02/27 13:37 | OK
2 | Project Name | Comment | F:\Develop\csm\Cases\ | Project Name_Module Name_s001e001v002.usc | 2009/02/27 13:37 | OK
3 | Project Name | Comment | F:\Develop\csm\Cases\ | Project Name_Module Name_s001e001v003.usc | 2009/02/27 13:37 | OK
4 | Project Name | Comment | F:\Develop\csm\Cases\ | Project Name_Module Name_s001e001v004.usc | 2009/02/27 13:37 | OK
Variable Selection
Scenario Setup
Topology
Notes

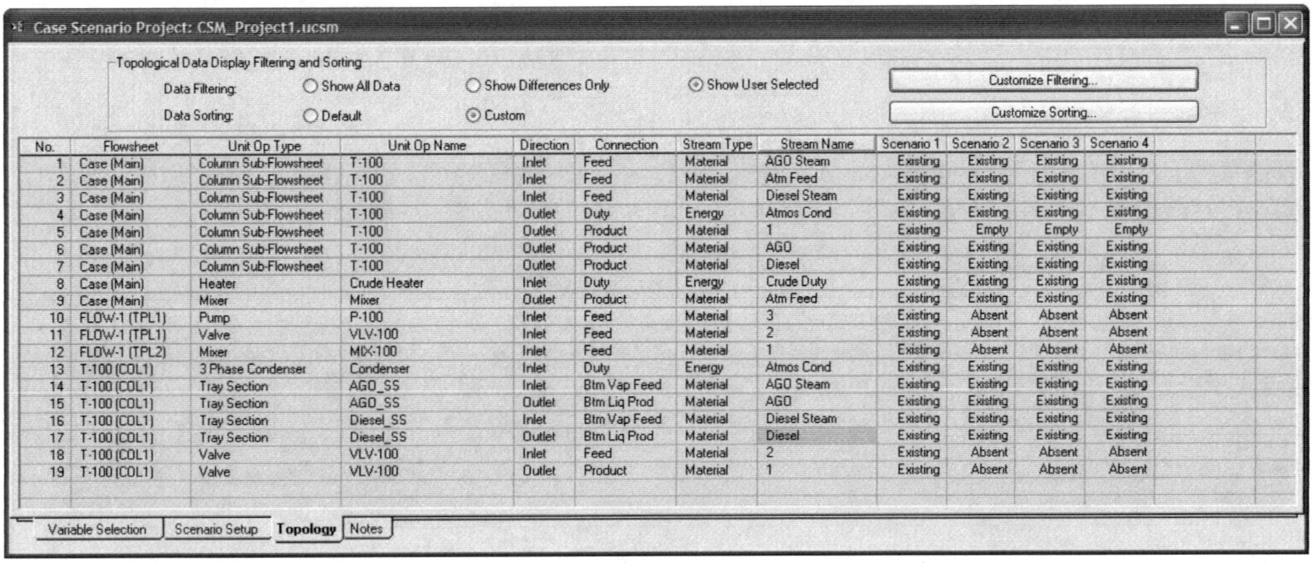

UniSim Excel Interface Tool (USE-IT)

This new feature of UniSim Design was built on the successful experience of UOP for case studies. It is particularly useful to process engineers and operational staff who are not necessarily familiar with process simulation tools, but have the real needs to use simulation models to perform "what-if" studies for troubleshooting and debottlenecking when making operational decisions. Through **UniSim Excel Interface Tool**, users can perform scenario analysis and case studies to see the impact of changes in key process variables on the plant performance without even seeing the model flowhseet. On the other hand, the model created by simulation experts can be properly maintained and updated without the risk of being compromised by others who use the model for their own applications.

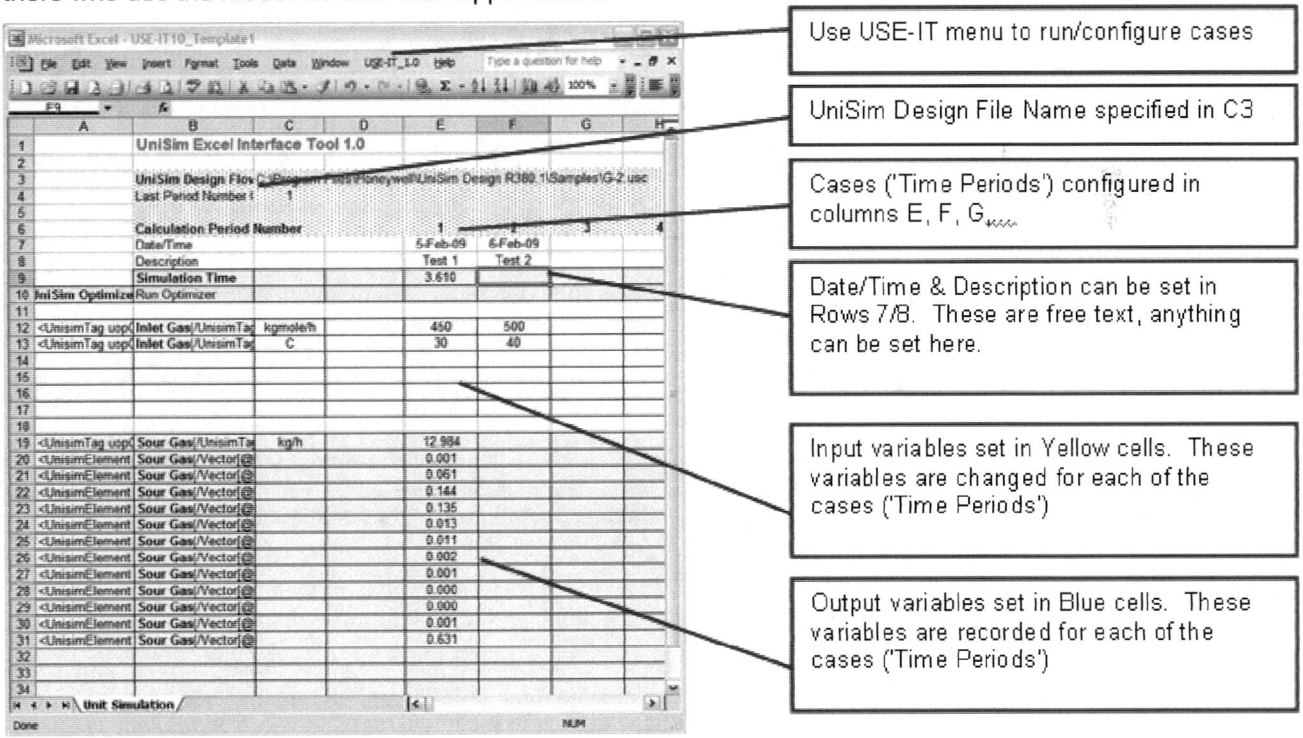

SmartPlant Foundation

Once a project enters the detailed design phase, it often needs an information management system to facilitate the engineering workflow. Such a system will provide support for concurrent/simultaneous engineering, multiple design alternatives, revision of the flowsheet, as well as multiple operating cases e.g. summer, winter, high production, turndown... At the center of such a system is the SmartPlant Foundation from Intergraph.

SmartPlant Foundation is the world's most proven, industry-standard engineering information management solution in the process, power, and offshore industries. SmartPlant Foundation ensures an open, independent data storage system to protect plant information for the life of the plant. It addresses all of the design, build, and operation requirements of a facility throughout the entire plant life cycle.

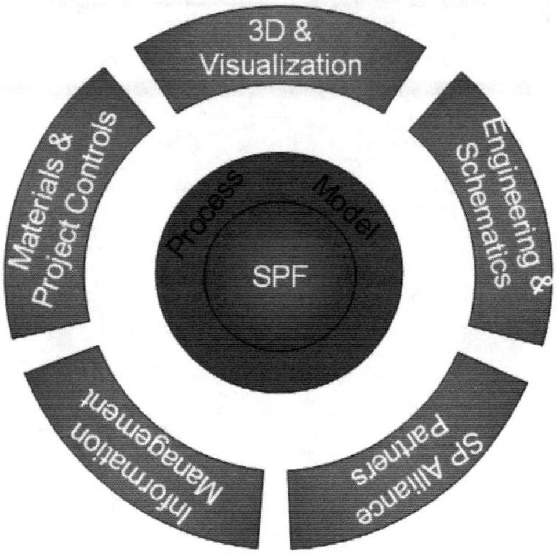

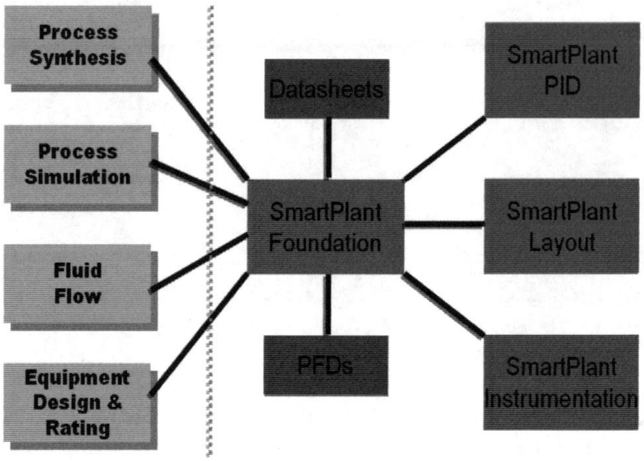

This means that once the process simulation information is linked to SmartPlant Foundation, it will become available to other tools used in the engineering workflow, such as SmartPlant PID, SmartPlant Layout, and SmartPlant Instrumentation.

Integration of Honeywell's UniSim Design and SmartPlant Enterprise from Intergraph

With the same vision and determination to facilitate the engineering workflow, Honeywell and Intergraph have undertaken joint effort to develop a seamless link between UniSim Design and SmartPlant Foundation. The screenshots below are taken from the prototype where the process data from UniSim Design are exported to SmartPlant Foundation and used to generate reports in different formats. The process flowsheet diagram from UniSim Design can also be recreated in Smart Sketch.

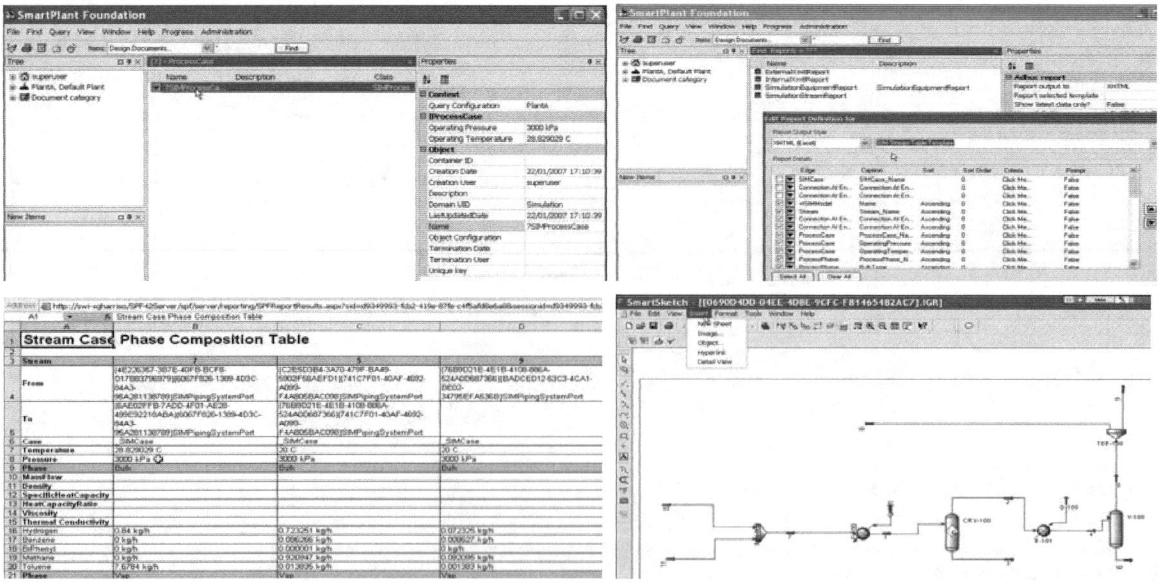

The outcome of this joint project will become available to users at the release of SmartPlant Process Engineering.

To close the information gap with an expanded scope, it is also highly valuable to integrate the live process data from operations.

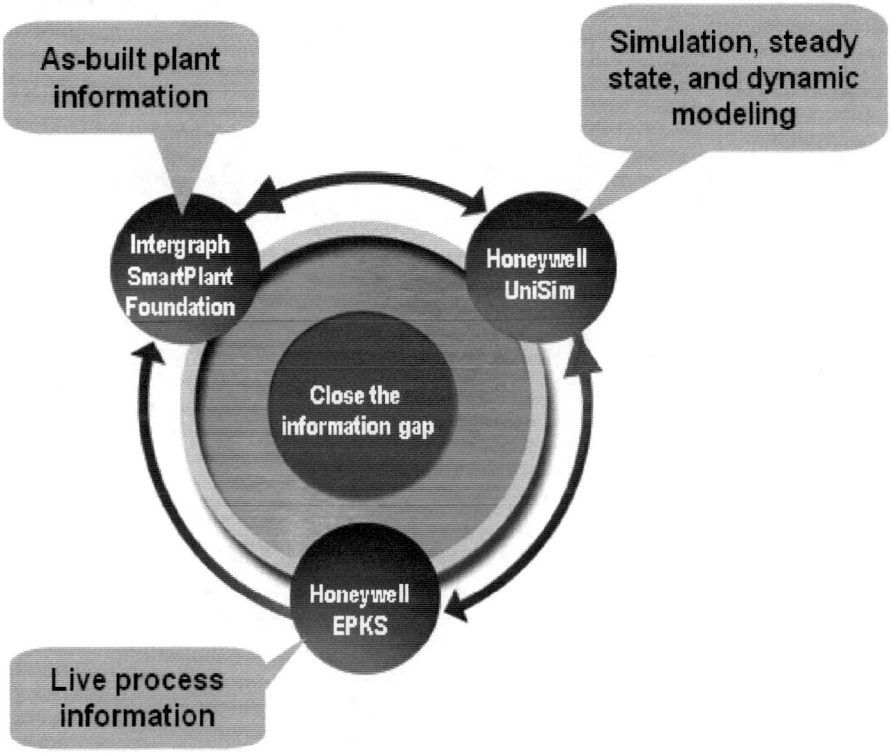

The best practice in facilitating engineering workflow through the Integration of Honeywell's Simulation and Process Information Management System with SmartPlant Enterprise from Intergraph offers tremendous benefits, including

- Optimizes the…
  - Project cycle
  - Collaboration phase
  - Validation phase
  - Commissioning phase
  - Startup phase
- Provides cost effective site-specific training (earlier in the process)
- Optimizes the Life-Cycle (operate/maintain)
- Solution to support both front-end as well as operational phase

In conclusion, there is tremendous value in enabling process simulation tool to track and manage changes in data (scenarios) in such way that it can be seamlessly used throughout the process design lifecycle without any rework and data entry mistakes.

Process Integration - Tutorial

AIChE Spring National Meeting, Tampa, FL
April 26-30,2009

Kirtan Trivedi
ExxonMobil Research and Engineering Company,
Fairfax, VA

ExxonMobil
Taking on the world's toughest energy challenges™

Outline

- Energy Outlook
- Process Design & Pinch Technology
- Fundamentals of Pinch Technology
- Grassroot Designs
- Retrofit Designs
- Utility Placement
- Total Site Analysis

ExxonMobil
Taking on the world's toughest energy challenges™

The Outlook for Energy – A view to 2030

Global Economics and Energy

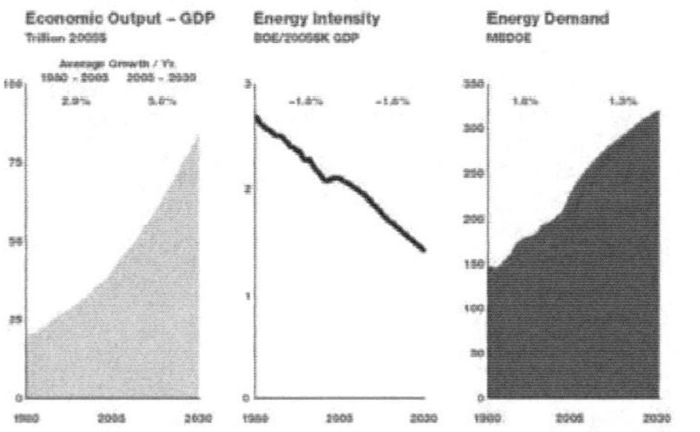

Ref: The Outlook for Energy – A view to 2030
http://www.exxonmobil.com/Corporate/energy_outlook.aspx

ExxonMobil
Taking on the world's toughest energy challenges™

The Outlook for Energy – A view to 2030

World Energy Demand and Supply

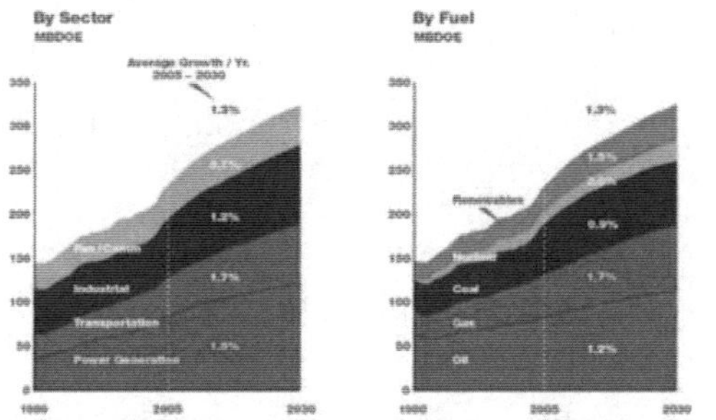

Ref: The Outlook for Energy – A view to 2030
http://www.exxonmobil.com/Corporate/energy_outlook.aspx

ExxonMobil
Taking on the world's toughest energy challenges™

The Outlook for Energy – A view to 2030

World Energy Demand - Primary Energy Supplies

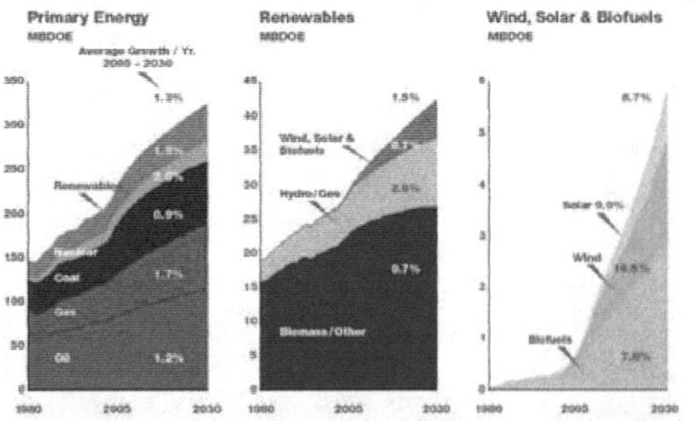

Ref: The Outlook for Energy – A view to 2030
http://www.exxonmobil.com/Corporate/energy_outlook.aspx

ExxonMobil
Taking on the world's toughest energy challenges™

The Outlook for Energy – A view to 2030

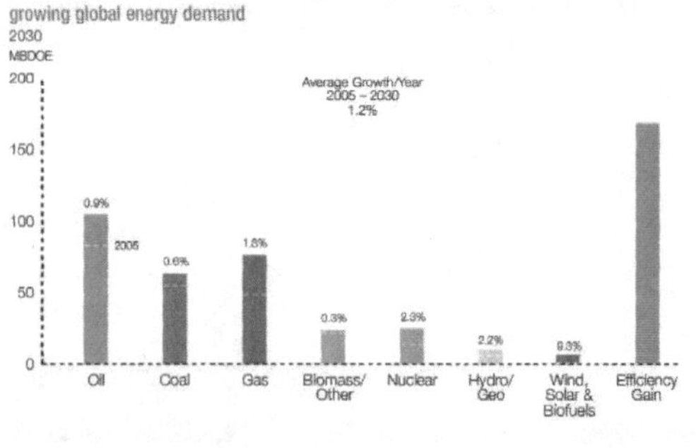

Ref: The Outlook for Energy – A view to 2030
http://www.exxonmobil.com/Corporate/energy_outlook.aspx

ExxonMobil
Taking on the world's toughest energy challenges™

The Outlook for Energy – A view to 2030

- Global energy demand is projected to grow at 1.3 percent per year
  - Underpinned by economic and population growth
  - Mitigated by efficiency gains
- Power generation represents more than 40 percent of demand increase
  - Growth concentrated in developing countries
  - Coal maintains highest fuel share, though gas has the largest growth
  - Nuclear power contribution grows, primarily post -2020
- Transportation sector demand growth slows over time
  - Increase in new light duty vehicle fuel economy helps offset impact of growing fleet
  - Biofuels supply increases but limited by cost and scale considerations
- Oil, natural gas and coal are indispensable to meeting this energy demand, even with rapid growth in renewables
- Demand growth and fuel mix will lead to increased CO_2 emissions
- Significantly impacting CO2 emissions requires global participation, step changes in energy efficiency, technology gains and massive investment

Ref: The Outlook for Energy – A view to 2030
http://www.exxonmobil.com/Corporate/energy_outlook.aspx

ExxonMobil
Taking on the world's toughest energy challenges™ 7

Process Design

- Goal
  - Optimal Design
- Subject to
  - Client Requirements
  - Process Design Constraints
  - Time Constraints

ExxonMobil
Taking on the world's toughest energy challenges™ 8

Typical Optimization

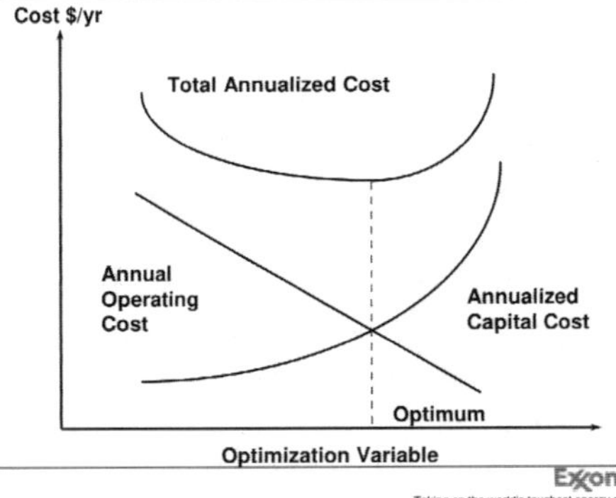

ExxonMobil
Taking on the world's toughest energy challenges™ 9

The 'Onion' Diagram

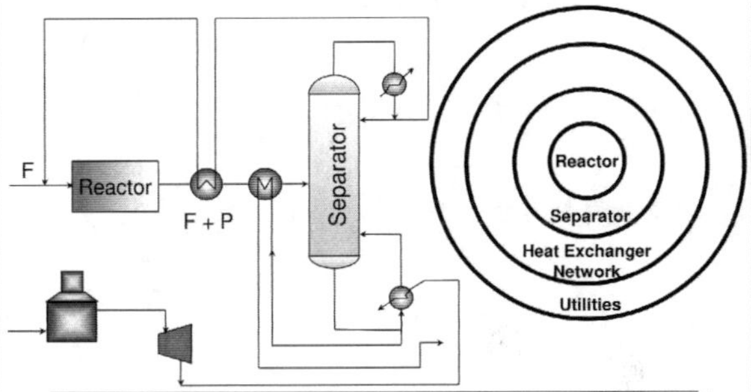

ExxonMobil
Taking on the world's toughest energy challenges™ 10

Process Modifications

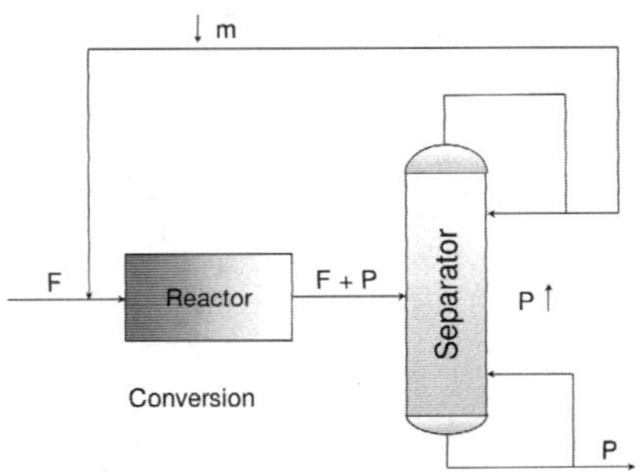

ExxonMobil
Taking on the world's toughest energy challenges™ 11

SUMMARY

- Designs based on experience
- Time consuming
- No real analysis of Capital-Energy trade-off
- End of the day - "It's the best we can do"

ExxonMobil
Taking on the world's toughest energy challenges™ 12

Beat the Learning Curve

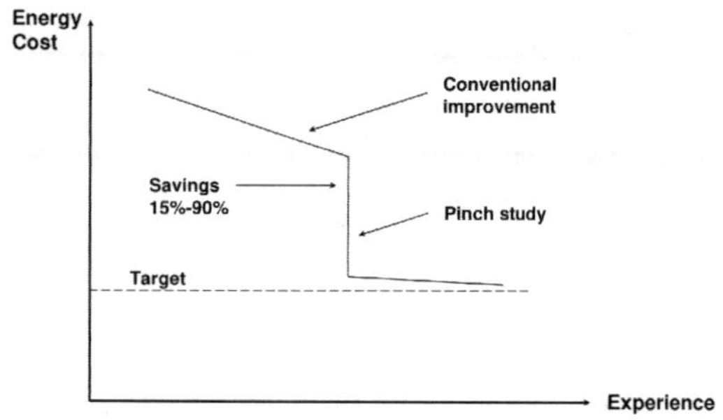

ExxonMobil

Taking on the world's toughest energy challenges™

13

Pinch Technology Track Record Achieved Energy Savings

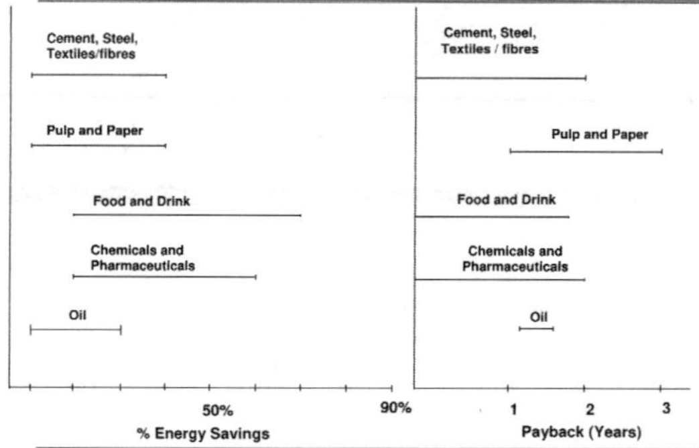

ExxonMobil

Taking on the world's toughest energy challenges™

14

"Hot" & "Cold" Streams

- "Hot" Streams Require Cooling

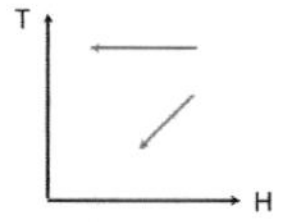

- "Cold" Streams Require Heating

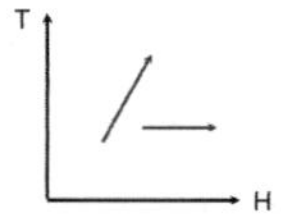

ExxonMobil

Taking on the world's toughest energy challenges™

15

Definitions

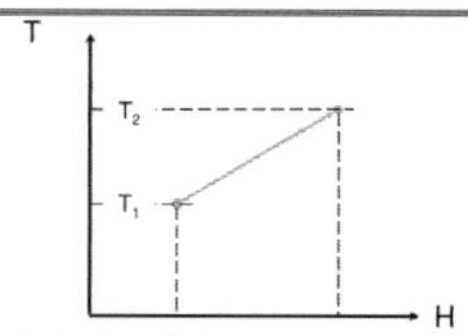

- H = Stream Enthalpy (MW)
- CP = m * Cp = $\Delta H / \Delta T$
  = Heat Capacity Flow Rate (MW/°C)
  = Stream Flow Rate * Specific Heat Capacity
  Kg/sec (MJ/Kg °C)
- ΔH = CP * ΔT

ExxonMobil

Taking on the world's toughest energy challenges™

16

ΔT_{min} "Minimum Permisible Temperature Difference"

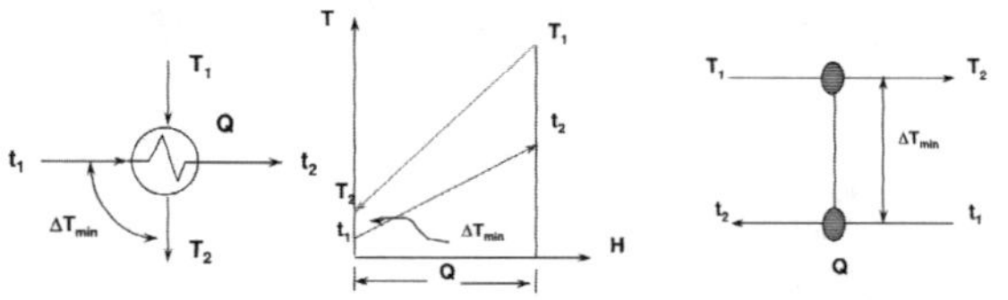

Conventional representation of an exchanger

Representation of an exchanger on the T-H diagram

Grid diagram representation of an exchanger

ExxonMobil

Taking on the world's toughest energy challenges™

17

Philosophy of Pinch Technology

- Targets Based on Thermodynamic Principles
- "Birds-eye-view" of the Solution Space
- Identify the Optimum Design Parameter
- Initialize the Design with the Optimum Design Parameter
- Fine Tune

ExxonMobil

Taking on the world's toughest energy challenges™

18

Two Stream System

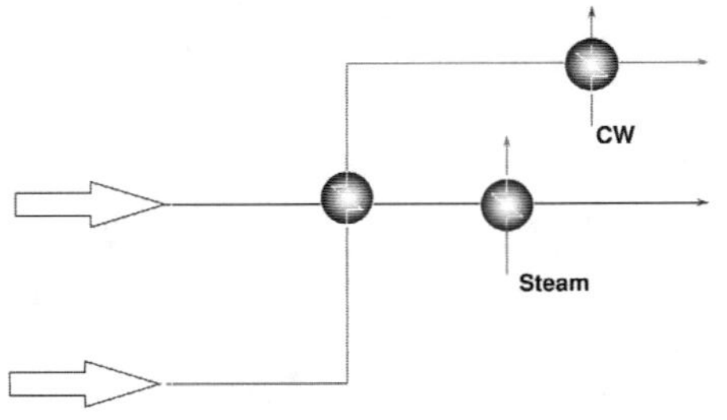

Energy Targets

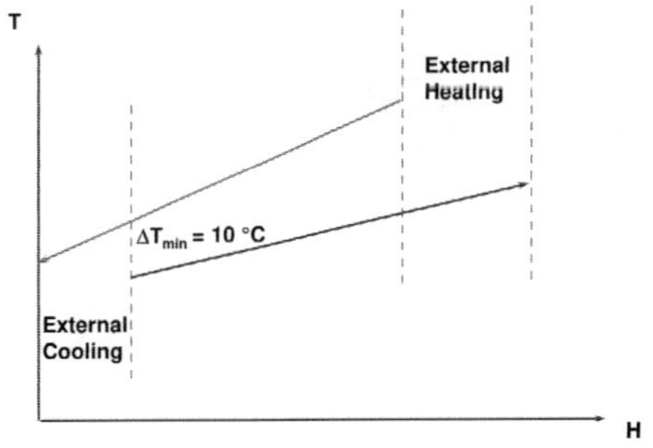

More in - More Out

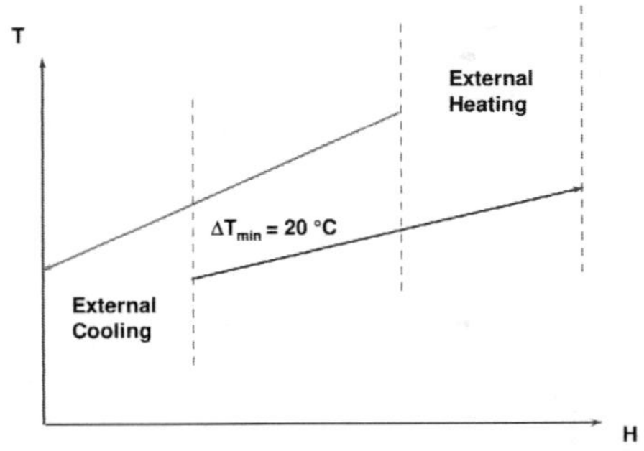

Area/Units Targets

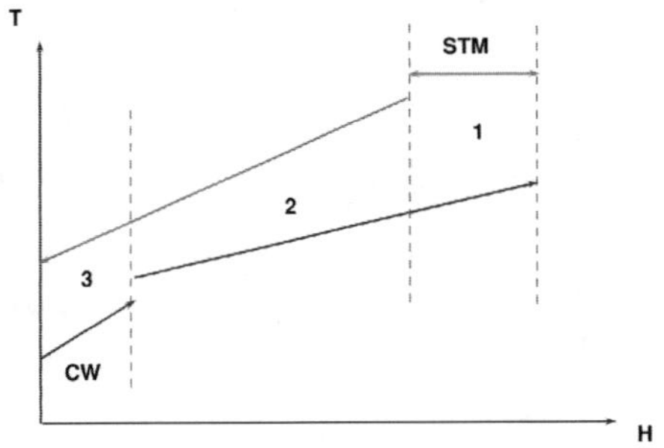

Shell Targets

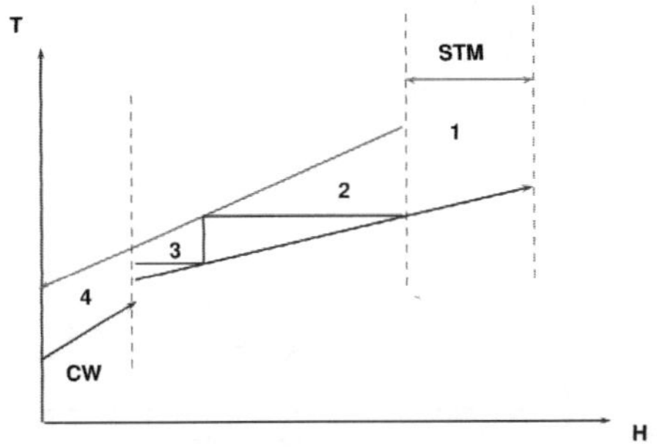

Total Annualized Cost Targets

- Energy targets ⇒ Operating cost
- Area target
- Units target
- Shell target

} ⇒ Capital cost

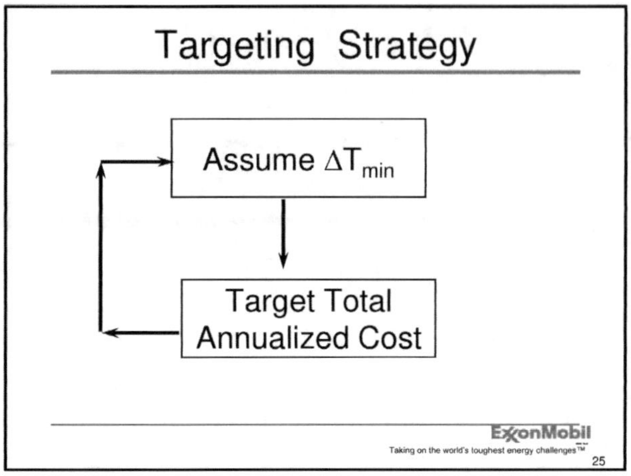
Targeting Strategy
Assume ΔT min
Target Total Annualized Cost
ExxonMobil
Taking on the world's toughest energy challenges™
25

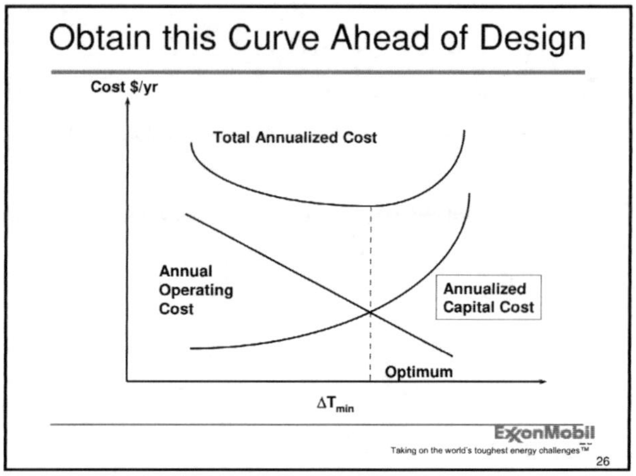
Obtain this Curve Ahead of Design
Cost $/yr
Total Annualized Cost
Annual Operating Cost
Annualized Capital Cost
Optimum
ΔT min
ExxonMobil
Taking on the world's toughest energy challenges™
26

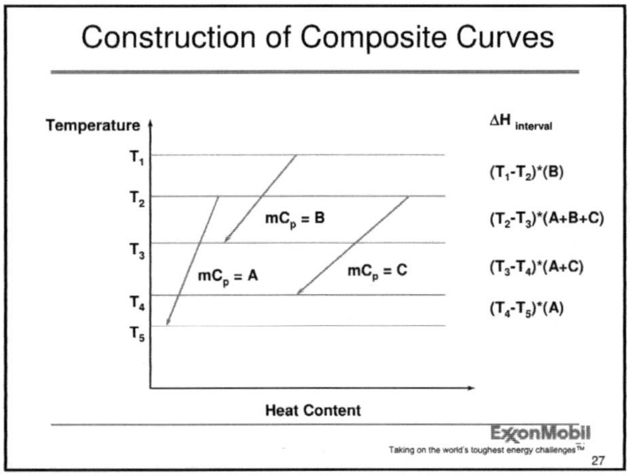
Construction of Composite Curves
Temperature
T1
T2
T3
T4
T5
mCp = B
mCp = A
mCp = C
ΔH interval
(T1-T2)*(B)
(T2-T3)*(A+B+C)
(T3-T4)*(A+C)
(T4-T5)*(A)
Heat Content
ExxonMobil
Taking on the world's toughest energy challenges™
27

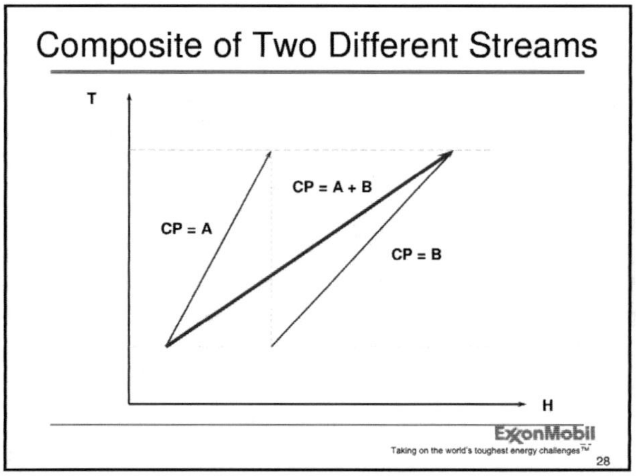
Composite of Two Different Streams
T
CP = A + B
CP = A
CP = B
H
ExxonMobil
Taking on the world's toughest energy challenges™
28

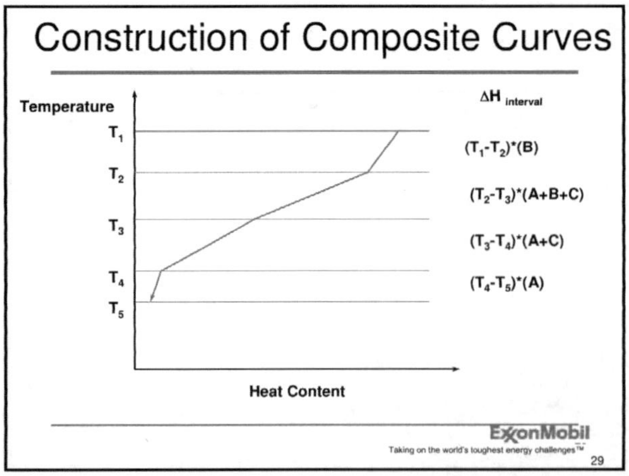
Construction of Composite Curves
Temperature
T1
T2
T3
T4
T5
ΔH interval
(T1-T2)*(B)
(T2-T3)*(A+B+C)
(T3-T4)*(A+C)
(T4-T5)*(A)
Heat Content
ExxonMobil
Taking on the world's toughest energy challenges™
29

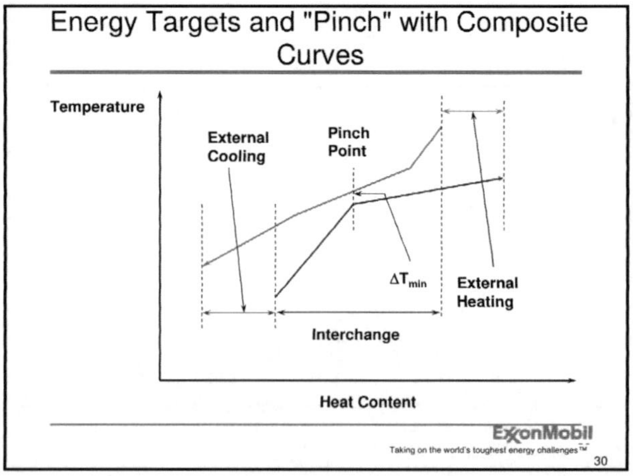
Energy Targets and "Pinch" with Composite Curves
Temperature
External Cooling
Pinch Point
ΔT min
External Heating
Interchange
Heat Content
ExxonMobil
Taking on the world's toughest energy challenges™
30

Pinch Point

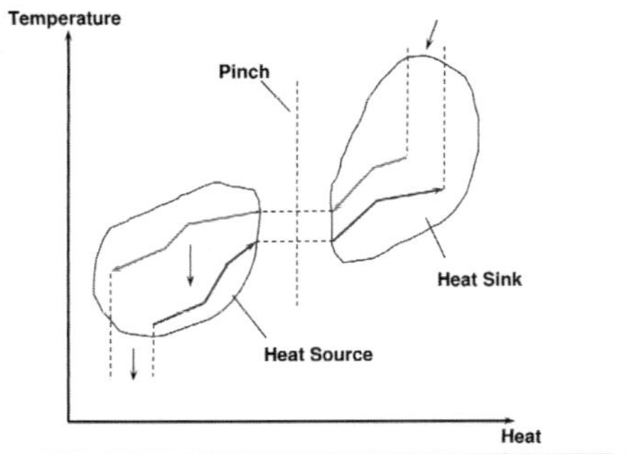

Intersystem Heat Transfer

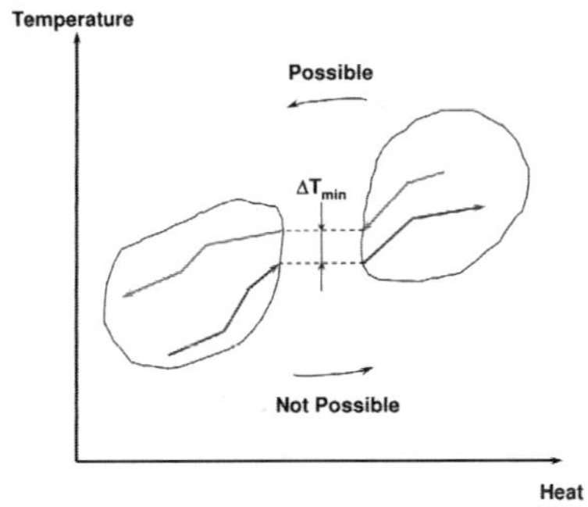

Heat Balance

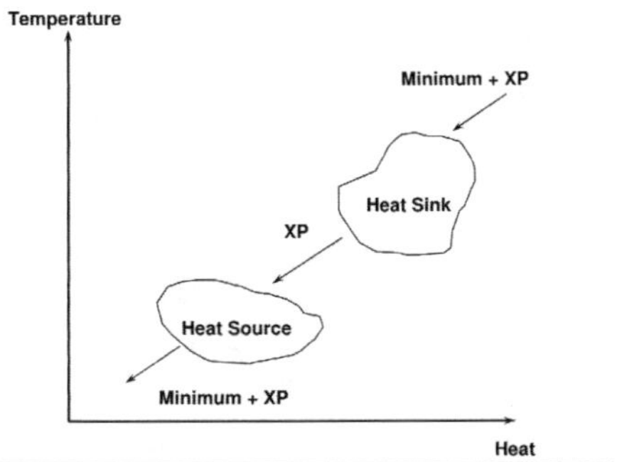

Approach

Keep **ABOVE** and **BELOW** separate and :

- do not transfer **ACROSS**
- do not import **BELOW**
- do not export **ABOVE**

Typical Plant Flowsheet

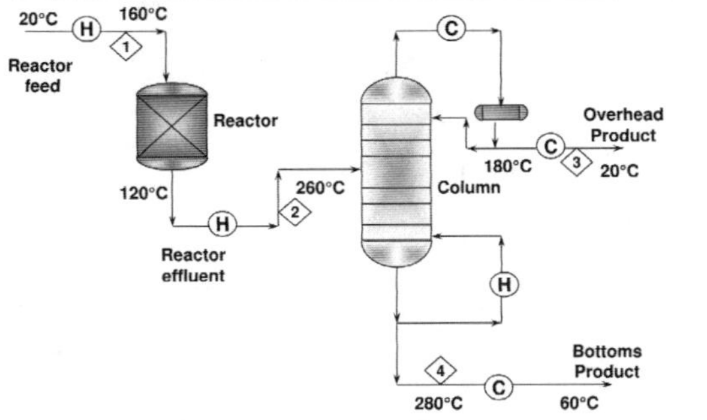

Stream Data

Stream Number	Stream Name	Stream No	T_s °C	T_t °C	CP (kW/°C)
1	Reactor Feed	1	20	160	40
2	Reactor Effluent	2	120	260	60
3	Overhead Product	3	180	20	45
4	Bottoms Product	4	280	60	30

Cost Data

Heat Exchangers

Installed cost per shell ($) = 10,000 $A^{0.6}$ (m^2)

Utility Data

Cost of using hot oil	=	68 ($/kW.yr)
Cooling water	=	2.5 ($/kW.yr)
Temperature of hot oil	=	320 °C - 310 °C
Temperature of cooling water	=	10 °C - 20 °C

Plant Data:

Interest rate	=	10 %
Lifetime	=	5 years
Operation Time	=	8000 hours/yr

Targeting Plots

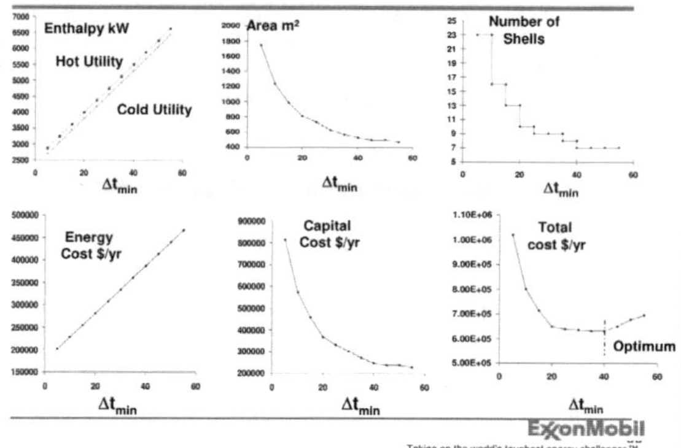

Final Flow Diagram for the Plant

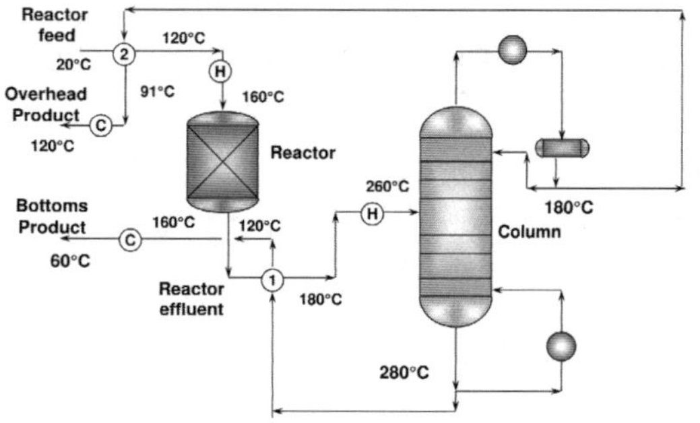

Retrofit Design Philosophy

- Set Targets Before Design
- Preliminary Design
- Detailed Design

Area Targets and "Vertical" Heat Transfer

$$A_{min} = \sum_{i}^{intervals} \frac{1}{\Delta T_{LMi}} \left[\sum_{j}^{streams} \frac{q_j}{h_j} \right]_i$$

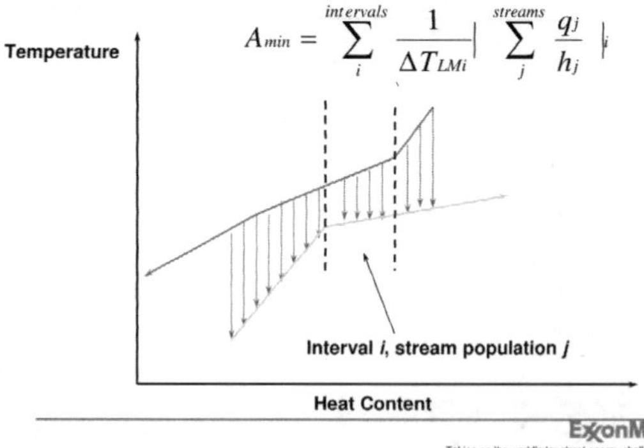

"Criss-Cross" Mode of Heat Transfer

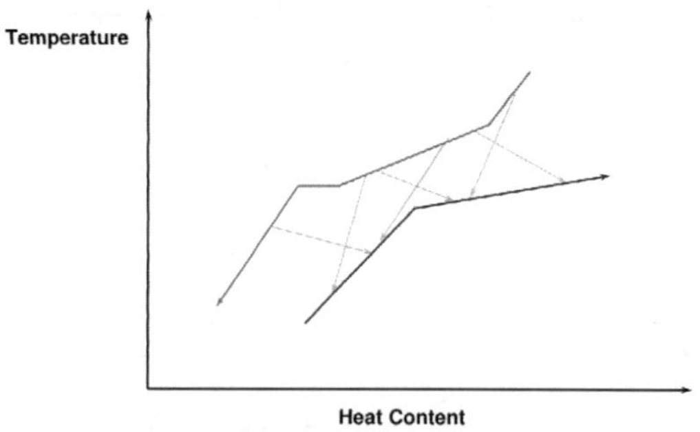

Area - Energy Plot

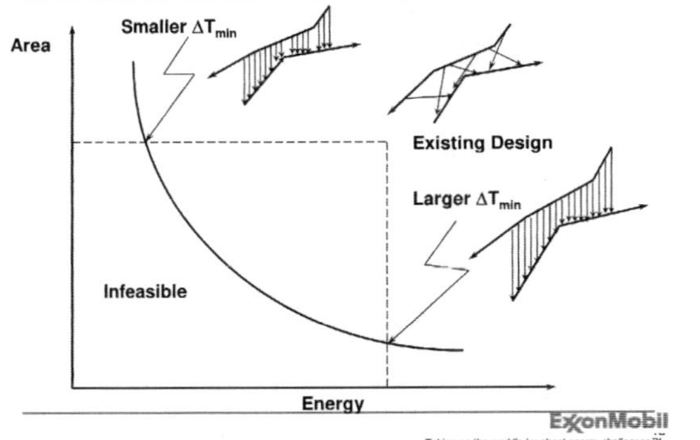

Area - Energy Plot For Revamps

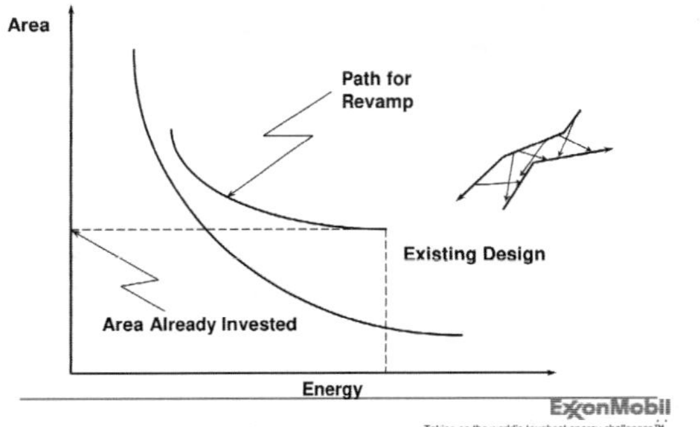

Area Efficiency

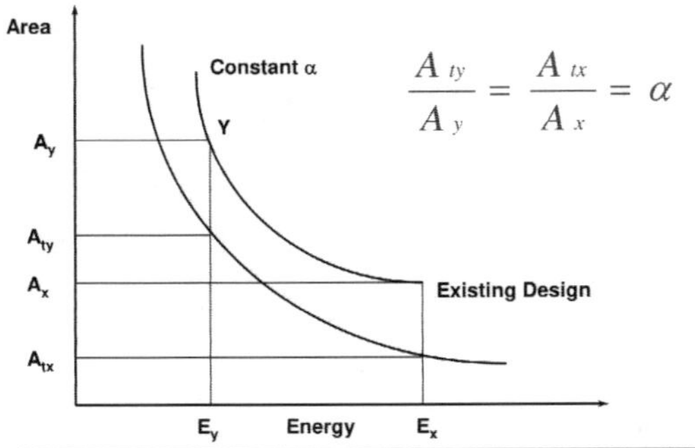

Economic Targets for Retrofit Design

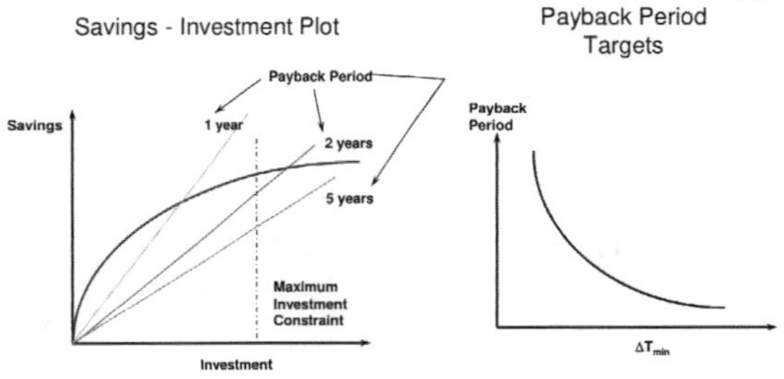

Summary

- Payback Period Constraints
- Investment Constraints
- Target Value of ΔT_{min}
- Correct Initialization of Design

Retrofit Design

Are the Existing Exchangers Properly Placed in The Network ?

ΔT Plot

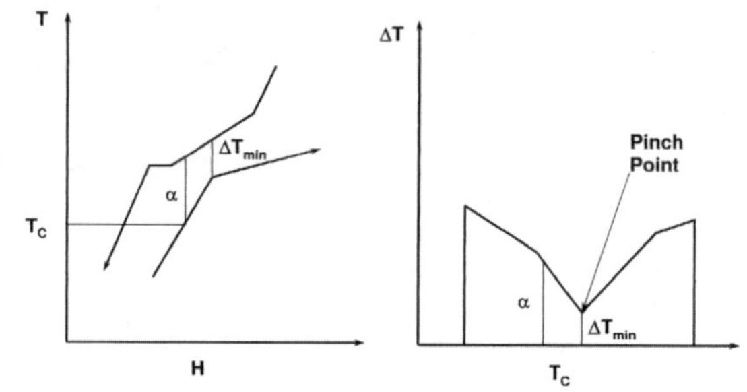

ExxonMobil
Taking on the world's toughest energy challenges™
49

Using ΔT Plot

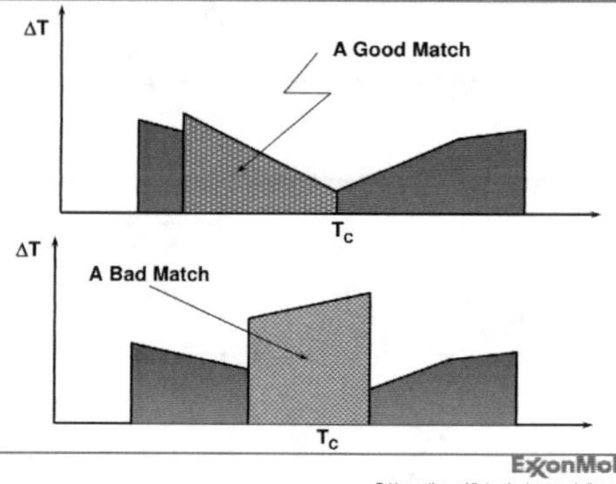

ExxonMobil
Taking on the world's toughest energy challenges™
50

Design Procedure

- Performance Test
- Target For ΔT_{min}
- Identify Bad Matches/Cross-Pinch Matches
- Eliminate Bad Matches/Cross-Pinch HX
- Use Pinch Design Method To Complete The Network
- Try To Reuse The Removed Exchangers

ExxonMobil
Taking on the world's toughest energy challenges™
51

Alternate Revamp Procedure

- Network Pinch Method
  - Identify the pinching exchanger in the network that limit energy savings and the location of these exchangers creates a 'network pinch'
  - Pinching exchanger can be determined by increasing the area of that exchanger till the ΔT_{min} of that exchanger reaches a practical minimum (LP problem)
  - Identify ways to modify the pinching exchanger (MILP problem)
    - Resequencing the exchanger
    - Repiping the exchanger
    - Add a new exchanger
    - Splitting a stream
  - Repeat the procedure till the target energy saving is obtained (Search tree)

ExxonMobil
Taking on the world's toughest energy challenges™
52

Utility Placement Scope

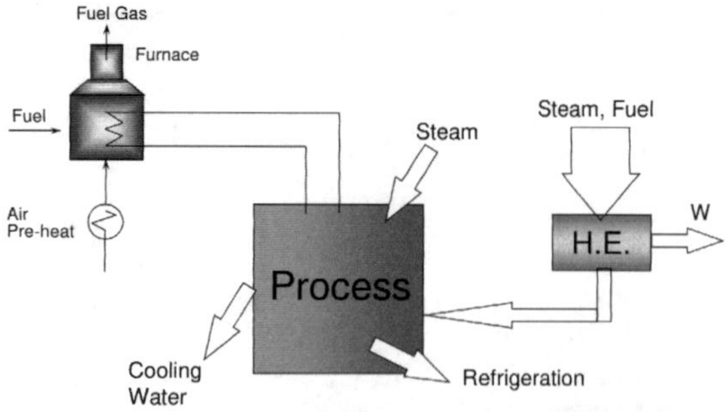

ExxonMobil
Taking on the world's toughest energy challenges™
53

Utilities on Composite Curves

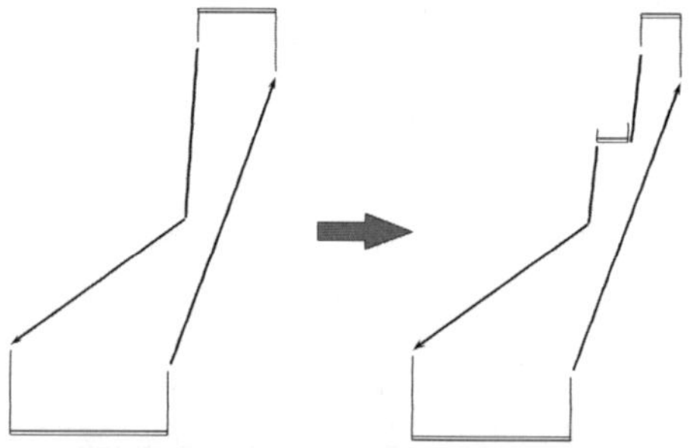

ExxonMobil
Taking on the world's toughest energy challenges™
54

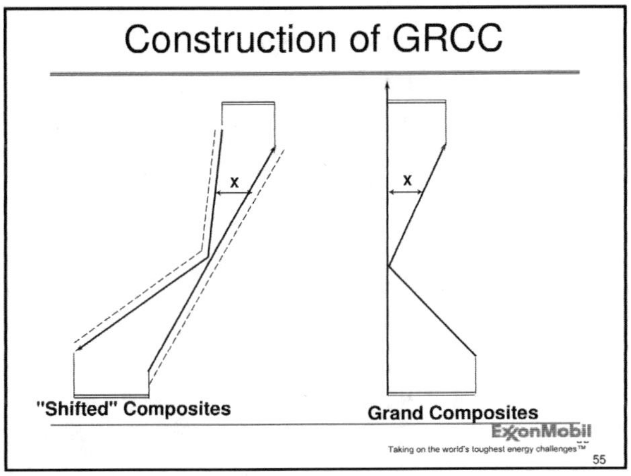
Construction of GRCC
x
x
"Shifted" Composites
Grand Composites
ExxonMobil
Taking on the world's toughest energy challenges™
55

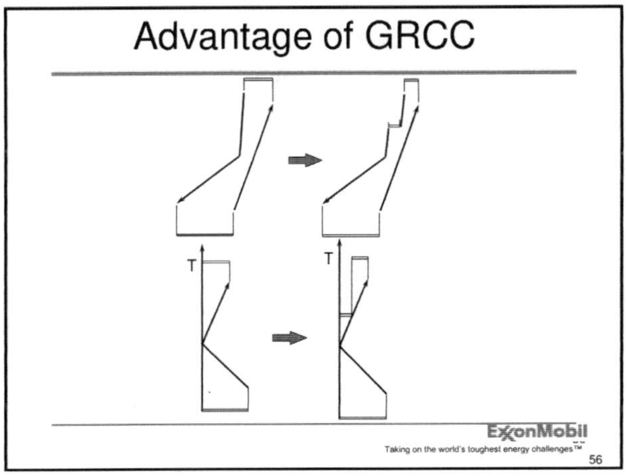
Advantage of GRCC
T
T
ExxonMobil
Taking on the world's toughest energy challenges™
56

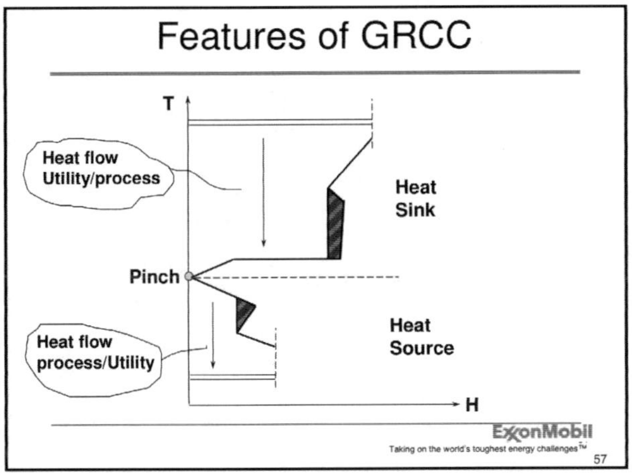
Features of GRCC
T
Heat flow
Utility/process
Heat
Sink
Pinch
Heat
Source
Heat flow
process/Utility
H
ExxonMobil
Taking on the world's toughest energy challenges™
57

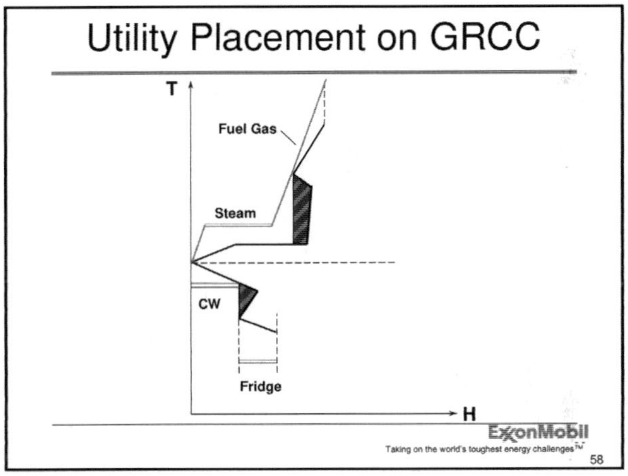
Utility Placement on GRCC
T
Fuel Gas
Steam
CW
Fridge
H
ExxonMobil
Taking on the world's toughest energy challenges™
58

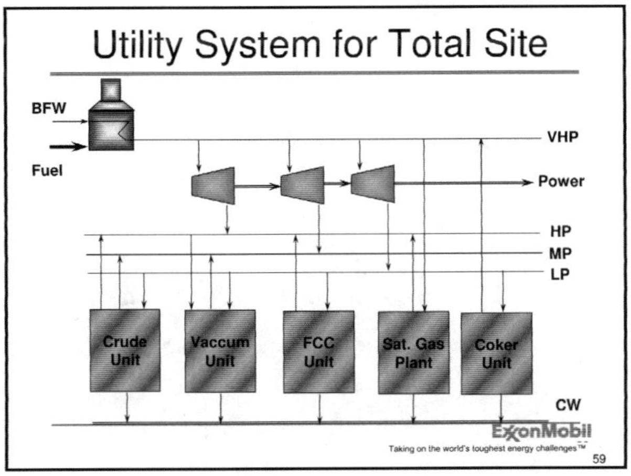
Utility System for Total Site
BFW
Fuel
VHP
Power
HP
MP
LP
Crude Unit
Vaccum Unit
FCC Unit
Sat. Gas Plant
Coker Unit
CW
ExxonMobil
Taking on the world's toughest energy challenges™
59

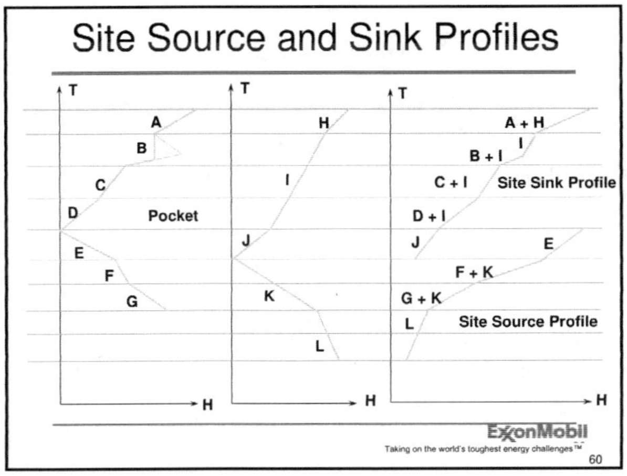
Site Source and Sink Profiles
T
A
B
C
D
Pocket
E
F
G
H
T
H
I
J
K
L
H
T
A + H
I
B + I
C + I
Site Sink Profile
D + I
J
E
F + K
G + K
L
Site Source Profile
H
ExxonMobil
Taking on the world's toughest energy challenges™
60

Site Pinch

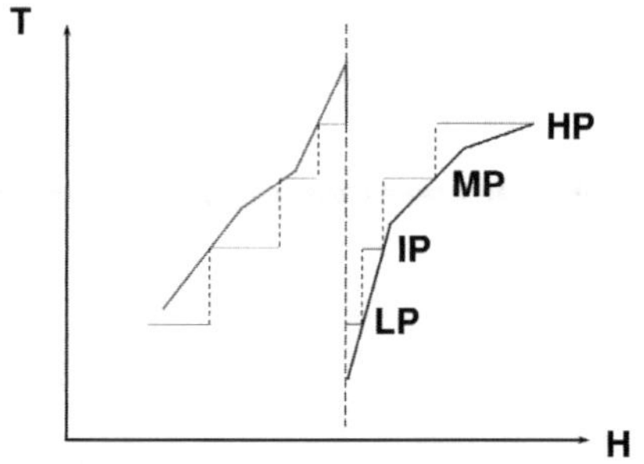

Site Pinch

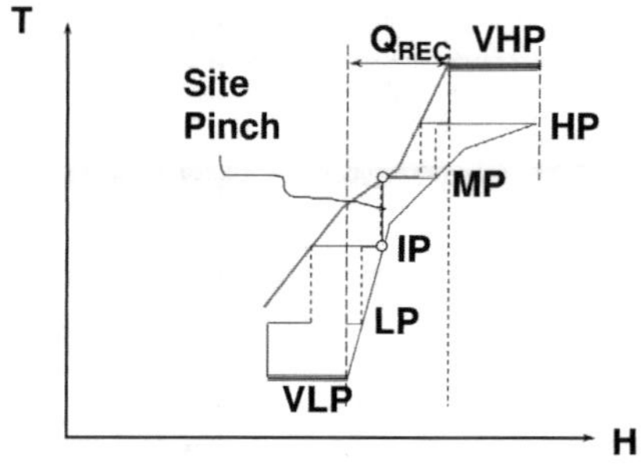

Site CC & GRCC

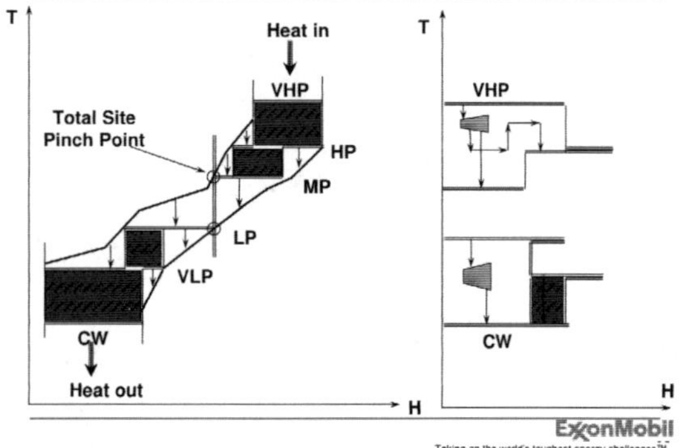

Other Process Integration Techniques

- Top level analysis and Optimization of combined Heat and Power systems
- Water Pinch – optimizing water use
- Hydrogen Pinch – optimizing hydrogen use
- Distillation Integration

Further Reading

- Trivedi, K. K., "Process Energy Efficiency: Pinch Technology," CRC Handbook of Energy Efficiency, editors F. Kreith and R. West, (1996).
- Smith, R, "Chemical Process: Design and Integration," 2005
- Linnhoff B., Townsend D.W., Boland D., Hewitt, G. F., Thomas BEA, Guy A.R., and Marsland R.H., "A User Guide on Process Integration for the Efficient Use of Energy," IChemE, Rugby, UK (1982)
- Kemp, I. C., "Pinch Analysis and Process Integration: A User Guide on Process Integration for the efficient Use of Energy," Second Edition, 2007

ENERGY CONSERVATION PROGRAM REDUCES COSTS AND CO2 EMISSIONS IN REFINERIES

Helmy Andrawis, Joe Ferrall
WorleyParsons
Monrovia, CA

INTRODUCTION

In addition to product supply and cost uncertainties, future regulations on CO2 emissions for operators of refineries and petrochemical plants represent another challenge. If legislation on CO2 emissions is enacted or other programs to limit CO2 emissions are realized, energy efficiency in plants will become increasingly important to maintain profitability. Typical energy costs to operate a refinery can be the largest single operating costs; with fuel and power costs expected to escalate, the effect of energy use costs to overall operating costs can be expected to increase.

Recent crude oil fluctuations from \$40 to near \$150/bbl have added significant uncertainty to the viability of energy investments. However, it is generally accepted that in the future, the costs of feedstock to a refinery will escalate over time. Figure 1 shows conceptually how production efficiency (= energy output/energy input x 100%) decreases as the cumulative production of the fossil resource is consumed including crude oil, shale and tar sands. As efficiency of production decreases, the cost of the recovered fuel resource should increase.

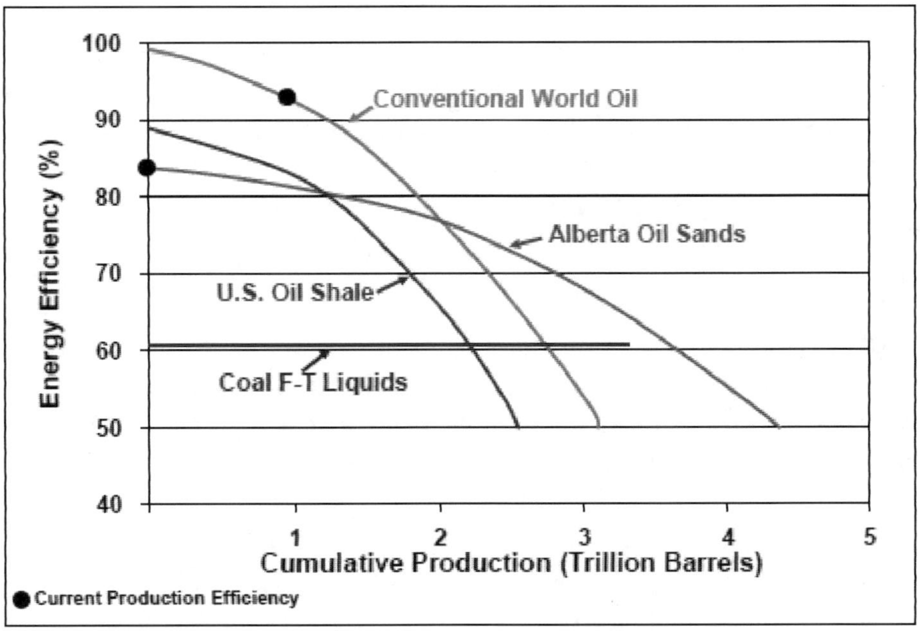

Figure 1 Energy Resource Recovery Efficiency (2006) (1)

In addition to fuel costs increasing, the prospect of a "carbon tax" or something similar to fuel or energy used in a refinery will make the energy use contribution even more significant to profitability. By planning and executing a well-organized and objective Energy Conservation Program, opportunities for improving both the refinery profitability and the environmental emissions can be simultaneously achieved (2).

This paper discusses typical refinery energy usage and the equivalent CO2 emissions attributable to fuel consumption. The impact of reducing energy use on cost savings is presented to include the effect of fuel and potential CO2 emission costs that may be passed on to the refinery.

A phased Energy Conservation Program is discussed showing what can be expected from implementing an energy program. Analysis tools that accurately depict the energy consumption and emissions picture can pinpoint opportunities available by targeting economic energy consumption and investment payback.

More importantly, the Energy Conservation Program needs total management commitment and operations need to continue operating the plant as efficiently as possible. In today's market, this can only be achieved using information management that translates energy usage to data points on the operator's screen just like a flow rate or a product specification. Operations can then be held accountable for meeting specific energy and performance targets including energy as well as CO2 emissions.

REFINERY ENERGY EFFICIENCY AND CO2 EMISSIONS

Energy consumption in a refinery can be classified into two major categories:

1. Thermal Energy Consumption. Examples include fired heaters, incinerators, boilers and coke burn in FCC units.

2. Electrical Energy Consumption. Examples include motors, lighting and power supply to instrumentation.

Table 1 summarizes energy consumption results of a site survey of four refineries. The survey results use energy units measured as Fuel Oil Equivalent Barrels (FOEB) to compare the total energy consumed at each of the sites with the total crude oil throughput. An FOEB is a well-established measure representing the volume of fuel consumed in a refinery. A unit of FOEB is equivalent to 6.05 MMBtu.

TABLE 1 TYPICAL REFINERY ENERGY CONSUMPTION

REFINERY	CAPACITY	THERMAL ENERGY (1)		ELECTRICAL ENERGY (2)		TOTAL ENERGY AS % OF CAPACITY
	BPSD	FOEBPD	MMBTU/HR	FOEBPD	MWe	
A	120,000	7,200	1815	1,300	32.8	7.1
B	160,080	6,600	1664	1,760	44.4	5.2
C	98,080	8,480	2138	1,510	38.1	10.2
D	140,000	9,560	2410	1,630	41.1	8
Nominal Average	130,000	8,000	2017	1,550	39.1	
Percent of Total Energy		84%		16%		100%

Notes

1 1 FOEB = 6.05 MMBtu

2 Heat rate = 10,000 Btu/kWhr

This table shows that thermal energy averages about 84 percent of the total energy consumed in a refinery and electrical energy represents about 16 percent of the total energy consumed. It is assumed that these values are representative of most refineries. The results suggest that thermal energy has the highest contribution to the energy consumption and should be the target of any conservation program. Also, reducing thermal energy consumption directly affects the CO2 emissions at the refinery.

Refineries typically generate and use fuel gas for internal use and can generate power for internal use. This paper only considers the imported fuel and power for all of its energy usage; CO2 emission sources as a result of refinery operations would be as indicated in Figure 2.

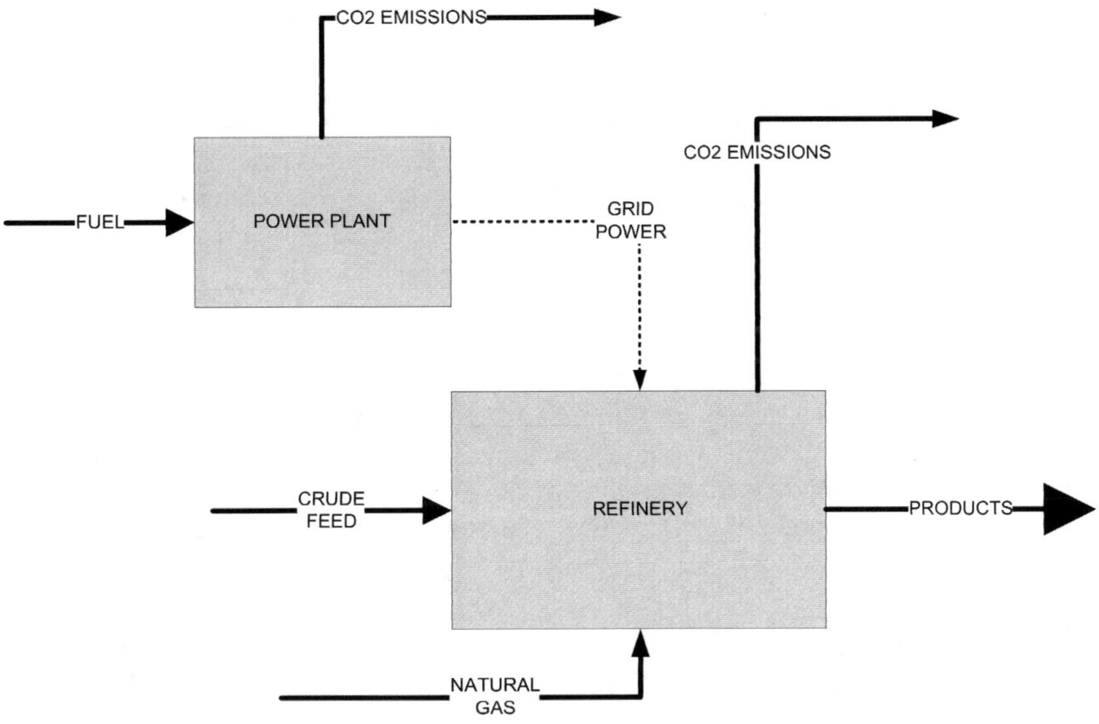

Figure 2 Generic CO2 Emissions Sources from Refinery Operations

Using EPA emissions equivalent factors (3), Table 1 shows over 3,000 metric tons per day of CO2 emissions due to refinery energy use. While the emissions from thermal energy use would occur at the refinery site, the emissions from power use would occur at the power plant as schematically shown in Figure 2.

Table 2 Estimated CO2 Emissions from Typical Refinery Energy Usage
(Crude feed rate = 130, 000 TPD)

	ENERGY CONSUMPTION	UOM	CO2 EMISSION FACTORS	UOM	CO2 EMISSIONS Metric tons/day
THERMAL ENERGY	2,017	MMBtu/hr	0.05	metric tons CO2/MMBtu	2,420
ELECTRICAL ENERGY	39.1	MW	7.12E-04	metric tons CO2/KwH	668
TOTAL					3,088

If only 10% of the combined energy and fuel usage can be saved, 310 tons/day CO2 emissions would be reduced for an average refinery.

POTENTIAL FUEL AND CO2 CREDIT COST SAVINGS

From a detailed refinery energy savings analysis of several refineries performed by WorleyParsons personnel (4), economically viable projects were identified that would result in 12-14% savings in total energy consumption; viability at that time was defined to include a three-year payback and very low crude oil costs ($12/bbl). With crude oil prices escalating to significantly higher prices, it is likely that if analyses were to be performed today, projects could be justified resulting in higher energy reductions.

A "what-if" analysis is presented here showing cost savings potential for the typical refinery. Figure 3 shows potential refinery cost savings for energy reductions up to 30% and for natural gas costs at $6, $8, and $10/MMBtu; gas prices are in the range of predictions of an Energy Information Agency report for industrial delivered prices (5). At $8/MMBtu and 30% energy savings, a $40MM/year cost savings is possible for a 130, 000 bpd refinery.

In addition to fuel savings, savings in the form of CO2 emission credits could be realized if "cap and trade", CO2 tax, or some other program is implemented affecting US refineries. A study was performed for the National Petroleum and Refiners Association (6) on the effect of such a program; CO2 costs to a refinery could escalate ranging from $20/metric ton in the near-term to over $60/ton in the future. Figure 4 shows potential CO2 credit savings for up to 30% energy use reduction. At $20/metric ton and 30% energy use reduction, potential CO2 credits would be over $5 MM/year.

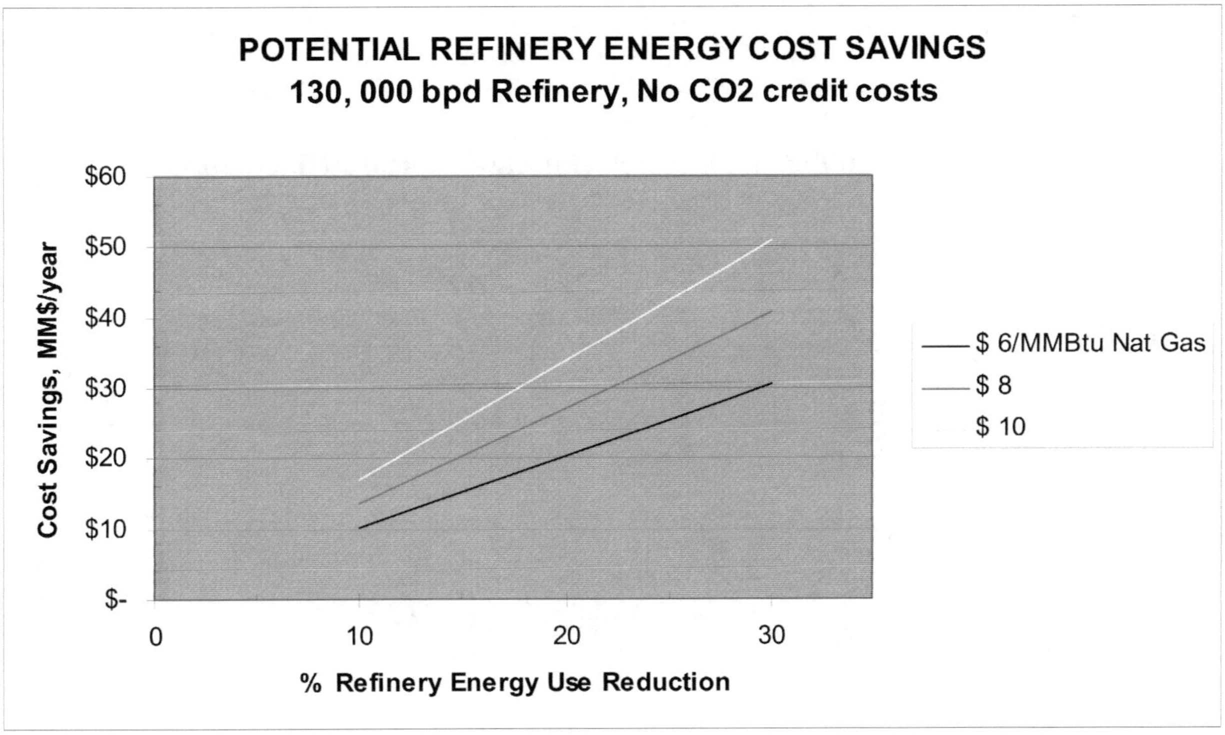

Figure 3 Potential Fuel Cost Savings and Refinery Energy Use Reduction

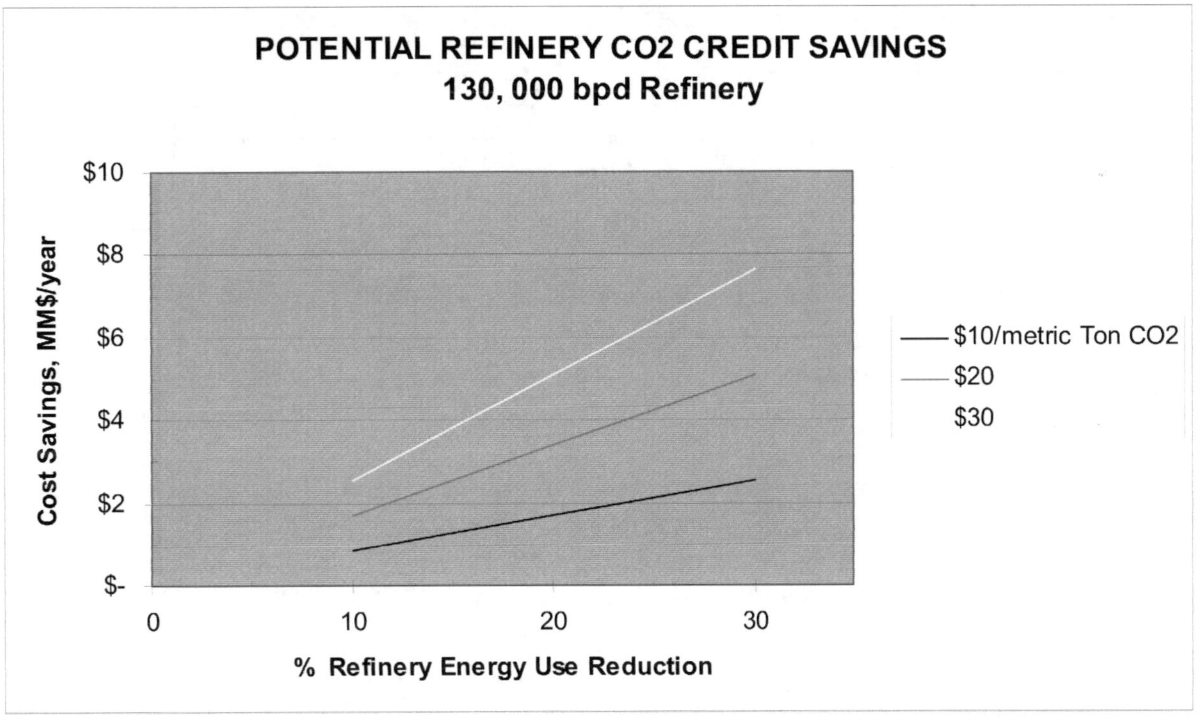

Figure 4 Potential CO2 Credit Cost Savings and Refinery Energy Use Reduction

The combined potential cost savings by reducing thermal energy use (energy and CO2 credits) are shown in Table 3. At $8/MMBtu, $20/metric ton CO2 and 30% refinery energy-use reduction, the refinery could save $46 MM/year.

Table 3 Total Thermal Potential Cost Savings for Typical Refinery Operation (Crude feed rate = 130,000 BPD, 30% Refinery Energy Use Reduction)

THERMAL ENERGY COST SAVINGS ($MM/YEAR)				
Fuel Costs	CO2 Credit Costs			
$/MMBtu	$0/metric Ton	$10/metric Ton	$20/metric Ton	$30/metric Ton
6	30	33	36	38
8	41	43	46	48
10	51	53	56	58

In addition to thermal energy reductions, power use reductions would also result in cost savings as indicated in Table 4 using a range of power costs from6 to 10 cents/kW-hr projected as a range typical for industrial users (7). However, the CO2 emissions credits would probably be credited to the power plant and not the refinery. At 8 cents/kW-hr and 30% power reduction, the refinery could save nearly $8MM/year.

Table 4 Potential Power Cost Savings for Typical Refinery Operation (Crude feed rate = 130,000 BPD)

POWER ENERGY COST SAVINGS ($MM/YEAR)			
Electricity Costs	Power Reduction %		
Cents / kW-hr	10%	20%	30%
6	2.0	3.9	5.9
8	2.6	5.3	7.9
10	3.3	6.6	9.8

ENERGY CONSERVATION PROGRAM

As discussed above, if a 30% energy use reduction is achieved, thermal and power savings of combined could amount to $54 MM/year. To achieve these savings, a rigorous energy reduction program must be adopted.

An energy conservation program has been developed in phases to offer a refinery an opportunity to review intermediate results of the program as it progresses as shown in Figure 5, Energy Conservation Program. The program is tailored for refiners operating one or more sites. The four distinct phases of the program are:

- Phase I, Site Survey and Energy Opportunity Identification
- Phase II, Energy Audit and Benchmarking
- Phase III, Energy Integration and Conceptual Design
- Phase IV, Project Implementation and Verification

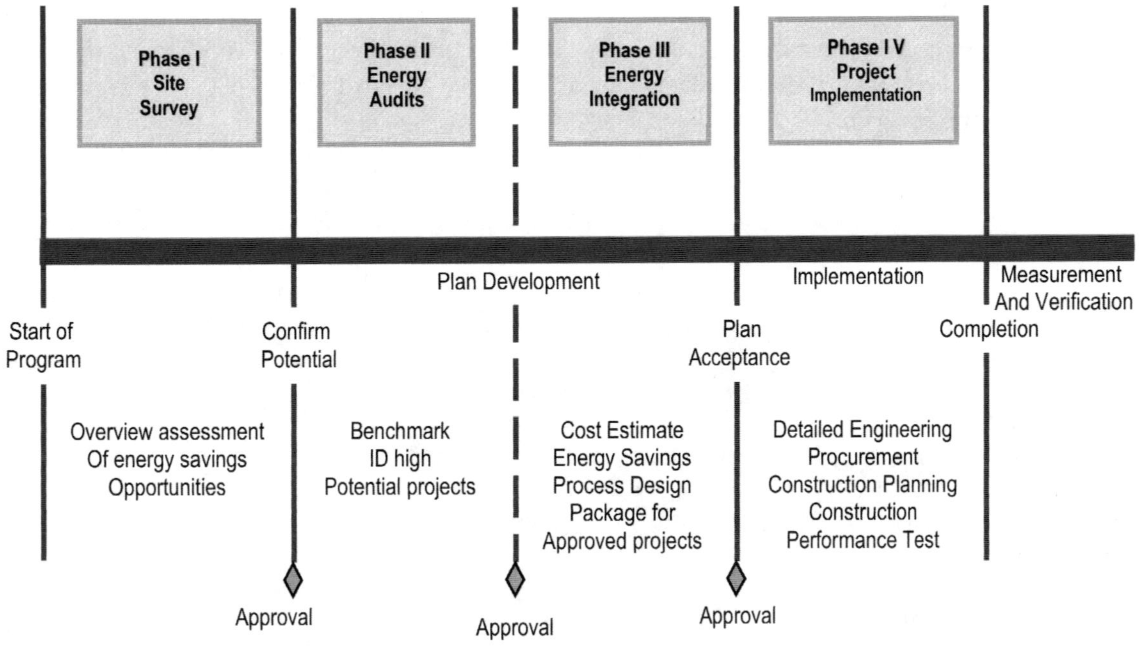

FIGURE 5 ENERGY CONSERVATION PROGRAM

PHASE I, SITE SURVEY: This phase starts with the most important step, Team Building. The teams for the energy program both the contractor and the owner are integrated into one Energy Team that is responsible for developing the following:

- Program goals, objectives and expectations
- Communicating the corporate objectives with each of the refineries
- Execution plans, schedules, costs and review cycles
- Follow-up, execution and implementation of the program
- Verify and measure results of the program

After the team has set the plans, check lists and needs lists are prepared for each site. The team visits and assembles energy teams at each site. The site teams support the Energy Team in collecting the necessary data and information and in conveying the messages to the operations teams.

The Energy Team visits each site to collect data and verify site conditions. Meetings with the site team establish common goals, plans, schedules and expectations.

The Energy Team reviews the site energy consumption using the data collected and compares it to a base energy consumption from typical consumption figures for each unit and for the site. An example of a thermal energy consumption survey by units is represented in Figure 6. The energy ratio comparing actual to base energy consumption guages the potential for energy savings. The higher the ratio the higher the potential. This procedure is used to prioritize the sites and units under evaluation and select the order of implementation. The Energy Team also solicits the refinery for actual energy projects that have been under consideration at each site. These projects include process improvements, reliability and energy projects.

The Energy Team reports the results of the survey in a final report. This report sets the basis for the second phase of the program and offers the refinery an opportunity to review the progress and direction of the program.

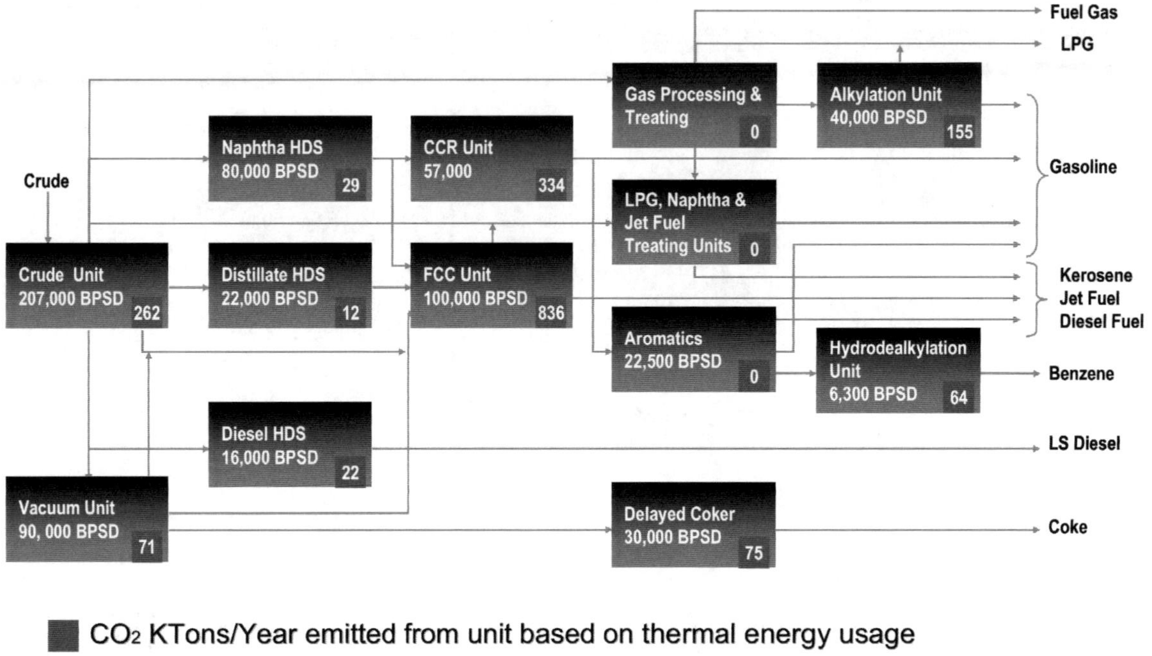

FIGURE 6 SURVEY EXAMPLE OF REFINERY UNIT EQUIVALENT CO2 EMISSIONS

PHASE II, ENERGY AUDIT This phase covers a more detailed analysis of the energy consumption at each site. The phase starts with developing Energy Audit Procedures and a Basis for the Audit. The Energy Team spends time in the refinery collecting more informaiton such as process flow diagrams, piping and instrumentation diagrams, operating data, DCS print-outs and major equipment data sheets. Performance testing and test results are collected from all major units.

The team uses the information to develop process simulations of the plant data and verifies and reconciles measured versus simulated data. The process models are used to calculate the actual energy consumption and establish energy consumption benchmarks for each unit and for the site.

The simulation results are used as a basis for developing the input to the Pinch Analysis tools. Specialized software are used to analyse existing systems and procedures such as Energy and Area Targeting, Utility Placements and Payback Analysis which are then used to establish design targets. A site-wide analysis establishes utility targets and opportunities for improving energy recovery and process integration.

Models for the steam system, which is one of the major energy consuming components in a plant, help identify losses and opportunities to improve energy recovery and optimize operating conditions.

A review of the performance and efficiencies of major equipment such as heaters, turbines and compressors completes the audit procedure.

In reviewing the results of the energy audit, project opportunities and improvements are identified that can be implemented at the site. The team use the audit results to develop conceptual project definitions that are completed and estimated to evaluate the expected economic paybacks as a result of implementing these projects. The projects are prioritized and reviewed with the refinery for their buy-in and agreement. The Phase II final report, which included energy benchmarks, saving projections, project economics, recommendations, project plans and schedules, sets the stage for phase III.

PHASE III, ENERGY INTEGRATION This phase develops the projects identified in Phase II to a higher level of detail and cost estimate to verify the economic and practical applications of the energy program. The projects that pass the technical and economic screens developed by the team are selected and are further developed into process design packages for project execution and installation.

This phase starts with a review of the project priority list. More data gathering takes place to verify as-built conditions at the site and update all project documents. Process engineering develops the projects and completes the design of the alternatives.

Tools such as Pinch Technology are used to verify the targets set in the audit phase and to improve heat exchanger network designs. Process integration and optimization is implemented in the designs of each of the identified projects. Utility systems are reviewed and modified to accommodate the designs in each of the units.

Process Design Packages complete with heat and material balances, process flow diagrams, marked P&ID's and equipment data sheets are prepared for the modified and new equipment. Cost estimates and budgetary costs as well as quotes are solicited from suppliers to confirm equipment costs. Constructability reviews take place with the refinery personnel; schedules are established with the refinery to plan project installations in accordance with the refinery shutdown and turnaround schedules. Plans for the Project Implementation along with schedules and design packages are issued for implementation. The packages are complete with costs estimates and economic anaylsis of the recommended designs.

PHASE IV, PROJECT IMPLEMENTATION In this phase , the finance plans are completed and the go-ahead decisions are made. The detailed design of the projects take place through a contracting stategy established by the Energy Team and the owner. An engineering contractor is selected to manage the detailed design, engineering, procurement and construction of the designs.

After the construction of the facilities are complete the team conducts performance tests to verify the plant performance and acceptance of the designs.

MEASUREMENT AND VERIFICATION As part of the program, the Energy Team implements a Measurement and Verificaiton program that ensures that the necessary instrumentation and control facitilties are installed.

Data is collected from the DCS and instruments and sends it to a dedicated computer system specifically designed for the Energy Management. This system is designed to run the plant linear program (LP) that completes the overall material balances around each of the units and for the site. The results of the material balance are reconciled using plant information extracted from each of the units. The LP results for each of the units drive the process simulation models developed for each unit. These models, developed during the initial design phases, are used to calculate the actual energy consumption. Reports can be generated, using models developed before and after the modificaitons, that compare the energy consumption before and after the modifications. This energy information is shared with the process control stations so that all operators see the energy targets and maintain an energy efficient plant operation. Critical equipment can be monitored for efficiency, fouling and performance. Performance targets are set to trigger maintenance requests for cleanning, testing and inspection.

SURVEY RESULTS OF AN ENERGY CONSERVATION PROGRAM

The results of an energy conservation program indicated that energy savings in the 12-14 percent of total energy consumption is achievable and that energy savings improves profitabilty and reduces emissions with the following assumptionms:

- 150 on-stream operating days/year
- $12/bbl crude oil
- 3-year investment payback
- Thermal energy reductions only

During the site survey it was recognized that savings can be enhanced by improving process operations through improvements such as feed tray locations, optimizing reflux rates, improving reactor performance and many other opporunities that the refinery operations and process studies can identify. Other areas for improvements includes reliability. Table 5 is representative of the classes of projects that can be found for energy improvement.

TABLE 5 CLASSES OF PROJECTS

PROJECT CLASS	PERCENT OF TOTAL
Energy Reduction Projects	51
Process Improvement Projects	42
Reliability Projects	7

Table 6 indicates representative equipment related projects where refiners could focus their efforts to improve energy consumption.

TABLE 6 EQUIPMENT FOR ENERGY SAVINGS

EQUIPMENT CLASS	PERCENT OF TOTAL PROJECTS
Heat Exchangers	22
Heaters and Boilers	23
Rotating Equipment	3
Process Operations	29
Steam System	23

CONCLUSIONS

Typical energy costs are the largest component of refinery operating costs. With fuel and power costs escalating over time, the effect of energy use costs to overall operating costs will increase. Reducing energy costs is necessary to refinery profitability.

In addition to fuel costs increasing, the prospect of a "carbon tax" or something similar to fuel or energy used in a refinery will make the energy use contribution even more significant to profitability.

Implementation of a phased Energy Conservation Program in refineries has the potential of significant energy cost savings. For a typical refinery (130,000 bpd, 30% energy use reduction), a cost savings of $54 MM/year could be realized as follows:

	$MM/year
• Thermal energy reduction fuel savings @ $8/MMBtu Natural Gas:	40
• Thermal energy reduction CO2 credits @ $20/metric ton CO2:	6
• Power energy electricity savings @ $0.08/kW-hr power cost:	8

TOTAL	54

REFERENCES

1. US Department of Energy, Office of Petroleum Reserves – Strategic Unconventional Fuels – Fact Sheet, http://www.fossil.energy.gov/programs/reserves/npr/Energy_Efficiency_Fact_Sheet.pdf

2. Andrawis, H, et al, "Energy Conservation Improves Refinery Margins", 2001 NPRA Annual Meeting, March 18-20, 2001

3. Environmental Protection Agency, www.epa.gov/cleanenerg/energy-resources/refs.html

4. Andrawis, H, et al, "Energy Conservation Improves Refinery Margins", 2001 NPRA Annual Meeting, March 18-20, 2001

5. "Annual Energy Outlook 2009", Energy Information Administration, Natural Gas Supply, Disposition, and Prices, **http://www.eia.doe.gov/oiaf/aeo/excel/aeotab_13.xls**

6. Harrison, D., "Impacts of Potential Climate Change Policies on the Refining and Petrochemical Sectors", April 2008 www.npra.org/files/NERA_Full_Report_S._2191.pdf

7. "Annual Energy Outlook 2009", Energy Information Administration, Electricity Supply, Disposition, Prices, and Emissions, http://www.eia.doe.gov/oiaf/aeo/excel/aeotab_8.xls

52c Energy Efficient Design In the Continuous Process Industries: From Nuisance to Necessity

Bill Townsend, Petrochemicals & Advanced Chemical Engineering, M W Kellogg Ltd., Kellogg Tower, Greenford Road, Greenford, UB6 0JA, United Kingdom

A retrospective view of design for energy efficiency in the process industries reveals that there has been a fundamental change of thinking over the last 30 years. It is useful to review the change in order to understand its significance for process design in an energy-challenged future.

Chemical Engineers have always sought to use energy efficiently, but priorities in the investment boom up to the late 1970s in an energy-abundant world dictated an emphasis on scale, 'repeatability' and materials yield. The 'oil shocks' forced an abrupt change, and process energy efficiency, including revamps, received much greater attention. In this context, 'crossover' thinking from the power and cryogenics industries brought systematic thermodynamic analysis into general process design.

A key concept was to analyse process plants as integrated systems and to apply temperature interval analysis. From this flowed all the benefits of 'pinch' understanding in heat recovery systems, including the concepts of placing CHP and process units in relation to the pinch. Associated concepts were 'cross-the-boundary' thinking applied to adjacent units and even sites, and the inclusion of pinch design techniques within overall process synthesis methods. Systematic analysis of the capex-energy cost tradeoff in energy systems was introduced.

The drive for energy efficiency lost some impetus in the 1990s due to very cheap oil, but interestingly, improvements in process selectivity and in catalyst development brought reduced process energy consumption. Also, innovative separation technologies and new distillation systems including practical dividing wall columns were developed, improving energy demand. Meanwhile, computing power continued to grow and powerful optimisation methods allowed engineers to explore many more optimum process configurations.

Examples of energy efficiency improvement are given, in ethylene, ammonia, and refining. For the future, the systems approach is being applied to the broader requirement for sustainability in process design, and the systems approach continues to be a focus of design research.

52d Energy Efficiency - Future Directions

Robin Smith, Centre for Process Integration, The University of Manchester, PO BOX 88, Manchester, United Kingdom

Increasing energy prices and pressures to decrease the emissions of greenhouse gases demand significant improvements in process efficiency, both for existing and new processes.

The utility systems of most sites have evolved over many years without fundamental questions being addressed as to their design and operation with much redundancy in place. This redundancy allows the site to respond to changes in the operation across the site, for maintenance of equipment and for breakdown. However, this redundancy also presents opportunities to improve performance by managing its operation more effectively. Thus, utility systems offer the opportunity of substantial savings given their large investment requirements and operating costs.

Investigating improvements in energy integration of individual processes should begin by understanding the utility system that services the process. Only in this way can the true incentives to improve the energy performance of individual processes be established. Thermodynamic techniques for heat exchanger network design and retrofit were developed in the 1980's. These have proved to be extremely successful industrially. However, these methods also have significant drawbacks, which is particularly true when improving the performance of existing processes in retrofit. This problem was difficult to address using the thermodynamic methods developed in the 1980's, since when they were applied to retrofit they effectively considered the problem as a pseudo new design. To be effective, whatever approach is used, should not attempt to move the design of an existing process towards an ideal new design, but to take account of what is existing.

This workshop will review state-of-the art technologies to improve energy efficiency in chemical processes.

New Sulfur Limits for Bunker Fuels: The Challenges and Opportunities for the Refinery Industry

Maoqi Feng, Chemical Engineering Department, Division of Chemistry and Chemical Engineering, Southwest Research Institute, San Antonio, TX 78238, email: mfeng@swri.org.

Abstract

The demand for bunker fuels (heavy fuel oils) is constantly increasing thanks to the booming shipping industry. Recently, the International Maritime Organization (IMO) approved the regulations for a decrease of sulfur limits of marine bunker fuels from 4.5% globally to 3.5% by 2012 and to 0.5% in 2020-2025. For the Emission Control area (ECAs) in the U.S., which includes California, the Pacific Coast, and other areas, the sulfur level requirements are more stringent: 1.5% until 2010, 1.0% after 2010, and a further reduction to 0.1% in 2015. The new regulations have raised concerns for the availability, cost and quality of lower sulfur bunker fuels, and this brings a lot of challenges for refinery industry. Since shifting to conventional diesel fuel for the shipping industry is an economic disadvantage, and after-treatment technology such as exhaust scrubbing can only solve part of the problems of meeting emission requirements, desulfurization of bunker fuels is becoming necessity in the refineries. Traditional diesel fuel desulfurization methods including hydrotreating might still be useful, but are less economically feasible than other technologies, e.g., hydrocracking with new catalysts and improved reactor design. This paper will review the technologies for low sulfur bunker fuels production, including hydrotreating and conversion technology, including coking, hydrocracking, visbreaking, oxidative desulfurization, and ultrasound treatment.

Introduction

On July 21, 2008, the U.S. officially signed the International Maritime Organization (IMO) "Marpol Annex 6" legislation, which requests cleaner marine diesel fuels and lower ocean-going ship engine emissions to cut ship pollution. Under the new IMO legislation, the sulfur limit will be reduced from the current level of 4.5% to 1.5% by 2010, to 1.0% after 2010, and a further reduction to 0.1% in 2015 for the Emission Control Areas (ECAs) [1], includingCalifornia, the Pacific Coast, and possibly other areas in the U.S.

Bunker fuel is also known as heavy fuel oil, No. 6 fuel oil, resid, Bunker C, blended fuel oil, residual fuel oil, furnace oil, etc. This fuel type is used to fuel ships, oil-fired power plants, and large-scale heating installations. Bunker fuel is the residuum derived from crude oil after naphtha-gasoline, No. 1 fuel oil (a light petroleum distillate, straight-run kerosene, consisting primarily of hydrocarbons in the range C9-C16), and No. 2 fuel oil (hydrocarbon chain lengths in the C11-C20 range) have been removed. No. 6 fuel oil is a heavy residual fuel normally a mixture of atmospheric and vacuum distillation residues cut back with kerosene or gas oil cutter stock to adjust for viscosity. The vacuum residuum has a boiling point range of 565°C and above. Bunker fuel is more complex in composition and impurities than distillate fuels, including polycyclic aromatic hydrocarbons and their alkylated derivatives, and metal-containing constituents. Limited comprehensive data are available on the composition of bunker fuel oil.

Before being sold as bunker fuel, vacuum resid needs to be diluted to meet various sales specifications for trace metals, sulfur and/or viscosity. In bunker fuel, sulfur exists as both water-soluble contaminants, e.g., contaminant-metal sulfates, sulfites, and sulfides, and oil soluble contaminants, such as polysulfides, thiols, mercaptans, etc. The normal practice for lowering sulfur content is to reduce the amount of sulfur in fuels that are sold by blending with low sulfur solvents.

New regulations are calling for a staged, step-by-step decrease in the sulfur content of this particular product. In the past, refiners have traditionally used the unregulated marine bunker fuel pool as a way to dispose of bottom-of-the-barrel refinery streams high in sulfur. The new regulation puts more restriction on refiners. Refineries may need to install processes, e.g. deep cracking and residue hydrocracking, to upgrade high sulfur bunker fuel to be legal to sell or to more valuable distillates such as gasoline and the middle distillates. This paper reviews the regulations for bunker fuel and its effect on the refining industry.

Bunker Fuel Specifications

The International Organization for Standardization (ISO) regulates the specifications of marine fuels in cooperation with marine and petroleum industries to meet the supply worldwide for use onboard ships. There are 19 categories of residual fuels available internationally; only four categories or grades are most frequently supplied and used by ships.

Industrial names for the four grades are IFO-180 (Intermediate Fuel Oil), IFO-380, MDO (Marine Diesel Oil) and MGO (Marine Gas Oil), which correspond to the ISO grades RME 25, RMG 35, DMB and DMA, respectively. Industrial nomenclature is based on the fuel viscosity at 50°C. IFO 180 indicates that the viscosity of the fuel is 180 cSt at 50°C [2], and IFO 380 fuel has a viscosity of 380 cSt at 50°C. MDO is a blend of gas oil and heavy oil. MGO is clear and not blended with residual oil.

IFO-180 (or RME 25) has about 6 to 7 % gas oil in it, while IFO-380 (or RMG 35) has about 3% gas oil. Because of the higher gas oil content, the price of IFO-180 is higher than the heavier IFO-380 grade. MDO or DMB is a blend of heavy fuel and gas oil and they are cheaper than pure MGO or DMA. Requirements are different for both the grades. In addition to gas oil content difference, requirements for 4 out of 11 parameters are different for the two grades. As you can see from the table, the maxima for viscosity, carbon residue, ash content and vanadium are different for the two grades.

The ISO specifications for bunker fuels are listed in Tables 1 and 2. Here the fuel types RMA, RMB, RMD, RME, RMF, RMG, RMH, RMK are used for heavy fuels.

Table 1. ISO 8217:2006(F) fuel quality standard for residual marine fuels.

Parameter	Limit	RMA 30	RMB 30	RMD 80	RME 180	RMF 180	RMG 380	RMH 380	RMK 380	RMH 700	RMK 700
Density at 15°C, kg/m^3	Max	960.0	975.0	980.0	991.0		991.0		1010.0	991.0	1010.0
Viscosity at 50°C, mm^2/s	Max	30.0		80.0	180.0		380.0			700	
Water, volume%	Max	0.5		0.5	0.5		0.5			0.5	
Micro Carbon Residue, weight%	Max	10		14	15	20	18	22		22	
Sulfur, weight%	Max	3.5		4.00	4.50		4.50			4.50	
Ash, weight%	Max	0.10		0.10	0.10	0.15	0.15	0.15		0.15	

Vanadium, mg/kg	Max	150		350	200	500	300	600	600
Flash point, °C	Min	60		60	60		60		60
Pour point, Summer, °C	Max	6	24	30	30		30		30
Pour point, Winter, °C	Max	0	24	30	30		30		30
Aluminum + Silicon, mg/kg	Max	80		80	80		80		80
Total Sediment Potential, weight%	Max	0.10		0.10	0.10		0.10		0.10
Zinc, mg/kg	Max	15							
Phosphorus, mg/kg	Max	15							
Calcium, mg/kg	Max	30							

Note: * - A sulfur limit of 1.5% m/m will apply in SOx Emission Control Areas designated by the International Maritime Organization, when its relevant Protocol comes into force. Theremay be local variations. See Reference [2].

Table 2. ISO 8217:2006(F) fuel quality standard for distillate marine fuels

Parameter	Limit	DMX	DMA	DMB	DMC	Test method
Density at 15°C, kg/m^3	Max	-	890.0	900.0	920.0	ISO 3675
Viscosity at 40°C, mm^2/s	Max	5.50	6.00	11.00	14.00	ISO 3104
Viscosity at 40°C, mm^2/s	Min	1.40	1.50	-	-	ISO 3104
Micro Carbon Residue at 10% Residue, weight%	Max	0.30	0.30	-	-	ISO 10370
Micro Carbon Residue, weight%	Max	-	-	0.30	2.50	ISO 10370
Water, volume%	Max	-	-	0.30	0.30	ISO 3733
Sulfur*, weight%	Max	1.00	1.50	2.00	2.00	ISO 8754 or ISO 14595
Total Sediment Existent, weight%	Max	-	-	0.10	0.10	ISO 10307-2
Ash, weight%	Max	0.01	0.01	0.01	0.05	ISO 6245
Vanadium, mg/kg	Max	-	-	-	100	ISO 14597
Aluminum + Silicon, mg/kg	Max	-	-	-	25	ISO 10478
Flash point, °C	Min	43	60	60	60	ISO 2719
Pour point, Summer, °C	Max	-	0	6	6	ISO 3016
Pour point, Winter, °C	Max	-	-6	0	0	ISO 3016
Cloud point, °C	Max	-16	-	-	-	ISO 3015
Calculated Cetane Index	Min	45	40	35	-	
Appearance		Clear & Bright		-	-	
Zinc, mg/kg	Max	-		-	15	IP 501
Phosphorus, mg/kg	Max	-		-	15	IP 501
Calcium, mg/kg	Max	-		-	30	IP 501

Note: * - A sulfur limit of 1.5% m/m will apply in SOx Emission Control Areas designated by the International Maritime Organization, when its relevant Protocol comes into force. There may be local variations. See Reference [2].

A simplified version is given in Tables 3 and 4. These are the principal properties that may be used to compare available bunker fuels. The following sections describe the refining operations that are employed to meet these specifications.

Table 3. The specification for marine diesel fuel

Properties		Test Method
Specific gravity @ 60°F	≤0.840	ASTM D 1298
Pour point, °F	≤15	ASTM D 97
Sulfur, wt%	N/A	ASTM D 129
Diesel index	≥50	IP 21
Flash point (PM), °F	≥150	ASTM D 93
Viscosity @ 100 °F	32.0-43.9	ASTM D 88
Carbon residue (condradson) on 10% bottoms, wt%	≤0.2	ASTM D 189

Note: Adapted from [3]

Table 4. Specification for No. 6 fuel oil product

Properties		Method of test
Specific gravity @ 60°F	≤0.990	ASTM D 1298
Pour point, °F	≤65	ASTM D 97
Sulfur, wt%	≤1.5	ASTM D 129
Viscosity kinematic @ 122 °F, cS	≤370	ASTM D 445
Flash point (PM), °F	≤160	ASTM D 93
Ash, wt%	≤0.1	ASTM D 482
Water (by distillation), vol%	≤1.0	ASTM D 95
Sediment (by extraction), wt%	≤0.1	ASTM D 473

Note: Adapted from [3]

Bunker Fuel Upgrading by Visbreaking

Visbreaking is a mild form of thermal cracking; about 15% conversion of the boiling range atmospheric residue to gas oils and naphtha can be achieved, and a low-viscosity residual fuel is produced [4]. Visbreaking is a flexible tool for heavy oil conversion.

There are two types of visbreaking: short contact and soaker. In short contact visbreaking, the feed is heated to about 900°F (480°C) and sent to a reactor (soaking zone) at 140 - 300 psig. The elevated pressure allows cracking to occur while restricting coke formation. To avoid over cracking, the residence time in the soaking zone is only several minutes, which is much shorter than in a delayed coker (several hours). The hot oil is quenched with cold gas oil (to inhibit further cracking) and sent to a vacuum tower for product separation.

Soaker visbreaking keeps the hot oil at elevated temperature for a longer time to increase the yield of middle distillates. The low-viscosity visbreaker gas oil can be sent to an FCC unit or hydrocracker for further processing, or used as heavy fuel oil.

Unless cracked or sold as asphalt, the vacuum reduced crude (VRC) is usually sold as low price heating oil. These heavy fuel oils have viscosity and pour point specifications they must meet in order to be sold. It is usually necessary to dilute the VRC with lighter gas oils in order to meet the #6 or bunker fuel specifications. These gas oils are more valuable to the refinery for use as feedstocks for conversion processes, and it is desirable to minimize the amount of cutter stocks used [5].

The heavy nonvaporized portion of crude oil is broken down into smaller components to be used in gasolines and heating oils. A visbreaker uses heat to break down the long, heavy molecules remaining at the bottom of the distillation towers. The visbreaking products from two different residues are listed in Table 5 [6].

Visbreaker gas oil often goes to a catalytic cracker for further refining. The residual fuel oil, is the residue left after naphtha, gasoline and other fuel oils have been removed. The bottom from a visbreaker can be mixed with other light oils and used as bunker fuel.

Table 5. Visbreaking products from different residue feedstock

	Residue Feed: Kuwait Long	Residue Feed: Agha-Jari Short
Visbreaking products	Residue Yields, wt%	Residue Yields, wt%
Gasoline	2.5	2.4
Naphtha	5.9	4.6
Gas oil	13.5	14.5
Tar	78.1	78.5

Note: modified from Reference [6].

Residue Demetallization

Compared with distillate desulfurization units, residue desulfurization units are expensive to build and operate because of the severe operating conditions (higher temperature and pressure) and higher catalyst consumption. Thus, it is beneficial or necessary to increase the onstream operating factors such as run length, operating unit throughput, and resid conversion to distillate [7].

A major obstacle in residue desulfurization is the high metal content (mainly nickel and vanadium) in the feed, which rapidly deactivates the catalyst in the guard reactor, resulting in short run lengths, and limits the severity of operation of the unit. Consider an atmospheric residue desulfurization (ARDS) unit with Middle Eastern atmospheric resid as the feed. This material had API gravity = 13.5°, and sulfur = 4.2 wt%, Table 6 lists the operating conditions for the atmospheric residue desulfurization unit. To achieve higher on-stream conversion and longer run time, a guard reactor is used and operated at high temperature and its catalyst deactivates fast due to metal deposits. This reactor is replaced. Reactors downstream of guard reactor receive feed with much lower metal content, and therefore can achieve much higher conversion rates [8].

Table 6. Operating Conditions for the Atmospheric Residue Desulfurizer Unit

Operating parameters	Value
Reactor inlet temperature	360°C
Reactor outlet temperature	380°C
Catalyst bed weighted average bed temperature	370°C
1st reactor inlet pressure	1880 psig

H_2 partial pressure, inlet	1475 psig
HP hot separator pressure	1812 psig
HP hot separator temperature	380°C
HP warm separator pressure	1785 psig
HP warm separator temperature	260°C
HP cold separator pressure	1735 psig
HP cold separator temperature	40°C
LP hot separator pressure	340 psig
LP hot separator temperature	370°C
LP cold separator pressure	325 psig
LP cold separator temperature	40°C
Recycled gas rate	5060 scf/bbl
H_2 chemical consumtion	460 scf/bbl
Makeup H_2	771 scf/bbl
Purge ratio, purge to makeup	45 wt%
Fractionator	
Flash zone temperature	368°C
overflash	0.6 vol%
Column top pressure	21.0 psig
Gas oil/resid cut point	368°C

Adapted from [7].

In industry, catalyst in the guard reactor is replaced using a variety of designs: two fixed-bed guard reactors in series, continuous addition of fresh catalyst and withdrawal of spent catalyst in an ebullated type guard reactor. Typically up to 50% in vanadium content is reduced. Generally, nickel and molybdenum on macroporous alumina with pore volume (0.9 cm^3/g) is used as the catalyst [7].

Residue Feedstock Desulfurization

Special hydrogenation is widely used for residue desulfurization. Residue desulfurization is achieved partly as a result of thermal decomposition. Thus, in the absenceof the catalyst, hydrodesulfurization reaches 30–40% of the conversion that might be obtained in the presence of the catalyst [8, 9]. As an example, typical desulfurization reactions of dibenzothiophene are shown in Figure 1.

For the reactions listed in Figure 1, there is still disagreement for the reaction order. For an atmospheric residue with an initial boiling point of 343°C, the reaction order ranged between 1.9 and 2.3, and the energy of activation was about 132–145 kJ/mole [8].

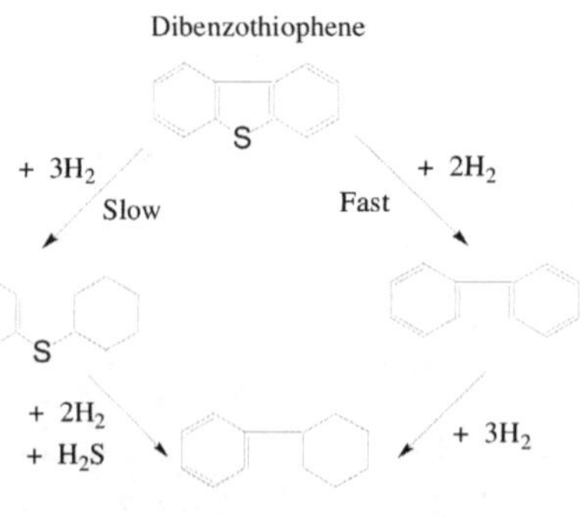

Figure 1. Typical desulfurization reaction by hydrodesulfurization [8]

The hydrodesulfurization of asphaltenes and of a residual fraction was reported [7]. The hydrodesulfurization process was accompanied by some cracking, which made the sulfur atoms in the previous more complex molecules accessible. Here Co-Mo/Al_2O_3 was used as the catalyst. The kinetics of converting the Conradson Carbon Residue (CCR) in catalytic hydroprocessing was also reported. A half-order kinetics was established with the activation energy being 277.58 kJ/mol. No dependence of the rate constant on hydrogen pressure was observed. It was found that the CCR is linearly related to the content of asphaltenes, hydrogen content, H/C atomic ratio, and residue content (350°C). Gas yield was also linearly related to the CCR content.

Catalytic hydrotreating and hydroconversion processes are extensively used in the petroleum refining industry for the upgrading of heavy oils and residues. Desulfurization of heavy oils is more difficult than that of distillates, including vacuum gas oil, because there are more contaminants present and very large, complex molecules are involved. The asphaltene fraction in heavy fuel oils forms coke readily, which deactivates the catalyst. Proper catalysts can overcome this, and other measures can be taken, such as the use of adequate hydrogen partial pressure [10].

For hydrotreating of the very heavy distillates and of residues, severe hydrogenation conditions are required in order to make accessible the heteroatoms contained within the molecule. For this reason, the hydrodesulfurization of such fractions is carried out in more severe hydrogenation conditions.

Recently hydrotreating of residues for desulfurization, denitrogenation, and demetallation has made significant improvement [9]. In combination with the catalytic cracking of the hydrotreated residues, hydrotreating is the most attractive methods for converting heavy residues to motor fuels with high yields and relatively low hydrogen consumption. Compared with the direct catalytic cracking of the residual fractions, hydrotreating has two major advantages: 1) avoid the poisoning of the catalyst by heavy metals such as Ni and V, and 2) produce gasoline that meets more severe specifications for sulfur content.

Three types of processes exist for the commercial implementation of the residues hydrotreating:

- Fixed catalyst bed systems
- Moving bed systems
- Ebullated bed systems

For the three systems, the pressure is of the order of 2600 psig, and an average temperature of 450°C [8].

A fixed catalyst bed system is shown in Figure 2. Four reactors connected in series are used with cooling between the catalyst bed and separator for the processing of distillates. In this design, the deposition of the metals on the catalyst is concentrated on the first portion of the contact zone, especially in the upper portion of the first reactor. Since the catalyst poisoned with metals cannot be regenerated on stream, the first reactor can be taken out of the process to replace the catalyst, without having to stop the operation of the unit.

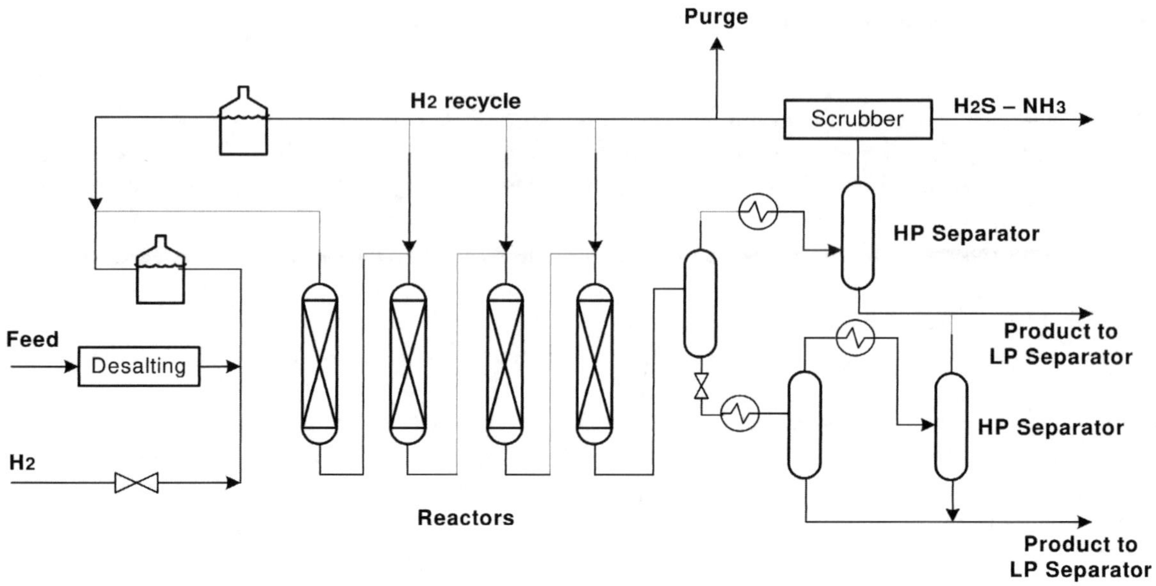

Figure 2. Fixed bed residue hydrotreating unit [8]

Generally two types of catalyst are used: the first oneis for breaking the asphaltene molecules and demetallation, and the second one is for hydrotreating. The proportion between the two catalysts depends on the properties of the feed and the quality requirements for the finished product. To protect the fixed bed catalyst, a two-step desalting of the crude oil is needed, followed by a filtration of the feed to the unit, in order to prevent plugging of the pores of the catalyst by metal salts.

A typical example for this system is the HYVAHL process developed bythe French Institute of Petroleum, which can process feed with high metal concentration (up to 500 ppm V and Ni). The system uses two interchangeable guard reactors located ahead of the main reactors: the first one containing demetallation catalyst, the other reactors contain both demetallation and desulfurization catalysts. Residfining process is also an example of catalytic fixed-bed process for the desulfurization and demetallization of residua. This process can also be used to pretreat residua to low contaminant levels prior to catalytic cracking.

The advantages of the process in fixed bed are the lack of catalyst attrition andlow risk of plugging of the bed, which can be of concern in system with moving bedor ebullating bed.

For systems with ebullated bed, the process flow diagram is shown in Figure 3.

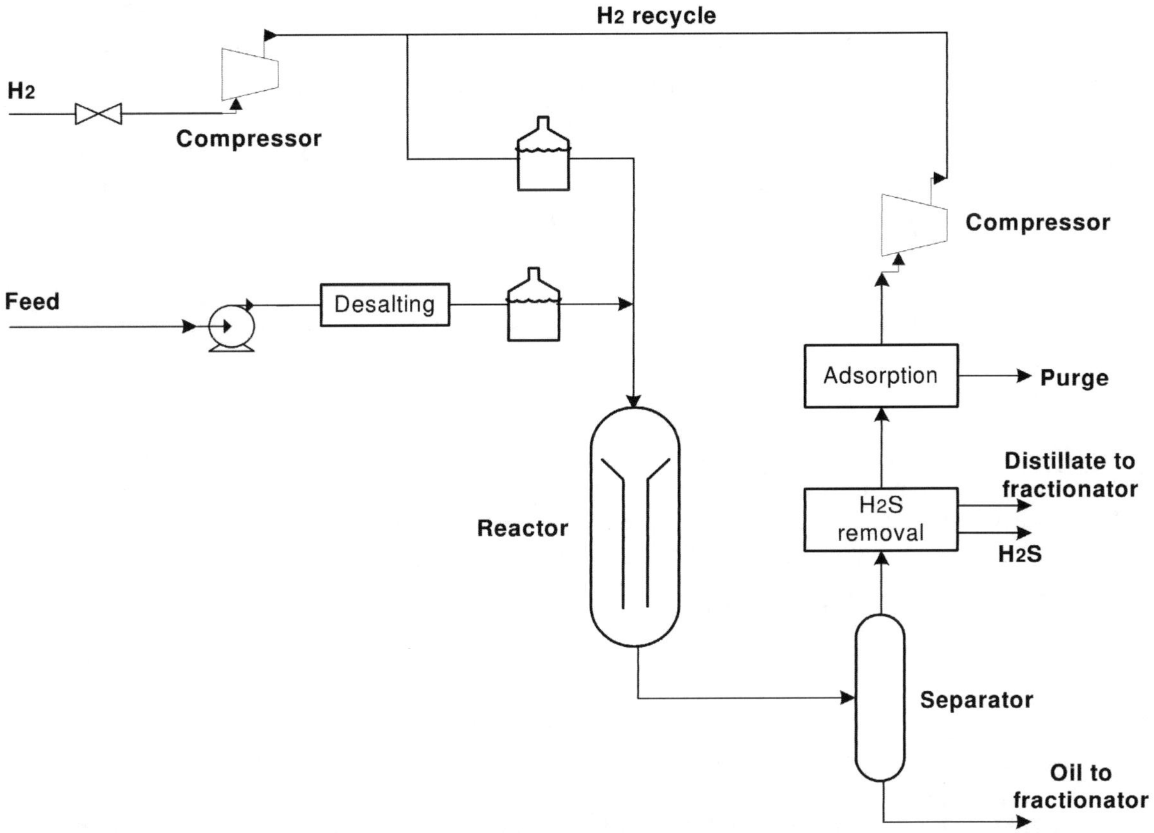

Figure 3. Process flow diagram for residue hydrotreating with one-step ebullated bed reactor. Adapted from Figure 10.28 in Reference [8].

There are three phases present in the reactor: moving gas, solid, and liquid phase. It is more difficult to achieve adequate contact for the gas phase with the liquid phase and solid phase. Currently, there are three examples for this process: H-Oil and L.C. Fining, with minimum differences in the design of the equipment. The typical operating conditions for such a unit are: temperatures of 420–450°C, pressures of 1000–2000 psig, space velocity 0.1–0.5 h^{-1}, the recycling coefficient inside the reactor is in the range of between 5 and 10 [8]. The recycled hydrogen is 3–4 times as the amount consumed in the reaction.

Deep thermal cracking takes residue conversion a step further, so that a larger amount of distillate product is being processed from the pool.

Hydrocracking

In hydrodesulfurization, residue desulfurization is achieved partly as a result of thermal decomposition [8]. Hydrocracking offers another method for residue desulfurization. Thelight distillate product produced from hydrocracking is used for blending with heavy esidue as bunker fuels.

The Mexican Petroleum Institute (IMP) has developed a new hydrocracking process for heavy crude oil and residua, which is 15–30% cost-effective than those of current commercial processes [11]. The two-stage process costs less since it operates at a lower pressure of ~100 kg/cm^2, compared with 200 kg/cm^2 for conventional methods. Typically the reaction temperature is 400–

430°C, and an alumina-supported nickel-molybdenum catalyst was used. In the tests, the process has reduced both the metals and sulfur content of Mexican Maya crude by about 80%. IMP has also developed a new catalyst for diesel desulfurization. The catalyst was 40% more active [12].

Idemitsu Kosan Co. and Petroleum Energy Center in Tokyo developed a new catalyst system for residue oil hydrodesulfurization (RDS) and residue oil fluid catalytic cracking (RFCC) processes for upgrading of heavy oils to boost gasoline and middle distillates to 64%, higher than the conventional RFCC process (60%) [13]. Instead of using conventional macroporous catalysts (100 nm), mesoporous catalyst (10 nm) was used for the RDS process and metals (V, Ni) removal yield was improved from heavy oils, and the desulfurization efficiency was also increased.

Biodesulfurization

Biotechnology can also be used for sulfur removal. The advantage is that bacteria work in mild ambient conditions, and thus saves a lot of equipment investment. Generally there are two types of biodesulfurization methods: aqueous phase and nonaqueous phase. In aqueous phase, the sulfur compounds are converted to sulfates, which dissolve in water, while in nonaqueous phase, the sulfur compounds are degraded to element sulfur for further removal by centrifuge or filtration.

One example of aqueous biodesulfurization is Energy Biosystems Corp. (EBC)'s biocatalytic desulfurization process, which uses genetically engineered bacteria to convert sulfur to sulfates and then removed from the fuel. In the process, water containing the bacteria is mixed with the fuel at ambient pressure and temperature, and air is sparged in to facilitate the conversion [14].

Tests in a 60-gal reactor in its pilot facility for diesel fuel desulfurization showed that the process decreased sulfur content in middle distillates from 2,000 ppm to 500 ppm [15]. A new biocatalyst was capable of increasing the conversion rate and lowering the sulfur content of the diesel fuel from 600 ppm to 300 ppm. The process is an alternative to conventional, high-pressure hydrodesulfurization and is expected to save 50% in capital costs and 20% in operating costs.

UOP is also developing a biodesulfurization process, called Thiopaq. This process was originally developed by Paques BioSystems B.V. [16].

Anaerobic biodesulfurization by sulfate-reducing bacteria has been applied in the desulfurization of model sulfur compounds such as thiols, sulfides, disulfides and benzothiophenes. The advantage of anaerobic biodesulfurization is that no oxygen is needed to introduce into the fuel, and may be capable of being directly applied to the oilfield reservoir. Thus there is no fuel stability problem, and this makes the process is safer [17].

For aqueous phase biodesulfurization of bunker fuel, high water content remaining in the bunker fuel can result in a number of operating difficulties [18]: 1) sludge formation; 2) tanks and pipelines corrosion; 3) erosion of injectors by cavitation, etc.

Nonaqueous biocatalytic desulfurization has been used in heavy oil viscosity reduction [19]. Nonaqueous biocatalysis reduces many concerns, including the design of expensive large volume bioreactors, and product recovery and separation by interfacing the non-aqueous biocatalytic system to existing chemical process reactors. Further developments for the implementation of this technology are: 1) more efficient biocatalysts, 2) more selective biocatalysts specifically for sulfur compounds in the feedstock, and 3) analytical systems for the bioprocessing systems monitor and control.

Only limited work has been done on the biodegradation of bunker oils, and many of the saturated, polycondensed components in the bunker fuel may be quite resistant to biodegradation [18].This technology is also limited by several factors: 1) some ancillary processes which are novel to petroleum refining would be needed, including a biocatalyst fermentor to reuse the bacteria; 2) The process is sensitive to environmental conditions, such as sterilization, temperature, and residence time of the biocatalyst; 3) some biocatalysts can not take high sulfur feedstock, thusthe process requires the existing hydrotreater to provide a lower sulfur feedstock to the unit.

Oxidative Desulfurization (ODS)

ODS is an alternative process to the traditional hydrodesulfurization (HDS) process and it has received much attention recently for deep desulfurization of high sulfur feedstocks. Hydrogen peroxide (H_2O_2) is generally used as the oxidant in organic acids (e.g. acetic acid or formic acid) medium for the ODS process (Equations 1-2). Compared to the HDS process, ODS process has two main advantages: 1) the ODS process can be carried out in the liquid phase and under very mild operating conditions; 2) the most refractory sulfur-containing compounds to the HDS process, e.g. dibenzothiophene (DBT) and its derivatives, show high reactivity toward the oxidation by this method. In the ODS process, the sulfur compounds are oxidized to their corresponding sulfoxides and subsequently sulfones [20]. Afterward, these highly polarized products can be removed by solvent extraction or adsorption.

$$\text{Sulfide} + 2H_2O_2 \xrightarrow[H^+]{75°C} \text{Sulfone} + 2H_2O \qquad (1)$$

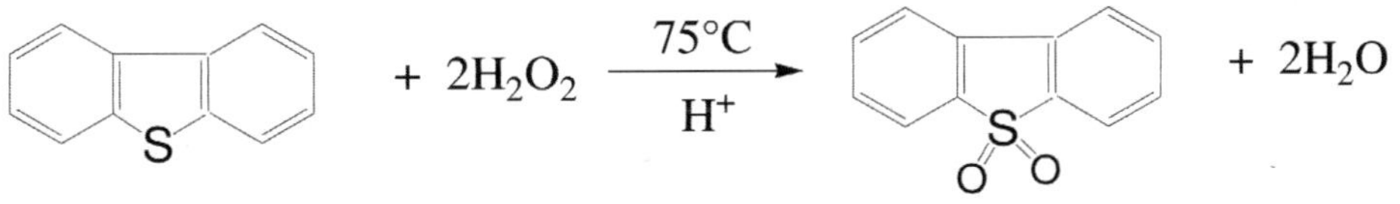

(2)

ODS process catalyzed by solid catalyst is of great interest recently. Activation of H_2O_2 had also been achieved with polyoxometallates with a Keggin structure, such as $H_3PM_{12}O_{40}$ (M = Mo(VI), W(VI)), and titanium silicalites (TS-1), which produce more effective and selective oxidants as polyoxoperoxo complexes, such as $PO_4[MO(\mu\text{-}O_2)(O_2)_2]_4^{3-}$, for the oxidation of sulfides [21].

ODS technology is being pursued by Unipure, Petrostar and Lyondell, etc [22].For the ODS process used by theses companies, the chemical oxidation desulfurization of diesel fuel is accomplished by first forming organic acid emulsion with the diesel fuel. In the emulsion, the sulfur atom is oxidized to sulfone by peracid formed from the reaction oforganic acid and H_2O_2. The sulfones hydrocarbon molecules are polar and hydrophilicand then move into the aqueous phase and removed.

Petrostar did a bench scale pilot plant test and they intended to demonstrate their technology next with a commercial demonstration unit. Unipure did a pilot plant test in 2003 [22]. Lyondell also developed an ODS process in 2003 which uses t-butyl hydroperoxide oxidant to convert sulfur species to sulfones; this eliminates the recycle of the organic acid. The oxidant was degraded to t-butyl alcohol after the sulfur species were converted to sulfones. The t-butyl alcohol byproduct can be used as fuel, and t-butyl hydroperoxide is not as corrosive as peroxyacetic acid. Lyondell had demonstrated pilot plant tests, which desulfurized 500 ppm diesel fuel to below 10 ppm.

Technically, it is possible to incorporate the ODS technology in bunker fuel desulfurization. Since stoichiometrically one mole of sulfur compounds needs two moles of H_2O_2, the economic feasibility of this process for bunker fuel desulfurization will depend on the price of H_2O_2 market.

Ultrasonic Processing

It has been reported that ultrasound can significantly improve the reaction efficiency under phase transfer conditions [23]. Ultrasound increases liquid–liquid interfacial area through emulsification, which is essential for viscous films containing gas-filled bubbles and cavitation bubbles. Cavitational bubbles spray solvent on the film that covers the pulsing gas bubble; therefore, the gas bubble is disrupted by the pulsing action, leading to highly dispersed emulsions.

Ultrasonic process has the advantages of low investment cost and without use of the expensive hydrogen for hydrotreating. The treatment of crude oil fractions, fossil fuels, and organic liquids by sonic energy with the presence of an oxidizing agent was reported [24]. The oxidative treatment with ultrasound reduced the concentrations of sulfur and nitrogen.

Sonochemical treatment of heavy gas oil (HGO) derived from Athabasca bitumen had alsobeen studied [26]. During the ultrasound treatment of HGO, low volatile hydrocarbon gases such as methane, ethylene, ethane, and propylene were produced, and the concentration of the lighter gases increased with increasing amplitude and sonofication time.

SulphCo, Inc. patented an ultrasound process for crude oil desulfurization. Field trials on a commercial scale gave promising results [25]. SulphCo is focusing on process optimization to minimize costs associated with the chemical package and the overal material and mass balance for future commercialization.

Blending with GTL Products

GTL products, such as Fischer-Tropsch (F-T) diesel fuel, have the characteristics of zero sulfur, zero nitrogen, and low aromatics, and high cetane number. Since time islimit for conventional refining to meet new bunker fuel standards, this could open the door for ultra clean blending feedstock, such as GTL products.

An ExxonMobil study shows that F-T diesel could help crude oil refinery use wider and higher margin refinery steams for future production of lower emission diesel fuels [27]. This can also be applied to bunker fuels.

One of the biggest challenges in bunker fuel desulfurization, besides dealing with the deteriorated crude oil supply, may be how to dispose unwanted refinery streams. High sulfur distillate cuts, which cannot be used as bunker fuel, have to be dumped if they cannot be used as high sulfur heating oils. Blending high sulfur residue fuels with F-T diesel as bunker fuel could boost the bunker fuel production and save a lot of waste disposal cost.

After-Combustion Scrubbing (Stack-Scrubbing) of Engine Exhaust

The new IMO emissions legislation opens the possibility of employing an abatement technology, such as seawater stack scrubbing, for 1.5% sulfur bunker fuel. Ship stack emissions controls could be achieved by after-combustion scrubbing for low sulfur bunker fuel. Stack-scrubbing technology could greatly reduce nitrogen oxides (NOx), particulates and other emissions.

For big ship owners, compared with the cost of low-sulfur fuels in a ship's lifetime, stack-scrubbing technology is much cheaper to be adopted by some ship operators. For small ship owners, the cost of scrubbers to remove the sulfur combustion products may force them to burn much cleaner and more expensive fuels.

Stack-scrubbing technology could avoid the need for refiner investments in desulfurization technologies for lower-sulfur marine fuel or more expensive low-sulfur crude. This technology appears very promising, however, large-scale and long-term deployment results are still very limited [28].

Refinery Challenges

The reduction of sulfur levels in bunker fuels from the current 4.5% to 1.5%, and then later to 0.5% and further to 0.1% in the future will require significant investments in catalytic hydrodesulfurization as the only method capable of declining the lowest sulfur concentration. In fact heavy fuel oil will not be sold without hydroprocessing in the future.Refiners need to make strategic investment decisions for high-sulfur residual fuel because of the new regulations. Overall conversion capacity requirements and deep residual conversion are projected to incease by more than 25% over the next 10 years [1]. New conversion projects will also need to focus on production of lower-sulfur and higher-cetane conversion products.

Currently, bunker fuel represents ~18% of the global residual market (10-11 million barrels/day) [1], and the increase is mainly in Asia pacific market. Bunker fuel refinery throughput in the region will increase by more than 5 million b/d over the next decade. The Asia-Pacific region, particularly China, has relied heavily on catalytic cracking of lower-sulfur residual streams. New bunker fuel regulation will challenge the region's residual processing capabilities.

Reduction in bunker fuel sulfur content will worsen the growing global sulfur surplus, which currently has a 2 million-ton/yr surplus. The new regulation on bunker fuels takes effect, it could increase the sulfur supply by 0.8 million tons. New applications of the over-supplied sulfur, such as fertilizer, building construction material and road construction, will need to be developed. One alternative developed by Southwest Research Institute was based on elemental sulfur and hydrocarbon binders called Sulphlex™.

Another possible scenario is that ships could massively abandon heavy residual bunker fuel and turn to marine gasoil/diesel. This will trigger big investments at oil refineries worldwide.

If it is less expensive to install and maintain marine exhaust after-treatment for conventional high-sulfur bunker fuel combustion, then it is not necessary to invest heavily in refinery desulfurization units. Instead, use low-sulfur crude, or blending with ultra-low sulfur diesel, for modest sulfur reduction in residual fuel.

In conclusion, compared with the mandatory ultra-low sulfur diesel (ULSD) fuel regulation, the goal of low-sulfur bunker fuel is much easier to achieve since the market is smaller. An EPA report [29] stated that the use of desulfurization techniques would be sufficient to meet the proposed ULSD sulfur standards.

References

[1] J. Vautrain, "New regs require lower bunker fuel sulfur levels", Oil & Gas Journal, Nov. 24, 2008, pp.46.

[2] R. Vis, "How do the four main fuel grades differ from each other?", http://www.bunkerworld.com/technical.

[3] D.S.J. Jones, P.R. Pujado, Handboook of Petroleum Processing, Springer, 2006, pp.712.

[4] P. R. Robinson, in Practical Advances in Petroleum Processing, Volume 1, edited By Chang S. Hsu and Paul R. Robinson, Springer, 2006, pp.22.

[5] L.Z. Pillon, Interfacial Properties of Petroleum Products, CRC Press, 2008, pp.101.

[6] J.H. Gary, G.E. Handwerk, Petroleum refining - technology and economics, 4th ed., Marcel Dekker Inc., New York, 2001, pp.159.

[7] S. Parkash, Refining Process Handbook, Elsevier, 2003, pp.55-58.

[8] S. Raseev, Thermal and Catalytic Processes in Petroleum Refining, Taylor & Francis, 2003, and the cited references therein.

[9] K. Al-Dalama, A. Stanislaus, "Comparison between deactivation pattern of catalysts in fixed-bed and ebullating-bed residue hydroprocessing units", Chemical Engineering Journal 120 (2006) 33–42.

[10] J.G. Speight, Chemistry and Technology of Petroleum, 4th ed., CRC Press, 2007, pp. 624.

[11] J.H. Gary, G.E. Handwerk, Petroleum Refining - Technology and Economics, 4th ed., Marcel Dekker Inc., New York, 2001, pp.159.

[12] G. Parkinson, "A new hydrocracking process, and an improved diesel-desulfurization catalyst", Chemical Engineering, Vol. 109, Jan 01, 2002.

[13] G. Ondrey, Chemical Engineering, "New catalytic system for heavy-oil reprocessing", 1999, Vol. 106, No.6, pp.39.

[14] G. Parkinson, Chemical Engineering, "Microbial desulfurization moves forward", 1997, Vol. 104, No.1, pp.50.

[15] U.S. Patent No. 5,232,854, assigned to Energy BioSystems Corp.

[16] Chemical Engineering, Dec. 1997, Vol. 104, pp. 32.

[17] S. M. Armstrong, B. M. Sankeyt, and G. Voordouw, "Evaluation of sulfate-reducing bacteria for desulfurizing bitumen or its fractions", Fuel 1997, 76, 223.

[18] Fossil Energy Biotechnology – A Research Needs Assessment, Final Report, Contract No. DE-ACO 1-ER30156, Nov. 1993.

[19] H. Song, X. Wang, and R. Bartha (1990). "Bioremediation potential of terrestrial fuel spills, Appl. Environ. Microbiol. 56:652-656".

[20] J.M. Campos-Martin, M.C. Capel-Sanchez, J.L.G. Fierro, "Highly efficient deep desulfurization of fuels by chemical oxidation", Green Chemistry 6 (2004) 557–562.

[21] V. Hulea, F. Fajula, and J. Bousquet, "Mild Oxidation with H2O2 over Ti-Containing Molecular Sieves—A very Efficient Method for Removing Aromatic Sulfur Compounds from Fuels", Journal of Catalysis 198, 179–186 (2001)

[22] H. G., Bernal, L. C. Caero, "Solvent Effects During Oxidation-Extraction Desulfurization Process of Aromatic Sulfur Compounds from Fuels", Intl. J. Chemical Reactor Engineering 2005, Vol.3, A28.

[23] H. Mei, B.W. Mei, T. F. Yen, "A new method for obtaining ultra-low sulfur diesel fuel via ultrasound assisted oxidative desulfurization", Fuel 82 (2003) 405–414.

[24] M. Cullen, U.S. Patent 20040035753.

[25] G. Ondrey, Chemical Engineering, "SulphCo announces positive results from commercial-scale field trials", Vol. 115, November 6, 2008.

[26] R. Gopinath, A. K. Dalai, and J. Adjaye, "Effects of Ultrasound Treatment on the Upgradation of Heavy Gas Oil", Energy & Fuels 2006, *20,* 271-277.

[27] J. Peckham, "FT diesel a valuable blendstock for marginal diesel streams", Diesel Fuel News, June 10, 2002.

[28] J. Peckham, Diesel Fuel News, "Bunker Fuel Desulfurization or Stack Scrubbing? New Study Seeks Answers", Nov. 24, 2008.

[29] EPA Report 420-R-04-014, "Summary and Analysis of the Highway Diesel Fuel 2004 Pre-Compliance Reports", Sept. 2004.

59b Benzene Management Options

David Shecterle[1], Richard Isherwood[2], Doug Becci[1], and Robert Madsen[1]. (1) UOP LLC, 25 E. Algonquin Road, Des Plaines, IL 60017, (2) UOP, 50 E. Algonquin Rd, Des Plaines, IL 60017

New benzene regulations have forced refiners to evaluate strategies to manage the amount of Benzene in their gasoline pools. The regulations enacted by the EPA under Mobile Sources Air Toxics Phase 2 (MSAT2) will require refiners in the U.S. to reduce benzene in their gasoline pool to a yearly average of 0.62 vol-percent by 2011. Refiners will have to optimize across multiple units in their refineries to estabilish their optimum strategy to meet the new standards with minimum capital expenditure.

This paper will discuss several options for reduction of benzene in gasoline pools such as: prefractionation, postfractionation, and benzene saturation. Some refiners may be able to meet the new regulations by adjusting process operations to minimize benzene production and using credits to off-set benzene levels higher than 0.62 vol%. Others will have to invest in revamp or new equipment to meet the new targets.

59c Application of Amine Solutions in Refinery

Pravin Sheth, Process Design, Sunoco Refining and Supply, 10 INDUSTRIAL HIGHWAY, Lester, PA 19029

This presentation will discuss the chemistry and application of various Amine solutions in Refinery Operation.

APPLICATION OF METAL SULFATES FOR REMOVING HYDROGEN SULFIDE FROM GAS STREAMS

Mahmood Torabi Angaji, University of Tehran, Iran*
Zahra Gholami, Islamic Azad University, Shahroud branch, Iran
Fatemeh Gholami, Islamic Azad University, Shahroud branch, Iran
Seyed Abolfazl Razavi Alavi, Islamic Azad University, Shahroud branch, Iran
Salim Salimi, Islamic Azad University, Shahroud branch, Iran

Abstract

Reactive absorption of hydrogen sulfide into aqueous ferric sulfate solution is a direct process for H_2S removal and Sulfur recovery. A distinct advantage of the process is that the reaction of H_2S with is so rapid and complete that there remains no danger of discharging toxic waste gas. In this study, the reactive absorption of hydrogen sulfide into aqueous ferric sulfate solution has been studied and design calculations for equipments have been done. Subsequently, diameter and height of absorption tower are 3.6 ft and 12.5 ft and pressure drop is 0.35 inH_2O/ft packing, oxidizer's height and diameter are 2.45 ft and 4.9 ft, for known gas flow rate (100000 kg/day sour gas, contain≈4% v/v H_2S). Effective operation parameters on this process considered. Results show that low temperature and high pressure are suitable for absorption reaction. Variation of hydrogen sulfide concentration and Fe^{3+} concentration with time in absorption reaction shown that the reaction of ferric sulfate and hydrogen sulfide is first order with respect to the both reactant. At low $Fe_2(SO_4)_3$ concentration the absorption rate of H_2S increase with increasing the $Fe_2(SO_4)_3$ concentration. At higher concentration, a decrease in the absorption rate was found as shown in figure 2. At higher concentration of $Fe_2(SO_4)_3$, the ionic strength and viscosity of solution increase remarkably resulting in a decrease of solubility, diffusivity and hence absorption rate.

Key words: Reactive absorption, Direct process, Ferric sulfate, Hydrogen sulfide

Introduction

Many natural and industrial gases contain H_2S. The presence of H_2S usually prohibits the direct use of these gases because of its toxic properties, the formation of SO_2 upon combustion (acid rain), and the problems it (usually) gives in downstream processing. This means that it is often necessary to remove H_2S from gas stream prior to use. Many processes have been developed to remove H_2S from gas streams [1]. Most of the process use gas-liquid contactors in which the H_2S is absorbed into a complexing reagent to give either another dissolved sulfide containing component and problems are the degradation of the solvent [2]. Iron (Fe) is an excellent oxidizing agent to convert H_2S to elemental sulfur (S). During the1960s, developmental work initiated in England focused on increasing the solubility of elemental Fe^{3+} in aqueous solutions [3].

In this study, an aqueous $Fe_2(SO_4)_3$ solution is used as H_2S absorbent. H_2S is absorbed and chemically oxidized to elemental sulfur, while Fe^{3+} reduced to Fe^{2+} [2]. Elemental sulfur is removed from the solution by a separator, and the reactant Fe^{3+} is regenerated from Fe^{2+} in a reactor. At the same time, Fe^{3+} is reduced to Fe^{2+} according to

$$H_2S + 2Fe^{3+} \rightarrow S^\circ + 2Fe^{2+} + 2H^+ \qquad (1)$$

Elemental sulfur is removed from the solution by separator and the reactant Fe^{3+} is regenerated from Fe^{2+} by oxidation in reactor to the following reaction:

$$Fe^{2+} + H^+ + 0.25O_2 \rightarrow Fe^{3+} + 0.5H_2O \qquad (2)$$

Depending on the gas flow rate and the efficiency required several types of absorbers are suitable, such as jet scrubbers, bubble-cap towers or packed towers. The sulfur separator can include filter pres, settler and sulfur melters, Depending on quality of elemental sulfur required. Apart from sulfur, only H_2O is generated in the process, and consequently, no waste treatment facilities are required [2].

A distinct advantage of this process is that the reaction of H_2S with $Fe_2(SO_4)_3$, is so rapid and complete that there remains no danger of discharging toxic waste gas.

Calculations of process design

In this study assumed a packed tower used as absorber, continuous filter used as solid-liquid separator to separate sulfur from outlet liquid, and a back mixing reactor used as oxidizer to regenerate Fe^{2+} to Fe^{3+}. According to these assumptions and design calculations have been done for known gas flow rate (100000 kg/day sour gas, contain ≈ 4%v/v H_2S).

Calculations have been done for a packed tower with 2 inches ceramic raschig rings as packing and 0.1 molar $Fe_2(SO_4)_3$ concentration. Onda's method used to calculate k_G and k_L, gas and liquid side mass transfer coefficient. The height of liquid and gas-film transfer units (H_G, H_L) and overall gas-phase transfer unit (H_{OG}) are obtained. And after calculations, tower's height (Z) calculated.
Subsequently, Diameter and height of absorption tower are 3.6 ft and 12.5 ft and pressure drop is 0.35 inH_2O/ft packing [4, 5, 6].

In this process assumed that Oxidizer is a back mixing reactor. According to this assume, calculations have been done for oxidizer. Fe^{3+} is regenerated from Fe^{2+} according to equation 2 and it is relatively slow reaction. The reaction is first order with respect to oxygen and is 0.536 with respect to Fe^{2+}, as follow:

$$-r_{Fe^{2+}} = 4.39\times10^4 \exp\left(\frac{-23.3}{R.T}\right).C_{Fe^{2+}}^{0.536}.C_{O2} \qquad (3)$$

Consequently, Oxidizer's height and diameter are 2.45 ft and 4.9 ft [7, 8, 9].

Results and discussion

Influence of $Fe_2(SO_4)_3$ concentration on the absorption rate

Recent studies have resulted at low $Fe_2(SO_4)_3$ concentrations (up to 0.3 M) the absorption rate of H_2S increase with total $Fe_2(SO_4)_3$ concentration. At higher $Fe_2(SO_4)_3$ concentrations a decrease in the absorption rate was found. At higher concentration of $Fe_2(SO_4)_3$ the ionic strength and viscosity of the solution increase remarkably resulting in a decrease of the solubility, diffusivity and liquid side mass transfer coefficient of H_2S and hence absorption rate, as shown in Fig. 3. [2].

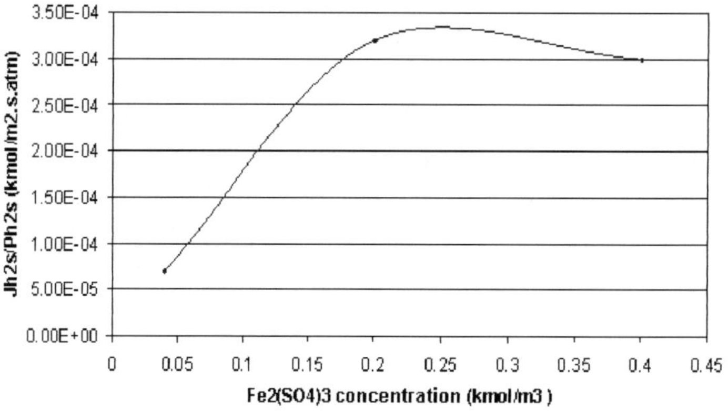

Fig. 1. Influence of $Fe_2(SO_4)_3$ concentration on the absorption rate of H_2S at 45°C

Calculations for Dimension of absorption tower

Calculations have been done for known gas flow rate,100000 kg/day sour gas, contain≈4% v/v H_2S, and dimension of absorber obtained with different types of packing and different concentration of ferric sulfate. These results Shown in Figures 1 and 2.

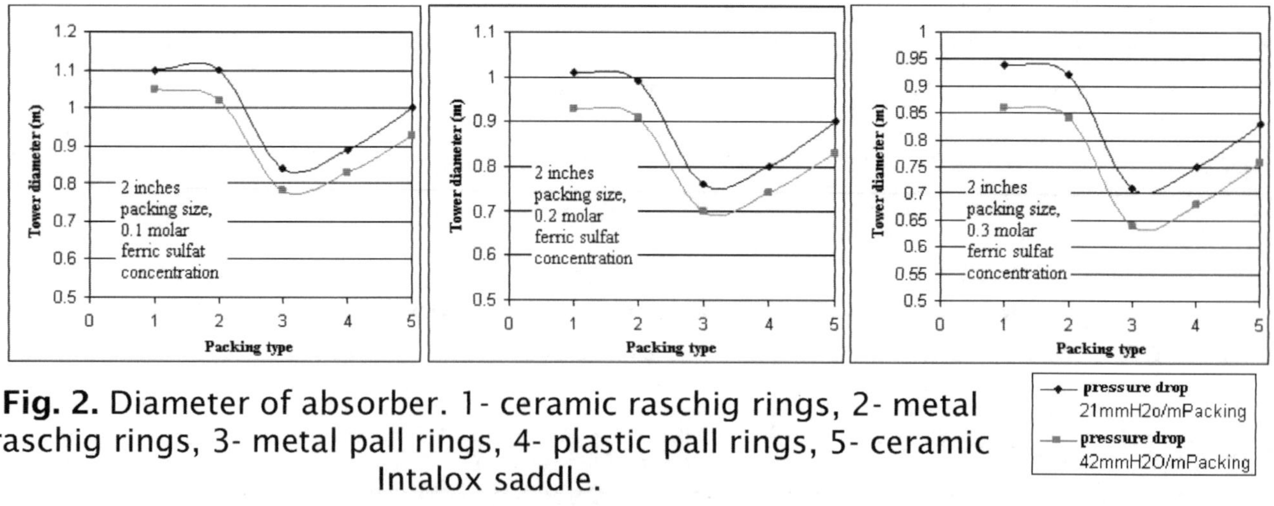

Fig. 2. Diameter of absorber. 1- ceramic raschig rings, 2- metal raschig rings, 3- metal pall rings, 4- plastic pall rings, 5- ceramic Intalox saddle.

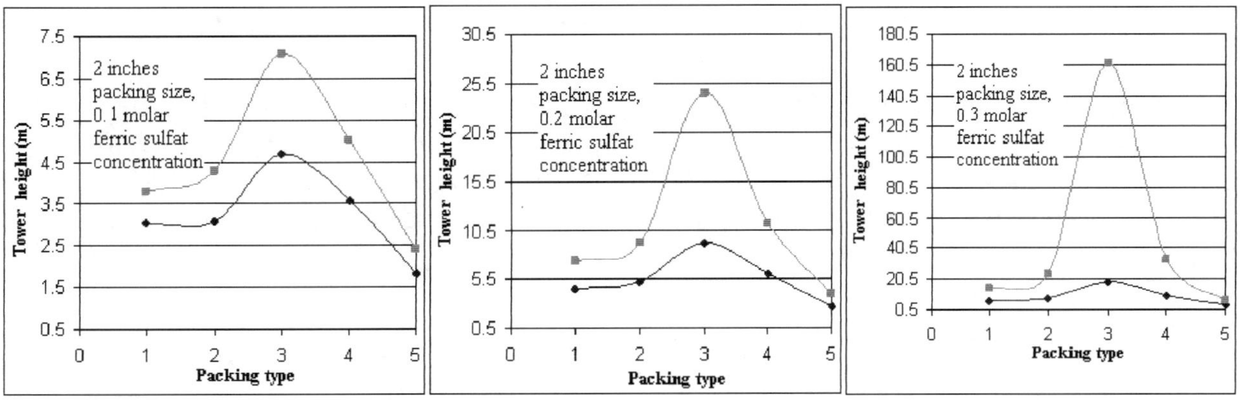

Fig. 3. Height of absorber. 1- ceramic raschig rings, 2- metal raschig rings, 3- metal pall rings, 4- plastic pall rings, 5- ceramic Intalox saddle.

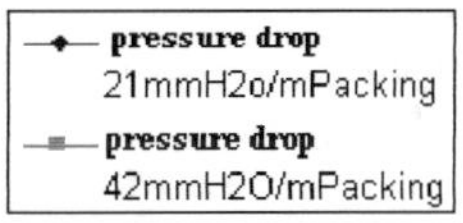

Results shown that whereas ceramic Intalox saddle packing used, height and diameter of absorber is minimum, with metal and plastic pall packing decrease in the absorption diameter was found, but the height increase remarkably. Furthermore, results shown whereas ceramic and plastic raschig rings are used, desirable dimension found for the absorption tower.

Variations of Reactant Concentration

The absorption reaction of hydrogen sulfide by ferric sulfate solution appears to irreversibly and is first order with respect to both reactants [2].
Variations of concentration profile of the reactants calculated in recent studies. Absorption reaction kinetic is as follow:

$$r_{H2S} = 16.815\ C_{H2S}.C_{Fe3+} \quad (10)$$

These profiles had shown in Fig. 4 and Fig. 5.

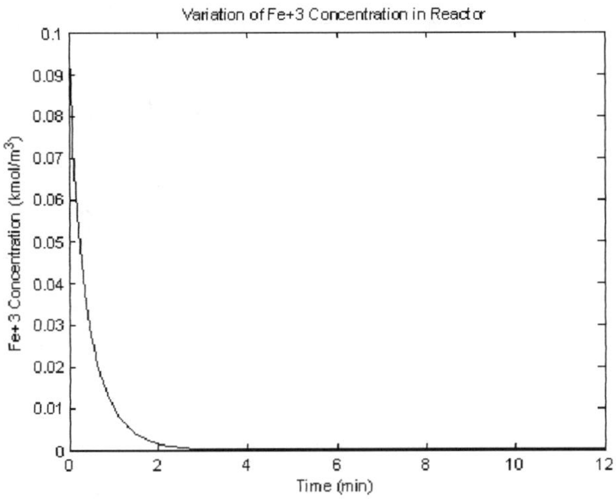

Fig. 4. Fe^{3+} Concentration profile

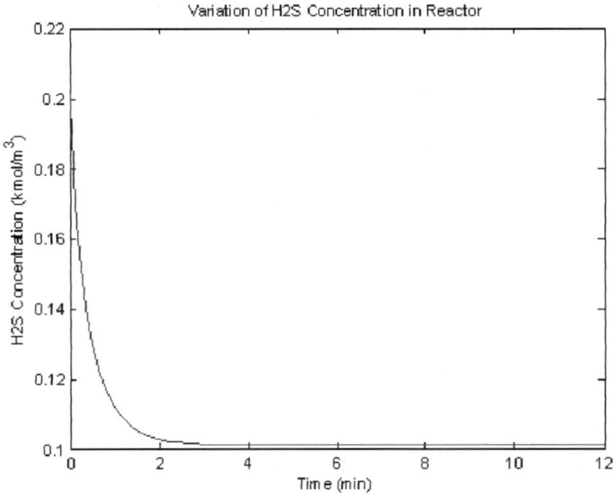

Fig. 5. H_2S Concentration profile

Conclusions

It is important to have an exact control on effective parameters in reaction to have high efficiency and low cost and energy consumption. H_2S removal from gas streams using aqueous ferric solution as absorbent has been studied theoretically and calculations for absorber and oxidizer have been done and their dimensions are obtained for a known sour gas flow rate.

The absorption reaction is first order with respect to both H_2S and ferric iron concentration and influence of temperature and pressure on H_2S absorption considered and shown that in high pressure and low temperature the process should have higher efficiency and optimum concentration of ferric sulfate should be used in the process due to increasing the ionic strength and viscosity of the solution at higher concentration of $Fe_2(SO_4)_3$, consequently, resulting in a decrease of the solubility, diffusivity and liquid side mass transfer coefficient of H_2S and hence absorption rate.

References

1. H. ter Maat, J. A. Hogendoorn, G. F. Versteeg (2005), "The removal of hydrogen sulfide from gas streams using an aqueous metal sulfate absorbent Part I. The absorption of hydrogen sulfide in metal sulfate solutions,"*Separation and Purification Technology* ,43, pp. 183-197.
2. S. Ebrahimi, R. Kleerebezem, M. C. M. van Loosdrecht, J. J. Heijnen (2003), "Kinetics of the reactive absorption of hydrogen sulfide into aqueous ferric sulfate solutions,"*Chemical Engineering Science*, 58, pp. 417-427.
3. D. L. HEGU and G. J. NAGL (2003), "Consider optimized iron-redox process to remove sulfur," *Hydrocarbon Processing*, January, pp. 53-57.
4. R. K. Sinnott (2005), "Coulson & Richardson's chemical engineering," vol. 6, Fourth edition, Elsevier Butterworth-Heinemann.
5. Nikolai Kolev (2006), "Packed Bed Column For absorption, desorption, rectification and direct heat transfer," First edition, Elsevier.
6. J. F. Richardson and J. R. Harker (2002), "Coulson and Richrdson's Chemical Engineering," Fifth edition, Butterworth-Heinemann.
7. Levenspiel. Octave (1999), "Chemical Reaction Engineering," Third edition, John Wiley & Sons.
8. Robert E. Treybal (1980). "Mass Transfer Operations," Third edition, Mc Grow-Hill.
9. Robbert H. Perry, Don W. Green (1997), "Perry's Chemical Engineers' Handbook," Seventh edition, Mc Grow-Hill Professional, New York.

THE EFFECT OF MURPHREE EFFICIENCY ON ABSORPTION OF CO2 AND H2S IN NATURAL GAS SWEETENING PROCESS

Somayeh Mirzaei ,Islamic Azad University, Shahroud branch, Iran*
Hasan Zare Ali Abadi,Islamic Azad University, Shahroud branch, Iran
Fatemeh Gholami ,Islamic Azad University, Shahroud branch, Iran
Zahra Gholami , Islamic Azad University, Shahroud branch, Iran

Abstract

Alkanolamine processes are used in the industry to remove acid gases, like CO_2, H_2S and other sulphur components, from natural gas and industrial gas streams. In this process the acid components react with the basic alkanolamine solution via an exothermic, reversible reaction in a gas/liquid absorber. The use of amine solvents for gas sweetening has been investigated using process simulation programs called HYSYS and ASPEN. We use Electrolyte NRTL and Amine Package and Amines(experimental) equation of state. The effect of murphree efficiency on the rate of absorption were studied . HYSYS software can not estimate murphree efficiency correctly and it apply the same contributions so in all diagrams related to HYSYS software, all the figures depending on variant efficiencies match to each other and in all cases efficiency of condenser and reboiler processes equals one. By improvement in murphree efficiency, maximum temperature of absorber decrease and the location of reaction teransfer to the stages of bottomn absorber and the absorption of acid gases increase. In stripper when murphree efficiency increase, the maximum temperature increases too and therefore the repel of acid gases increase.

Keywords: Simulation, gas sweetening, Diethanolamine, Methyldiethanolamine, Absorber

Introduction

Acid gases like CO_2 , H_2S and other sulphuric components are usually to som extent present in natural gas and industrial gases. They may have to be removed (selectively) from these gas streams for operational, economical or environmental reasons. One of the most commonly used processes for the removal of acid components is absorption in alkanolamine based solvents. In this process the acidic components react with an alkanolamine absorption liquid via an exothermic, reversible reaction in a gas/liquid contactor. In a following process step the acidic components are removed from the solvent in a regenerator, usually at low pressure and/or high temperature [6]. Industrially important alkanolamines for this operation are monoethanolamine (MEA), diethanolamine (DEA), di-isopropanolamine (DIPA) and N-methyldiethanolamine (MDEA) [9].

Aqueous diethanolamine (DEA) is a common chemical absorbent used in refineries to remove H_2S from refinery off gases [8]. Aqueous MDEA is used to accomplish selective removal of H_2S [10]. Besides MDEA, diisopropanolamine (DIPA) has also been reported to have a greater selectivity for H_2S over CO_2 than either MEA or DEA [9]. The sweet gas specification is ¼ grain H_2S /100 scf [4].

Process Chemistry

$$H_2S + Amine \Leftrightarrow [Amine]H^+ + HS^- \quad (1)$$

$$CO_2 + H_2O + Amine \Leftrightarrow [Amine]COOH^+ + OH^- \quad (2)$$

$$CO_2 + H_2O + R_2NCH_3 \Leftrightarrow R_2NCH^+4 \; + HCO_3^- \quad (3)$$

[4, 7, 11, 2, 3].

The following chemical reactions occure in an aqueous MDEA solution when CO_2 and H_2S are present :

$$CO_2 + H_2O \leftrightarrow HCO_3^- + H^+ \qquad HCO_3^- \leftrightarrow CO_3^{2-} + H^+$$

$$H_2O \leftrightarrow OH^- + H^+ \qquad RR'R'NH^+ \leftrightarrow RR'R'N + H^+$$

$$H_2S \leftrightarrow HS^- + H^+ \qquad HS^- \leftrightarrow S^{2-} + H^+$$

Where R corresponds to a methyl group and R' to an ethanol group [6].

Process Operating Parameters

Table 1. Typical operating data of amine-acid gas absorber

Parameter	Value (MDEA)	Value (DEA)
Inlet gas flow rate (SCMH)	173000	173000
Inlet liquid flow rate (M3/HR)	350	405
Inlet gas temperature ©	58	58
Inlet liquid temperature ©	58	58
Amine Concentration(%wt)	45	34
Gas in press.(PSIA)	1063	1063
L. Amine in press.(PSIA)	1100	1100
H_2S inlet gas composition (%mole)	3.588	3.588
CO_2 inlet gas composition (%mole)	6.459	6.459

Results and Discussion

We use Electrolyte NRTL and Amine Package and Amines (experimental) equation of state. The Electrolyte-NRTL model was originally proposed by Chen et al. [1, 12]. for aqueous electrolyte systems. It was later extended to mixed solvent electrolyte systems [12, 5].

Table 2. Simulation results of amine-acid gas absorber in HYSYS software

Parameter	MDEA	DEA
Outlet gas flow rate(kgmole/hr)	6687	6567
Outlet liquid flowrate(kgmole/hr)	1.283e+4	1.737e+4
Outlet gas temperature©	58.5	58.14
Outlet liquid temperature©	82.05	87.01
CO_2 Outlet gas composition(%mole)	1.7376e-2	4.3e-5
H_2S Outlet gas composition(%mole)	7.376908e-9	2.031665e-8
No. of stages	20	20

Table 3. Simulation results of amine-acid gas absorber in ASPEN software

Parameter	MDEA	DEA
Outlet gas flow rate(kgmole/hr)	6597.1692	6577.2
Outlet liquid flow rate(kgmole/hr)	12938.94	19276.6068
Outlet gas temperature©	34.27	58.16
Outlet liquid temperature©	61.79	82.07
CO_2 Outlet gas composition (%mole)	1.44981e-2	2.75279e-5
H_2S Outlet gas composition (%mole)	1.84811e-9	2.07944e-9
No. of stages	20	20

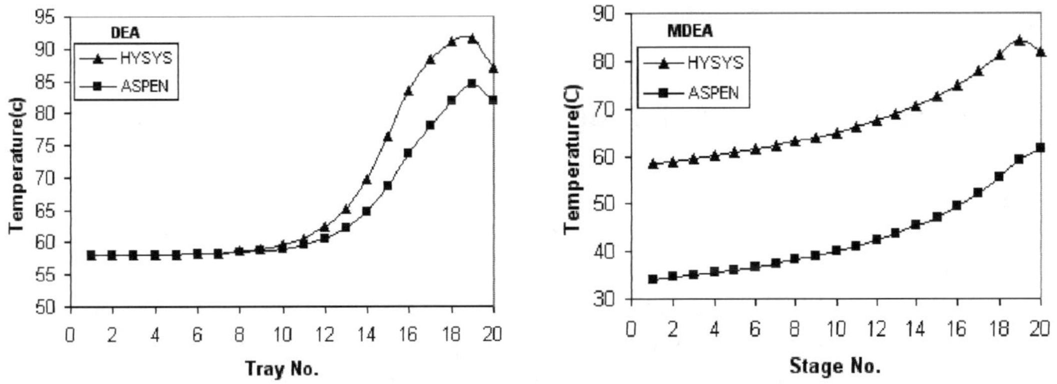

Fig. 3. Absorber temperature profile

From these profiles (Fig. 3) , it can be noticed that a temperature bulge occurs. This temperature bulge may be explained as follows. As the liquid flows down the tower, it continues to absorb acid gas. This absorption is accompanied by a heat of reaction, which causes the temperature of the liquid to continue to rise. The temperature drop at the bottom of the tower results from the cold gas entering the bottom and contacting the hot liquid flowing downwards. The cold gas absorbs heat from the hot liquid causing its temperature to decrease. This results in a temperature bulge at bottom of the tower. The heat of reaction for MDEA is less so temperature bulge is less.

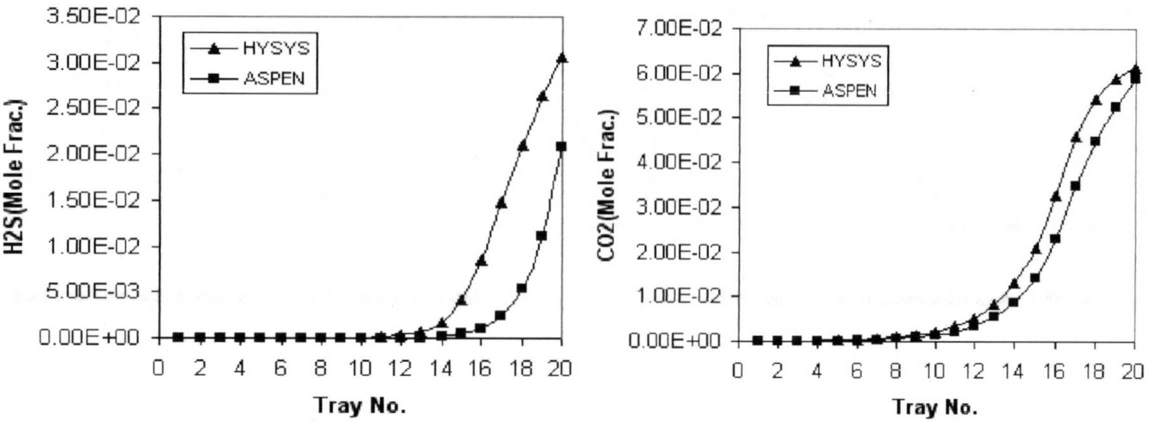

Fig. 4. Stage number vs. CO_2 and H_2S vapor composition profiles in the DEA absorber

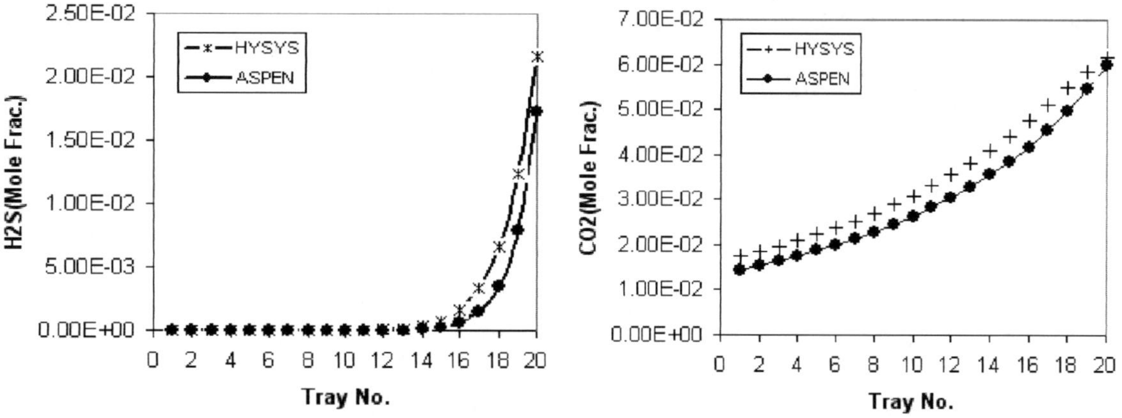

Fig. 5. Stage number vs. CO_2 and H_2S vapor composition profiles in the MDEA absorber

The less CO_2 will be absorbed by the MDEA solvent because aqueous MDEA is used very often for selective removal of H_2S from gas streams containing both CO_2 and H_2S. As can be seen (Fig. 4 and Fig. 5) , the concentration of H_2S drops down to almost zero at about stage 11(HYSYS software) and 14(ASPEN software) for DEA as compared to stage13 (HYSYS software) and 15(ASPEN software) for MDEA.

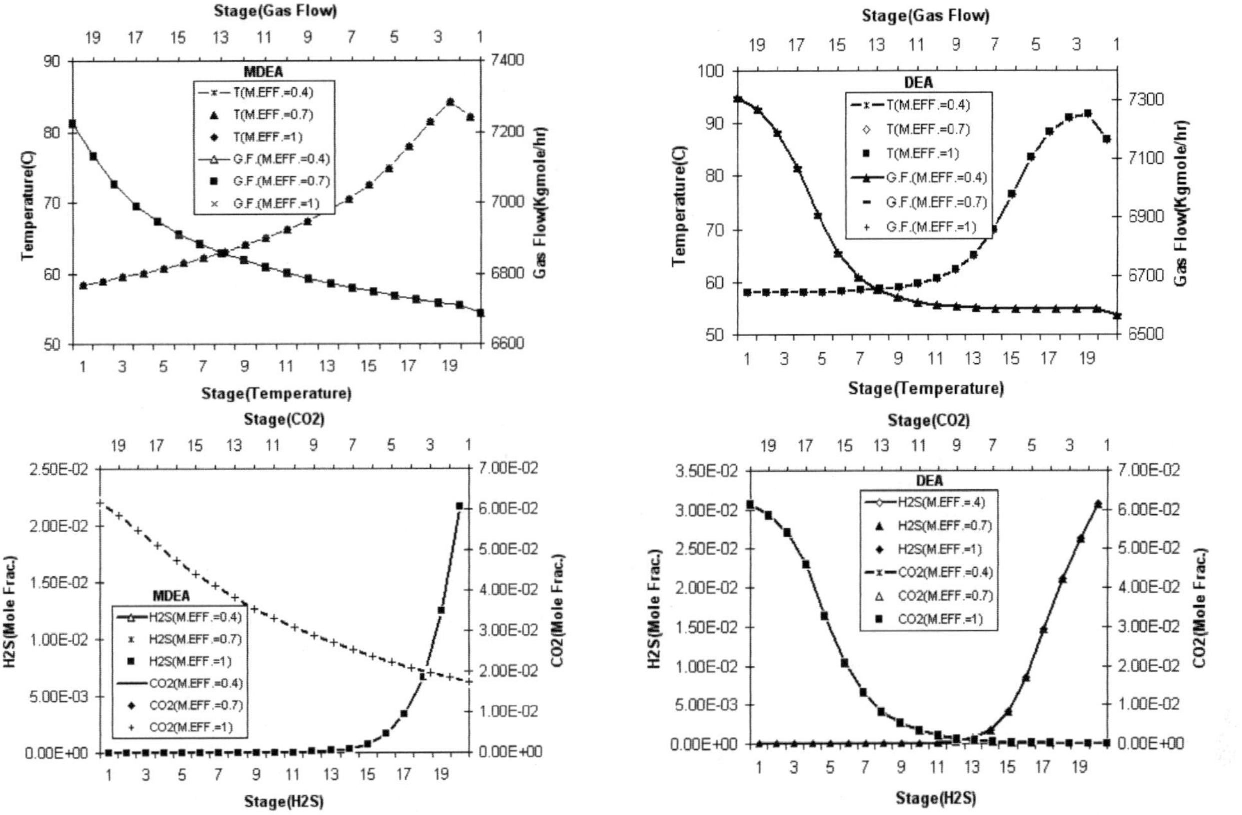

Fig. 6. Effect of murphree efficiency on temperature and gas flow and CO_2 and H_2S vapor composition in absorber in HYSYS software

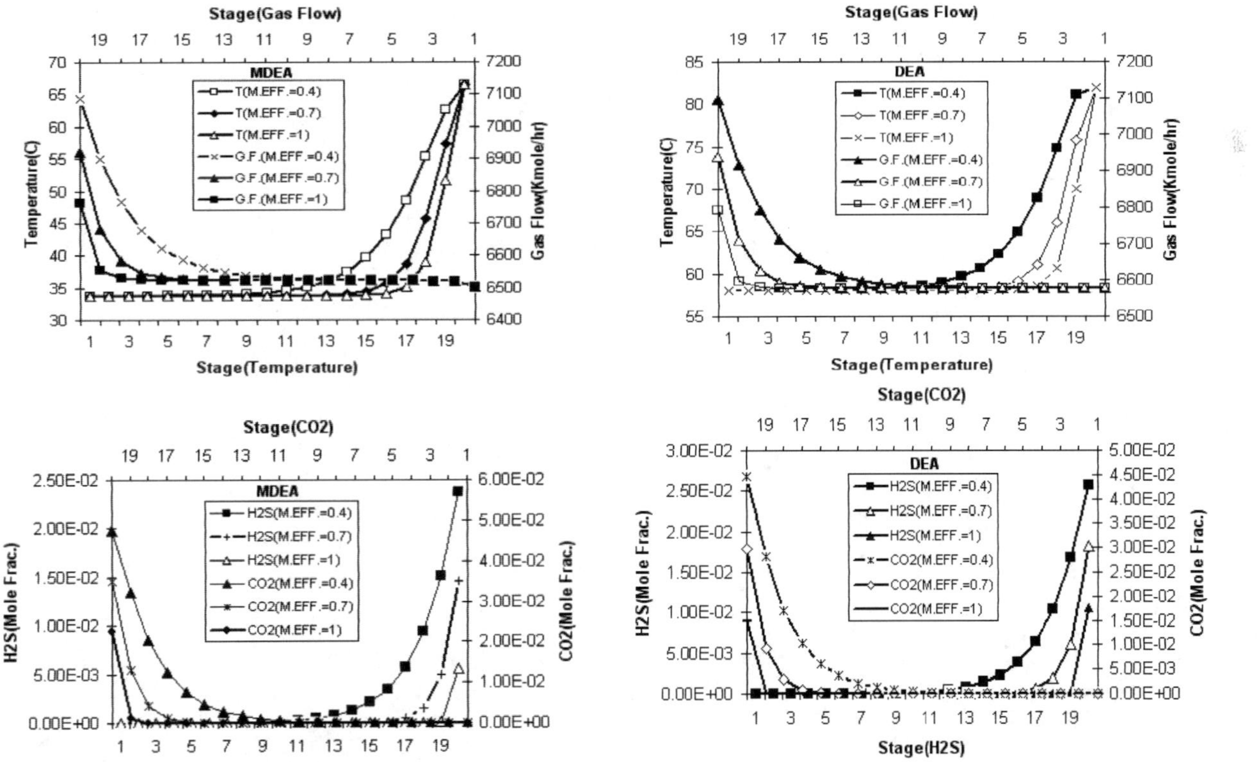

Fig. 7. Effect of murphree efficiency on temperature and gas flow and CO_2 and H_2S vapor composition in absorber in ASPEN software

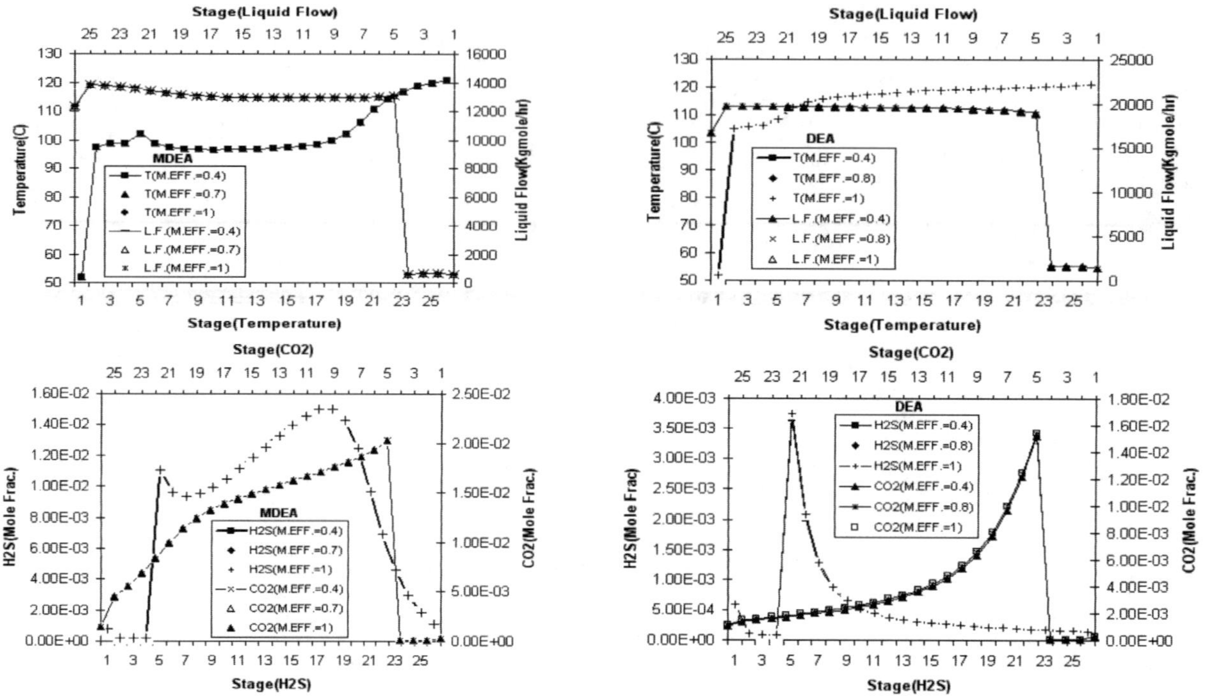

Fig. 8. Effect of murphree efficiency on CO_2 and H_2S liquid composition in stripper in HYSYS software

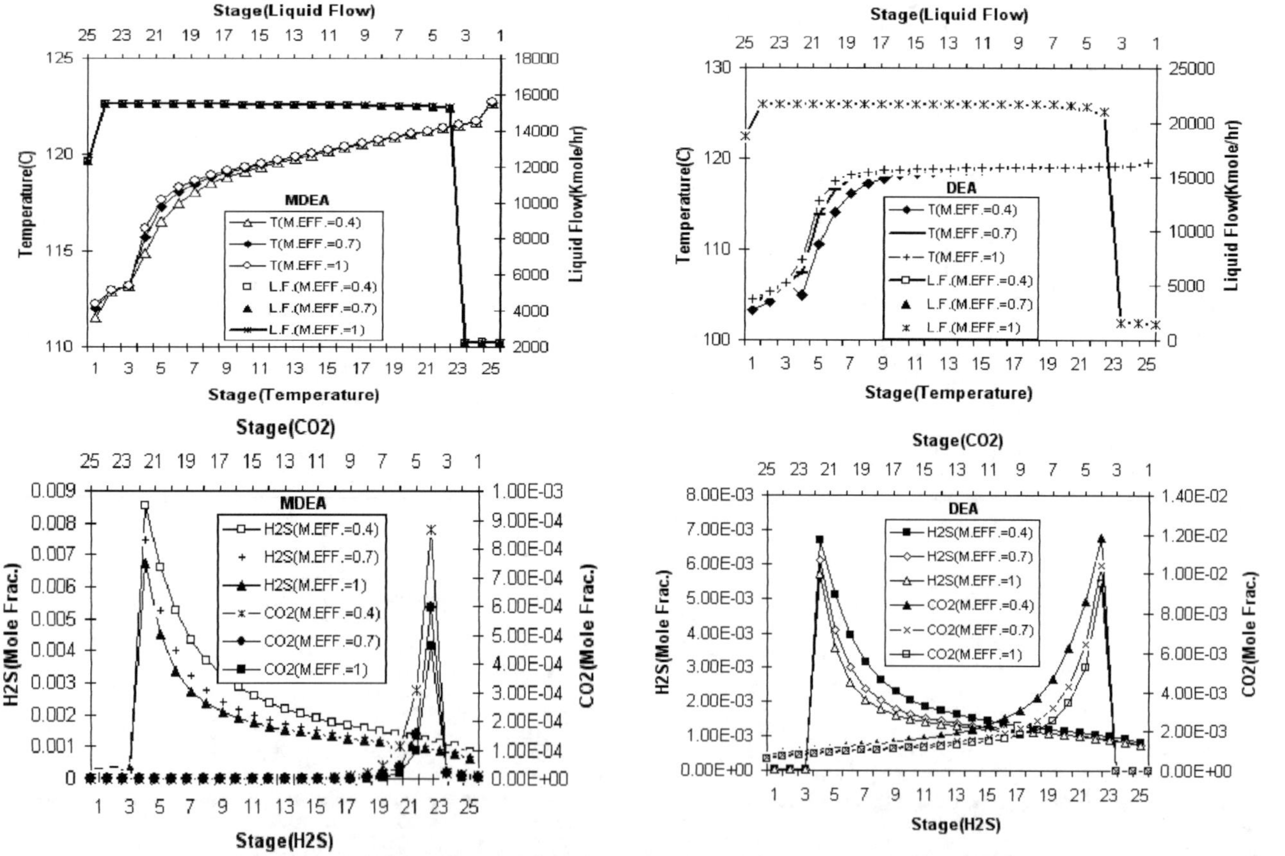

Fig. 9. Effect of murphree efficiency on temperature and liquid flow and CO_2 and H_2S liquid composition in stripper in ASPEN software

Conclusions

HYSYS software can not estimate murphree efficiency correctly and it apply the same contributions so in all diagrams related to HYSYS software, all the figures depending on variant efficiencies match to each other and in all cases efficiency of condenser and reboiler processes equals one. By improvement in murphree efficiency, maximum temperature of absorber decrease and the location of reaction teransfer to the stages of bottom absorber and the absorption of acid gases increase. In stripper when murphree efficiency increase, the maximum temperature increases too and therefore the repel of acid gases increase.

References

1. C.C. Chen, H.I. Brit, J.F. Boston, L.B. Evans (1982), " A local composition model for the excess Gibbs energy of aqueous electrolyte systems: part I: single solvent, single completely dissociated electrolyte system," *AICHE J.*, 4, pp. 588-596
2. Danckwerts, P. V. (1979), "The Reaction of CO_2 with Ethanolamines," *Chem. Eng. Sci.*, 34, pp. 443.
3. Savage, D. W., and E.W.Funk (1981), "Selective Absorption of H_2S and CO_2 into Aqueous Solutions of Methyldiethanolamine," *AIChE meeting*, Houston, Texas, April 5-9..
4. Douglas H. Mackenzie, and Christina A. Daniels (1987), "Design & Operation of a Selective Sweetening Plant Using MDEA," *Bryan Research & Engineering* , Inc.
5. G. Vallee, P. Mougine, S. julian, W. Furst (1999), "Representation of CO_2 and H_2S absorption by aqueous solutions of diethanolamine using an electrolyte equation of state," *Ind. Eng. Chem. Res.* 38, pp. 3473-3480.
6. Huttenhuis, P.J.G., Agrawal, N.J., Hogendoorn, J.A., Versteeg, G.F. (2007), "Gas solubility of H_2S and CO_2 in aqueous solutions of N-methyldiethanolamine." *Journal of Petroleum Science and Engineering* 55,122-134.
7. John C.Polasek, and Gustavo A.Iglesias-Silva (1992), "Using Mixed Amine Solutions for Gas Sweetening," *Bryan Research & Engineering , Inc.*
8. Kohl, A.L., Riesenfeld, F.C. (1985), "Gas Purification," 4th ed. Gulf Publishing Company , Houston.
9. Mandal, B.P., Bandyopadhyay, S.S. (2005), "Simultaneous absorption of carbon dioxide and hydrogen sulfide into aqueous blends of 2-amino-2-methyl-1-propanol and diethanolamine," *Chemical Engineering Science*,60, pp. 6438-6451.
10. Mandal, B.P., Biswas, A.K., Bandyopadhyay, S.S. (2004a), "Selective absorption of H_2S from gas streams containing H_2S and CO_2 in aqueous solutions of N-methyldiethanolamine and 2-amino-2-methyl-1-propanol", *Separation and Purification Technology* 35, pp. 191-202.
11. Maddox , R., (1977)."Gas and Liquid Sweetening," Second edition, Campbell.
12. Markus Bolhar-Nordenkampf, Anton Friedl, Ulrich Koss, Thomas Tork (2004), "Modelling selective H_2S absorption and desorption in an aqueous MDEA-solution using a rate-based non-equilibrium approach," *Chemical Engineering and processing* 43, pp. 701-715.

Process Integration Saves Capital and Energy

AIChE Spring National Meeting, Tampa, FL
April 26-30,2009

Kirtan Trivedi
ExxonMobil Research and Engineering Company,
Fairfax, VA

ExxonMobil
Taking on the world's toughest energy challenges™ 1

The Outlook for Energy – A view to 2030

Global Economics and Energy

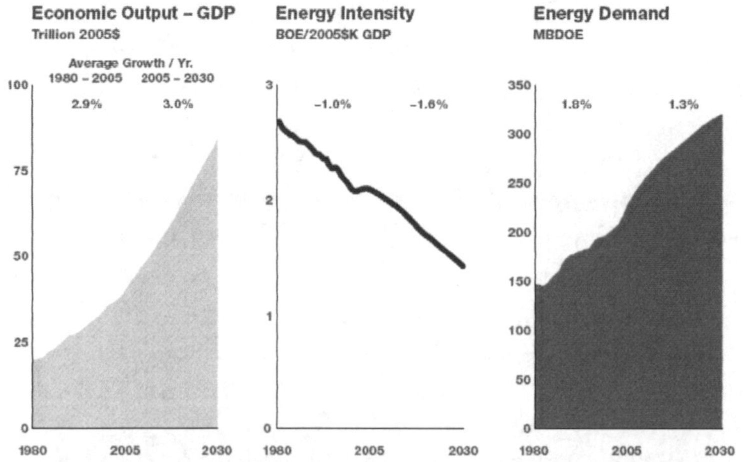

Ref: The Outlook for Energy – A view to 2030
http://www.exxonmobil.com/Corporate/energy_outlook.aspx

ExxonMobil
Taking on the world's toughest energy challenges™ 2

The Outlook for Energy – A view to 2030

World Energy Demand and Supply

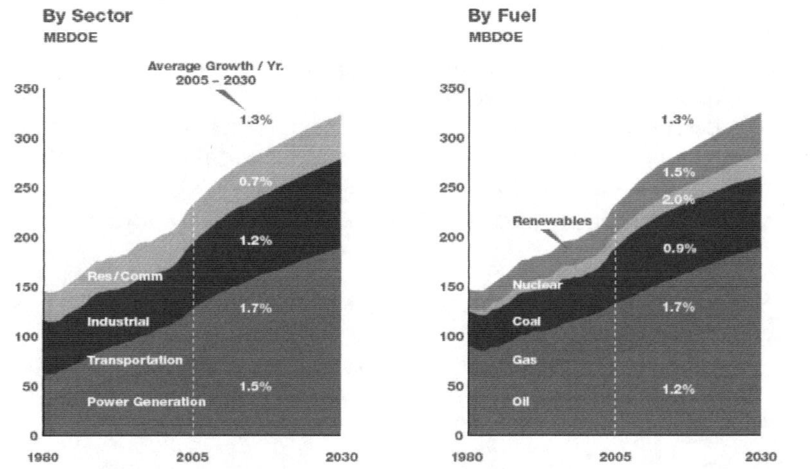

Ref: The Outlook for Energy – A view to 2030
http://www.exxonmobil.com/Corporate/energy_outlook.aspx

ExxonMobil
Taking on the world's toughest energy challenges™

3

The Outlook for Energy – A view to 2030

World Energy Demand – Primary Energy Supplies

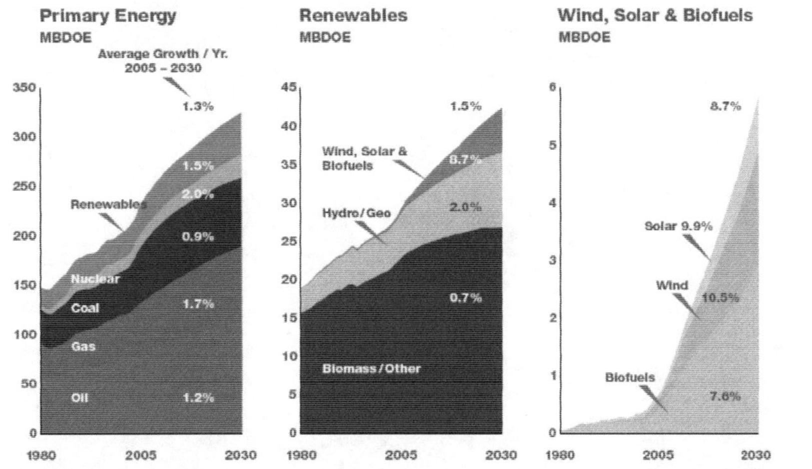

Ref: The Outlook for Energy – A view to 2030
http://www.exxonmobil.com/Corporate/energy_outlook.aspx

ExxonMobil
Taking on the world's toughest energy challenges™

4

The Outlook for Energy – A view to 2030

- Global energy demand is projected to grow at 1.3 percent per year
  - Underpinned by economic and population growth
  - Mitigated by efficiency gains
- Power generation represents more than 40 percent of demand increase
  - Growth concentrated in developing countries
  - Coal maintains highest fuel share, though gas has the largest growth
  - Nuclear power contribution grows, primarily post -2020
- Transportation sector demand growth slows over time
  - Increase in new light duty vehicle fuel economy helps offset impact of growing fleet
  - Biofuels supply increases but limited by cost and scale considerations
- Oil, natural gas and coal are indispensable to meeting this energy demand, even with rapid growth in renewables
- Demand growth and fuel mix will lead to increased CO_2 emissions
- Significantly impacting CO2 emissions requires global participation, step changes in energy efficiency, technology gains and massive investment

Ref: The Outlook for Energy – A view to 2030
http://www.exxonmobil.com/Corporate/energy_outlook.aspx

ExxonMobil
Taking on the world's toughest energy challenges™

5

Common Processes and Systems

OIMS
Operations Integrity Management System

Controls Integrity Management System

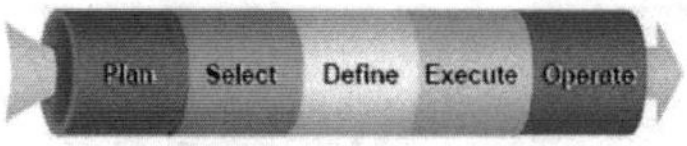

ExxonMobil Capital Project System

GEMS
Global Energy Management System

GRS
Global Reliability System

Ref: ExxonMobil 2008 Analyst Meeting
http://ir.exxonmobil.com/phoenix.zhtml?c=115024&p=irol-EventDetails&EventId=1746215

ExxonMobil
Taking on the world's toughest energy challenges™

6

Energy Security & Environmental Leadership

- Safely and reliably meeting energy demand
- Reducing environmental impact
  - Preventing spills and releases
  - Protecting biodiversity
- Reducing GHG emissions from energy production
  - Energy efficiency and cogeneration
  - Flare reduction
- Improving consumer use of energy

Protect Tomorrow. Today.

Operations Excellence

Disciplined Investments

Environmental Planning

Technology Applications

Ref: ExxonMobil 2008 Analyst Meeting
http://ir.exxonmobil.com/phoenix.zhtml?c=115024&p=irol-EventDetails&EventId=1746215

ExxonMobil
Taking on the world's toughest energy challenges™ 7

GEMS - A Global Approach

OBJECTIVES

- Establish a single, comprehensive global energy management system (G-EMS)
- Utilize a common methodology to identify performance gaps, implement closure plans, sustain progress, and continuously improve results

BUSINESS MODEL

- Operate existing facilities optimally and efficiently through application of best practices
- Identify economic investment opportunities above an optimized base for step-change improvement
- Implement strong management systems to sustain progress and drive continuous improvement

ExxonMobil
Taking on the world's toughest energy challenges™ 8

GEMS Design

ExxonMobil
Taking on the world's toughest energy challenges™ 9

GEMS Development

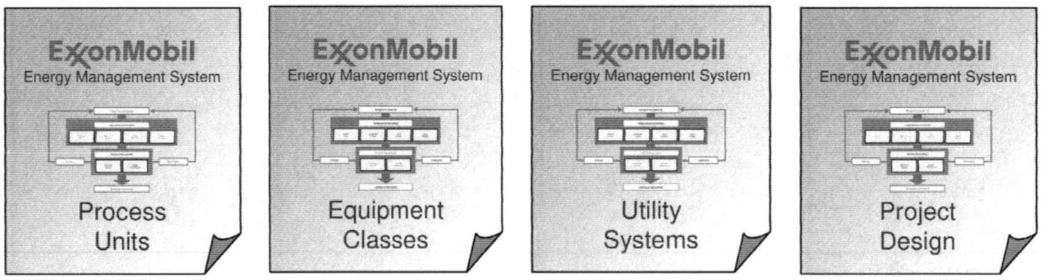

- Best practices documented in 12 volume set of manuals
- Contain 1200 pages and identify over 200 key energy variables
- Cover key process, equipment, and utility operations
- Also address energy efficiency in project design

ExxonMobil
Taking on the world's toughest energy challenges™ 10

Identify Best Practices

Provide Performance Measures and "How-to's"

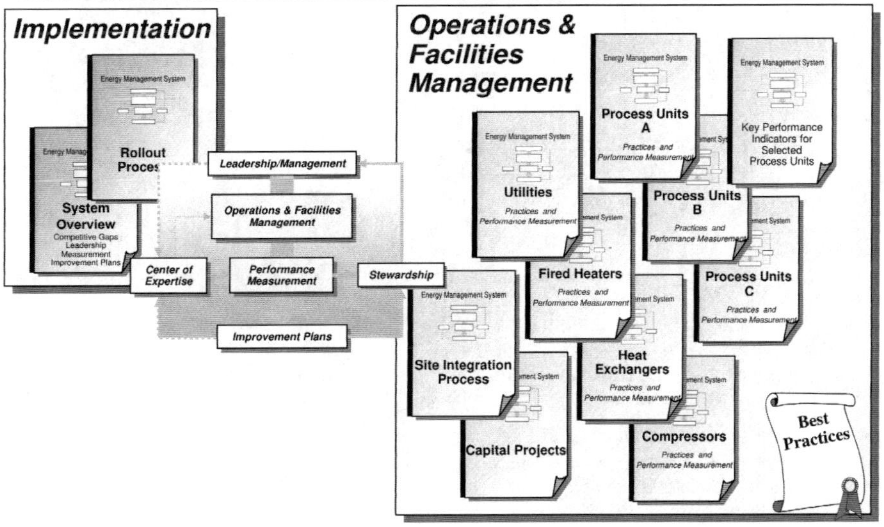

ExxonMobil

Taking on the world's toughest energy challenges™ 11

GEMS Measurement

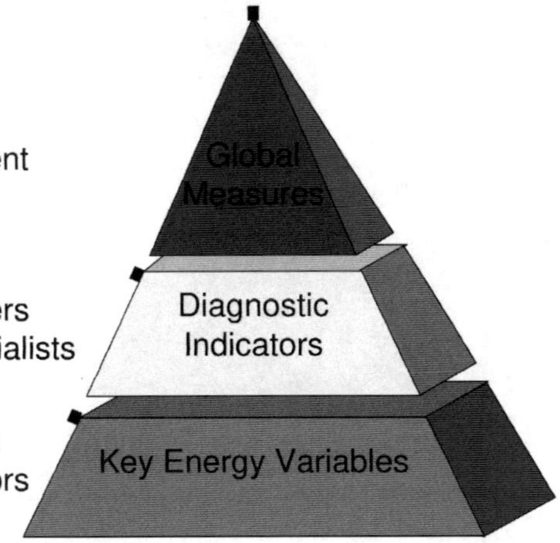

Headquarters &
Plant Management

Business Teams
Process Engineers
Equipment Specialists

Shift Supervision
Console Operators
Field Operators

<u>Plant / Site</u>
Energy Index
Energy Expense
Profitability

<u>Area / Complex</u>
Implementation
Lost Opportunity

<u>Unit / Equipment</u>
Targets & Handles
Directly Actionable

ExxonMobil

Taking on the world's toughest energy challenges™ 12

GEMS Rollout Results

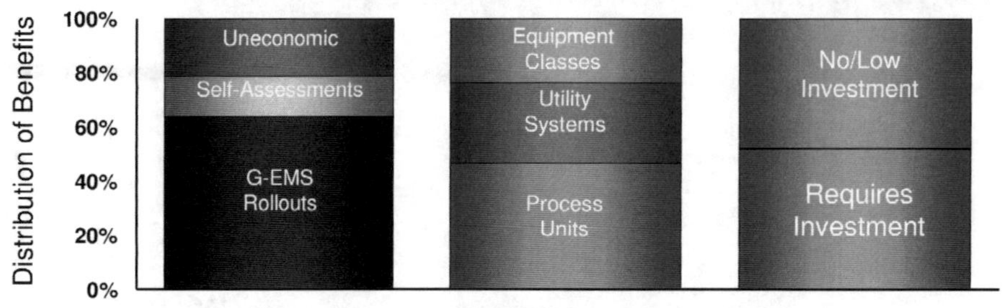

- Rollouts and Self-Assessments address 80% of total opportunity
  - Process Units total nearly 50% -- mainly investment related
  - Utility Systems account for over 30% -- mix of no/low and investment items
  - Equipment Classes total 20% -- many no/low investment options
- Potential savings equal to 15-20% of plant energy bill on average
  - About half of the benefits identified require little or no investment to capture
  - Potential to further reduce GHG emissions

ExxonMobil
Taking on the world's toughest energy challenges™ 13

Results from GEMS Implementation

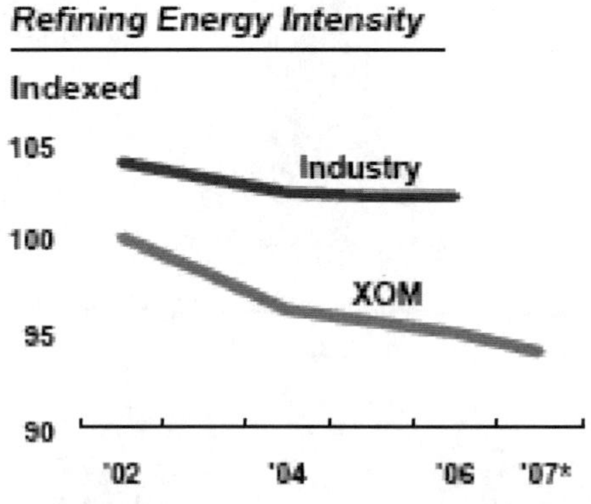

Source: Solomon
* 2007 Estimated – Solomon survey only prepared in even years

Ref: ExxonMobil 2008 Analyst Meeting
http://ir.exxonmobil.com/phoenix.zhtml?c=115024&p=irol-EventDetails&EventId=1746215

ExxonMobil
Taking on the world's toughest energy challenges™ 14

Project Background

- Clean fuels project
- Non-discretionary
- Integrated site
- Revamp of two Hydrotreaters to produce LSD (50 wppm)
- Two new furnaces initially determined to be required to meet throughput and product specifications in Hydrotreaters

ExxonMobil
Taking on the world's toughest energy challenges™ 15

Block Flow Diagram of the Refinery

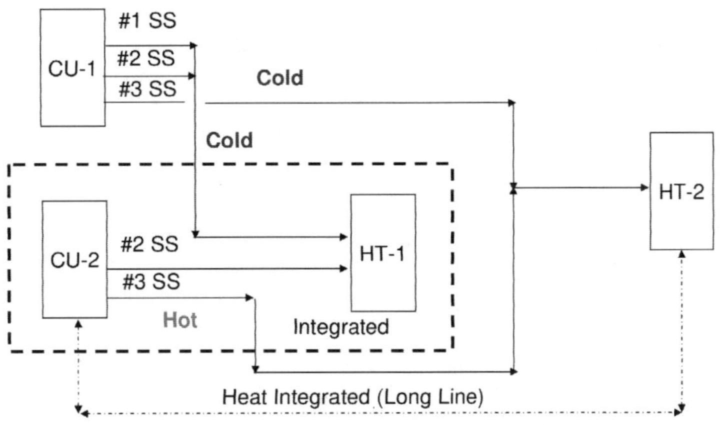

ExxonMobil
Taking on the world's toughest energy challenges™ 16

New Furnaces in HT-1 and HT-2

- Expensive
- Execution Issues
- HT-1
  - Construct the furnace near the actual location
  - During TA demolish the existing furnace
  - Slide the new furnace on the plot space for the old furnace
- HT-2
  - Relocation of road and other underground lines
  - Construct the new furnace on the new plot space
  - Leave the old furnace in place

ExxonMobil
Taking on the world's toughest energy challenges™ 17

Revamp Process Flow Scheme for HT-1

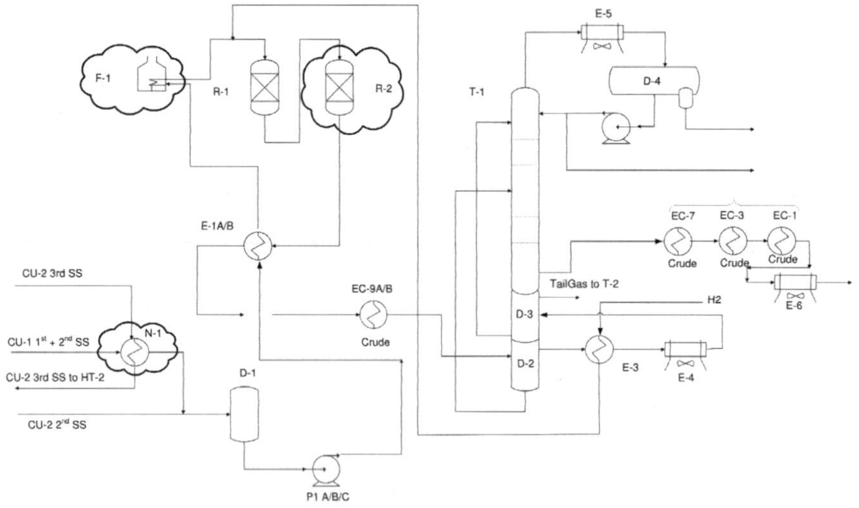

ExxonMobil
Taking on the world's toughest energy challenges™ 18

Challenges and Design Constraints

- Challenges
  - Cost reduction required by the project
  - New furnaces represented significant cost of revamp for the hydrotreaters
  - No obvious low cost solutions
- Design Constraints
  - CU-2 and HT-1 on the same plot space and heat integrated
  - CU-2 and HT-2 not on the same plot but heat integrated
  - Long distances between CU-2 and HT-2
  - Product routings from CU-2 changed
  - Pressure drop limitations in the treat gas circuit
  - Heat that can be recovered from the reactor effluent stream is fixed by the HP separator temperature and the heat required in the Crude Unit

ExxonMobil
Taking on the world's toughest energy challenges™ 19

Process Integration Principles Applied

- Composite Curves
- Classical revamp techniques
- Process Modification based on +/- principles and fundamental understanding of the process

ExxonMobil
Taking on the world's toughest energy challenges™ 20

Composite Curves

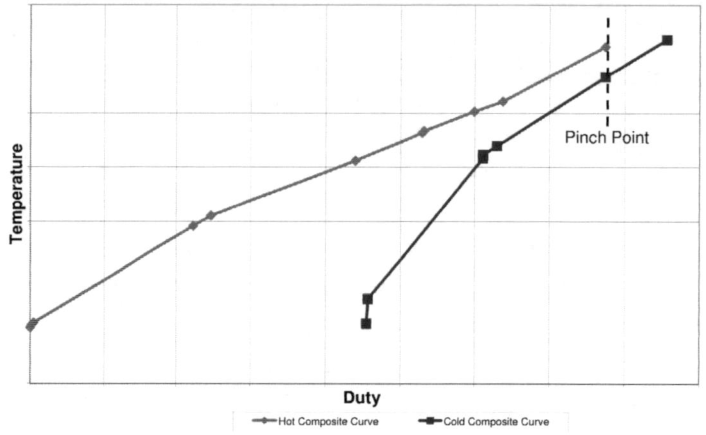

- Pinch is at the hot end
- Heat recovery can be improved by reducing ΔTmin
- Heat required by crude not included in the cold compsite curve

ExxonMobil
Taking on the world's toughest energy challenges™ 21

Grid Diagram – Heat Path

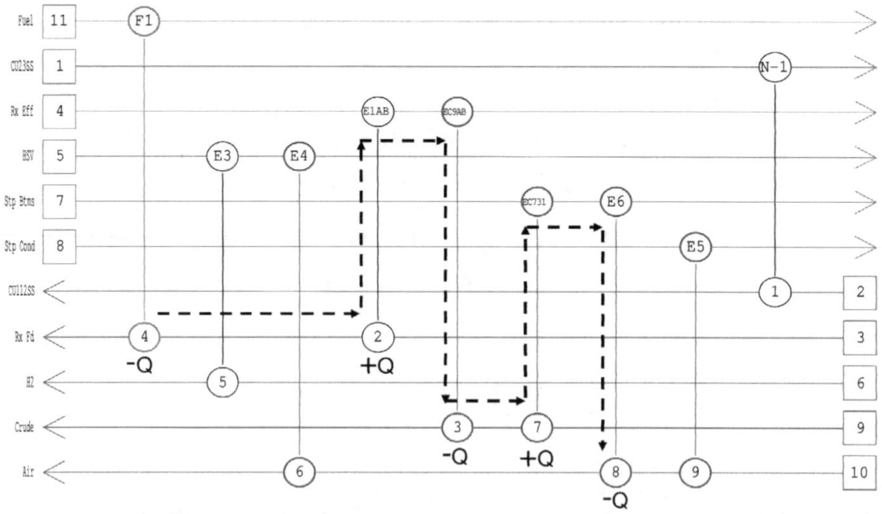

ExxonMobil
Taking on the world's toughest energy challenges™ 22

Heat Path Approach

- Increase area on E1AB to recover more heat from reactor effluent
- Decrease heat recovery in EC9AB
- Increase heat recovery in EC731
- Decrease the duty on E6
- Crude Unit HX train will have to significantly modified

Grid Diagram – Process Interactions

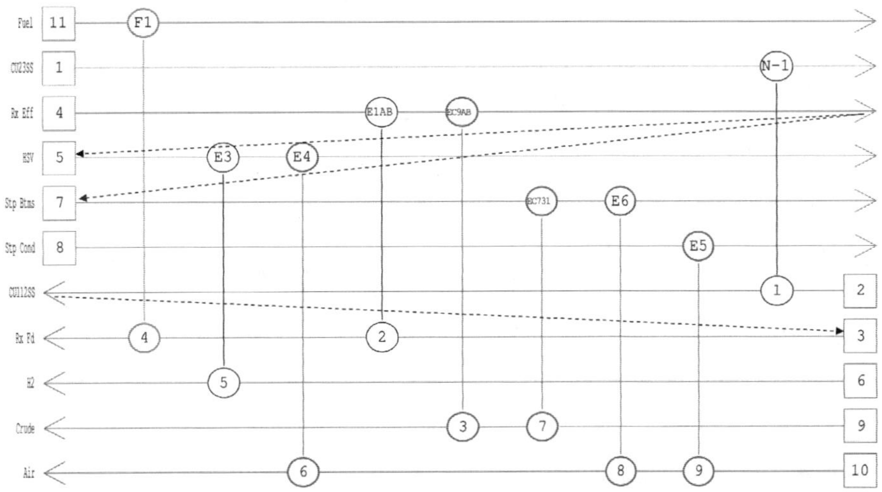

Options – with process modifications

- High ΔT available in N-1
  - Feasible to increase feed drum temperature
  - Decreases the heat recovery in the existing feed/effluent heat exchanger
  - New heat exchanger will be required to restore original heat recovery
  - Overall will increase the CIT to the furnace and decrease the duty
  - Energy savings required to avoid the new furnace will come from preheating the feed to avoid impacting the heat balance in the CU-2 unit
- Put additional Heat exchangers in the feed effluent circuit
  - Subject to pressure drop limitations
  - Move the Crude / Rx Effluent Heat exchanger
- HP Separator temperature increases
  - Possible to increase heat recovery in the treat gas/ HP vapor HX
  - Further decreases the furnace duty
  - Stripper bottoms is hotter
  - Additional heat at higher temperature can be recovered from the bottoms stream
  - Can be used in CU-2 to restore heat balance

ExxonMobil
Taking on the world's toughest energy challenges™ 25

Modified Flow Sheet

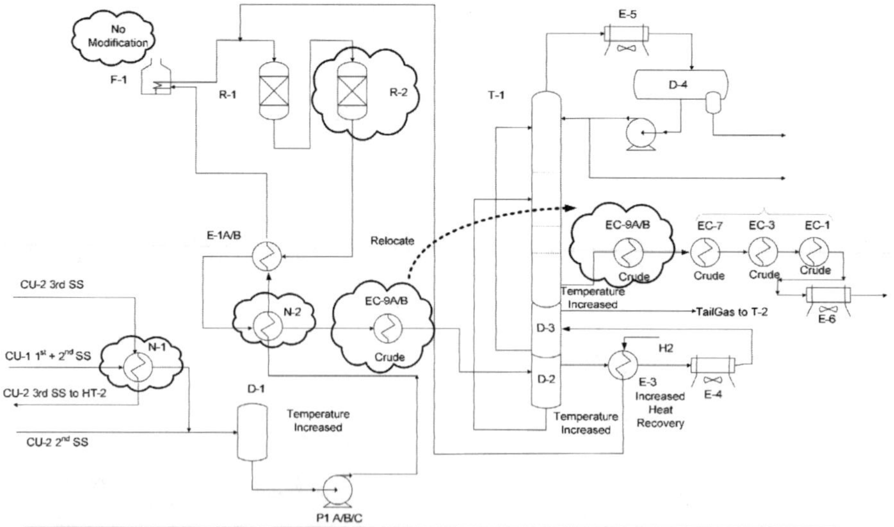

ExxonMobil
Taking on the world's toughest energy challenges™ 26

Hardware and plot plan

- Selection for the type of heat exchanger to minimize cost
- Plot plan constraints
- Maintainability and repairs
- Reuse of existing heat exchangers in different services
  - Crude/Reactor effluent HX reused as Crude/Stripper Btms HX and relocated
  - New reactor feed effluent exchanger put in place of the Crude/Reactor Effluent HX
    - Virtually no change in reactor effluent piping
    - Minimized the impact on pressure drop in the reactor effluent circuit
- Heat integration of CU-2 indicated no issues with the modifications and heat balance

ExxonMobil
Taking on the world's toughest energy challenges™ 27

Conclusions

- No new furnaces required for HT-1 and HT-2
- Significant savings in Capital and Energy
- Eliminated high risks involved with execution
- Process Integration techniques can help in generating ideas which may not be obvious
- PI techniques coupled with fundamental understanding of the process, design constraints, process interactions can yield significant savings
- Do not conduct pinch analysis in isolation, as the real value-added by these techniques may be missed

ExxonMobil
Taking on the world's toughest energy challenges™ 28

69b Opportunities for Energy Savings on Thermal Coupling CDU and DCU Grassroot Designs

Claudio L. M. Kuboski, Antonio V. S. Castro, Roberto A. Oliveira, and Washington Geraldelli. CENPES – PETROBRAS Research and Development Center, Av. Horacio Macedo, 950 - Ilha do Fundao., Rio de Janeiro, 21941-958, Brazil

Crude oil distillation unit (CDU) and Delayed coking unit (DCU) present significant energy consumption in a refinery, given the high flow rates involved on process streams. As refiners are increasing the process of opportunity crudes, heat integration on these units becomes more relevant.

Because DCU is essentially a semi-continuous process, heat integration involving its hot streams may impact considerably the CDU operation. However, thermal coupling between CDU and DCU can be explored through the evaluation of adequate DCU feed temperature. By increasing vacuum residue temperature to DCU, the DCU cooling demand increases while CDU cooling demand decreases. As pinch location on each unit is relatively high, little influence on heating demand is expected, unless the pinch is modified by variation on vacuum residue temperature to DCU. Capital costs are supposed to suffer variations as well, especially on DCU, due to the required area to preheat the feed and to cool the product streams.

This work presents the evaluation of DCU feed temperature relevance on both CDU and DCU overall costs and utility consumption, considering grassroot designs, based on Pinch Analysis approach in a conceptual design.

The paper also investigates the potential of medium pressure steam generation on each unit, discussing the impacts on both processes.

Low Cost Energy Savings Through Automation

Paper 69c
AIChE 2009
Spring National
Meeting

Doug White
Emerson Process
Management

Speaker

Doug White

Principal Consultant and Vice President, APC Services

Advanced Applied Technologies

Process Systems and Solutions

Emerson Process Management

EMERSON™
Process Management

Many years experience designing, justifying, installing and commissioning advanced real time computer applications in the process industries.

Natural Gas Prices

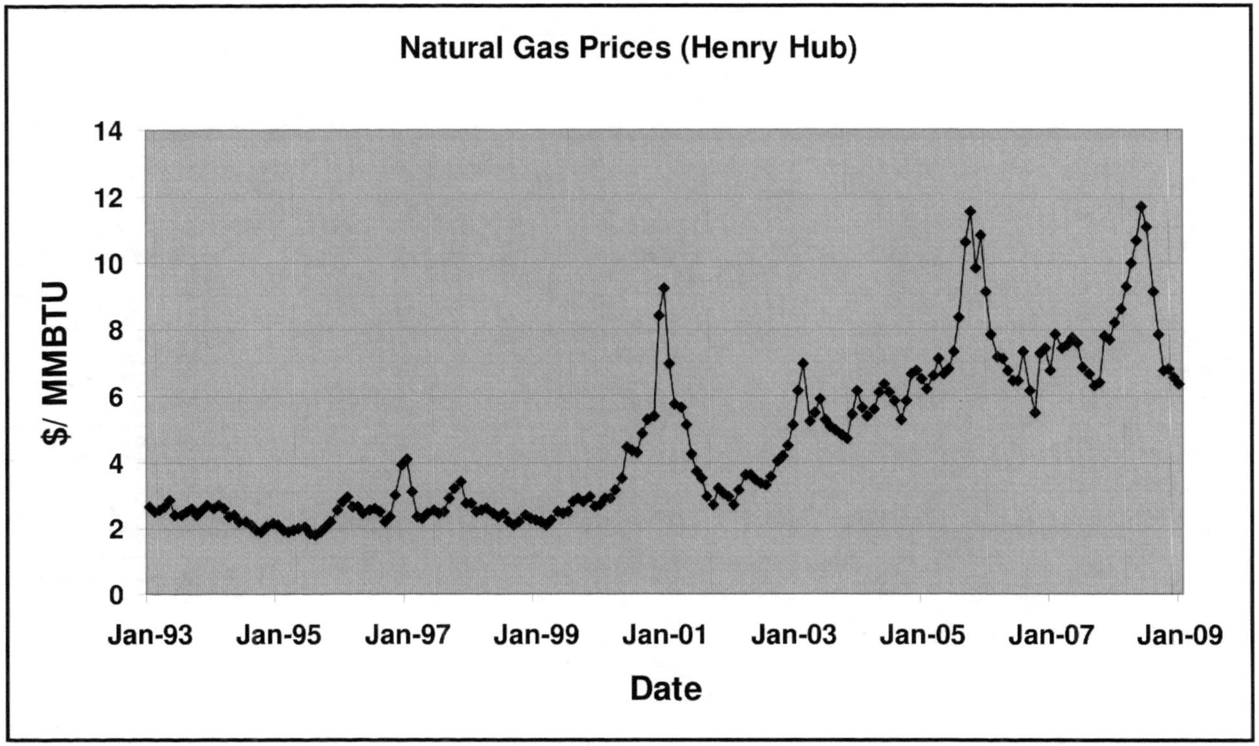

Process Energy Usage

	Process Energy; MM BTU/ Ton	Value; 10% Energy Reduction; $/ Ton $7/MMBTU
Petroleum Refining	4.4	3.1
Integrated Pulp/Paper Mill	29.0	20.3
Cement Production	7.9	5.5
Chemicals		
Ethylene	7.1	5.0
Polyethylene	6.7	4.7
EDC	9.4	6.6
PVC	4.0	2.8
EO	6.2	4.3
EG	7.1	5.0
Ethylbenzene	2.9	2.1
Styrene	38.8	27.2

Largest Controllable Cost in Most Plants!

AIChE 2009 Spring National Meeting

Presentation Objectives

Help you recognize the many ways that automation, advanced automation and asset management can save energy in today's plants

Agenda

→ How is energy used in plants?

→ How can automation help save energy?

General Process Site Energy Flow

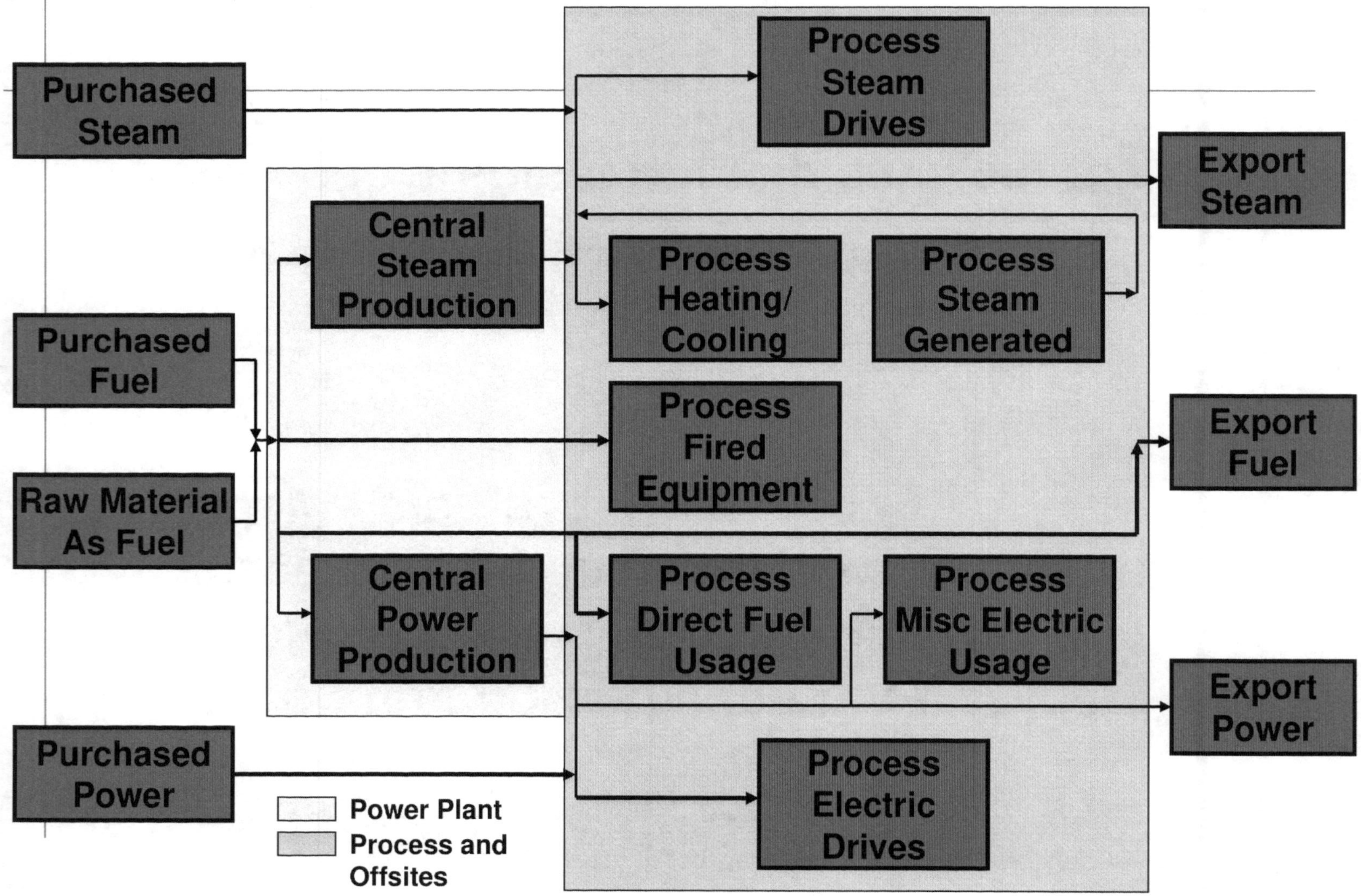

AIChE 2009 Spring National Meeting

Source - DOE EIA

"Average" Chemical Site Energy Flow

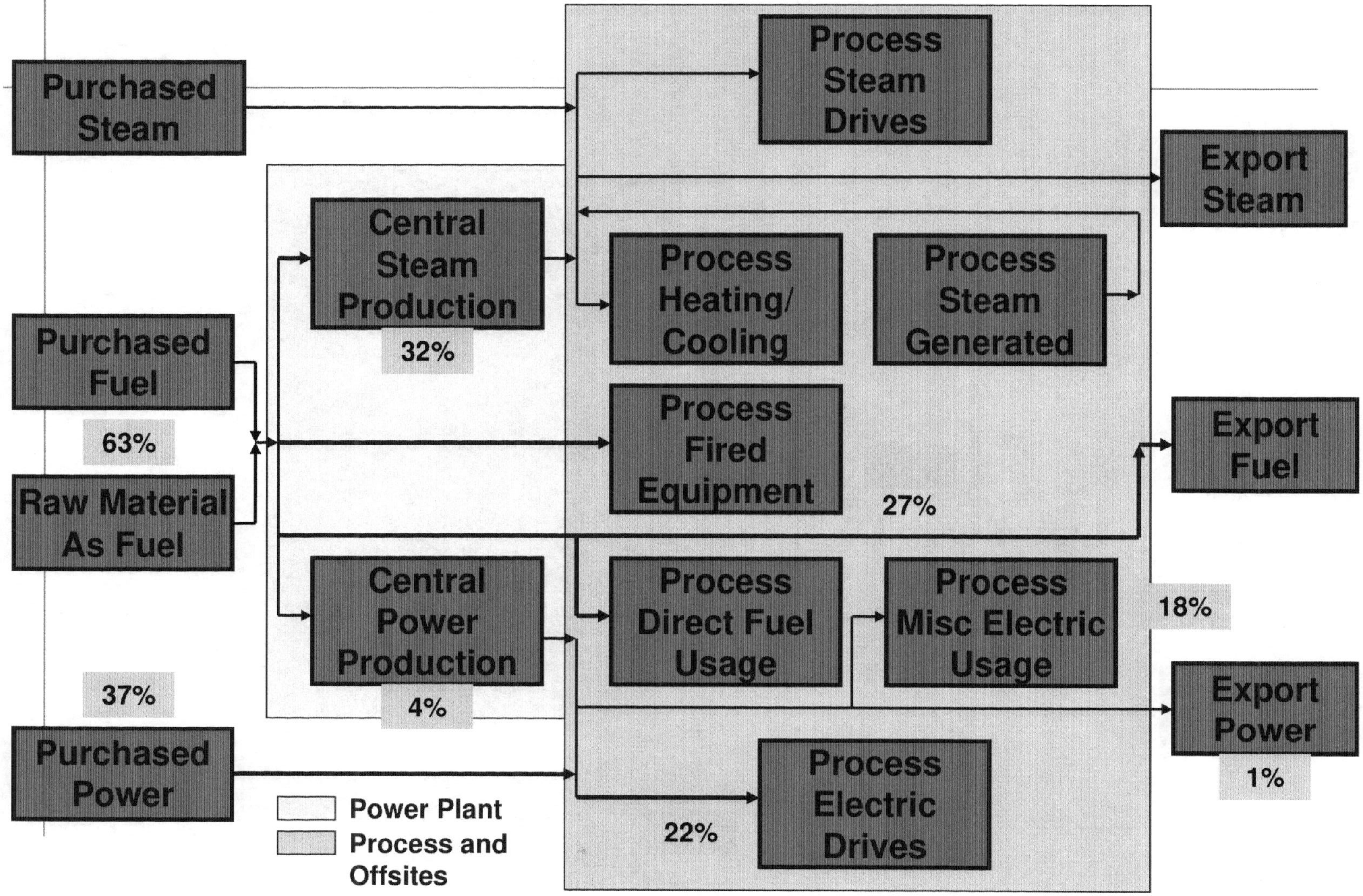

%-Equivalent BTU basis (Including losses) on total input

Source - DOE EIA

Integrated Pulping Paper Mill Energy Flow

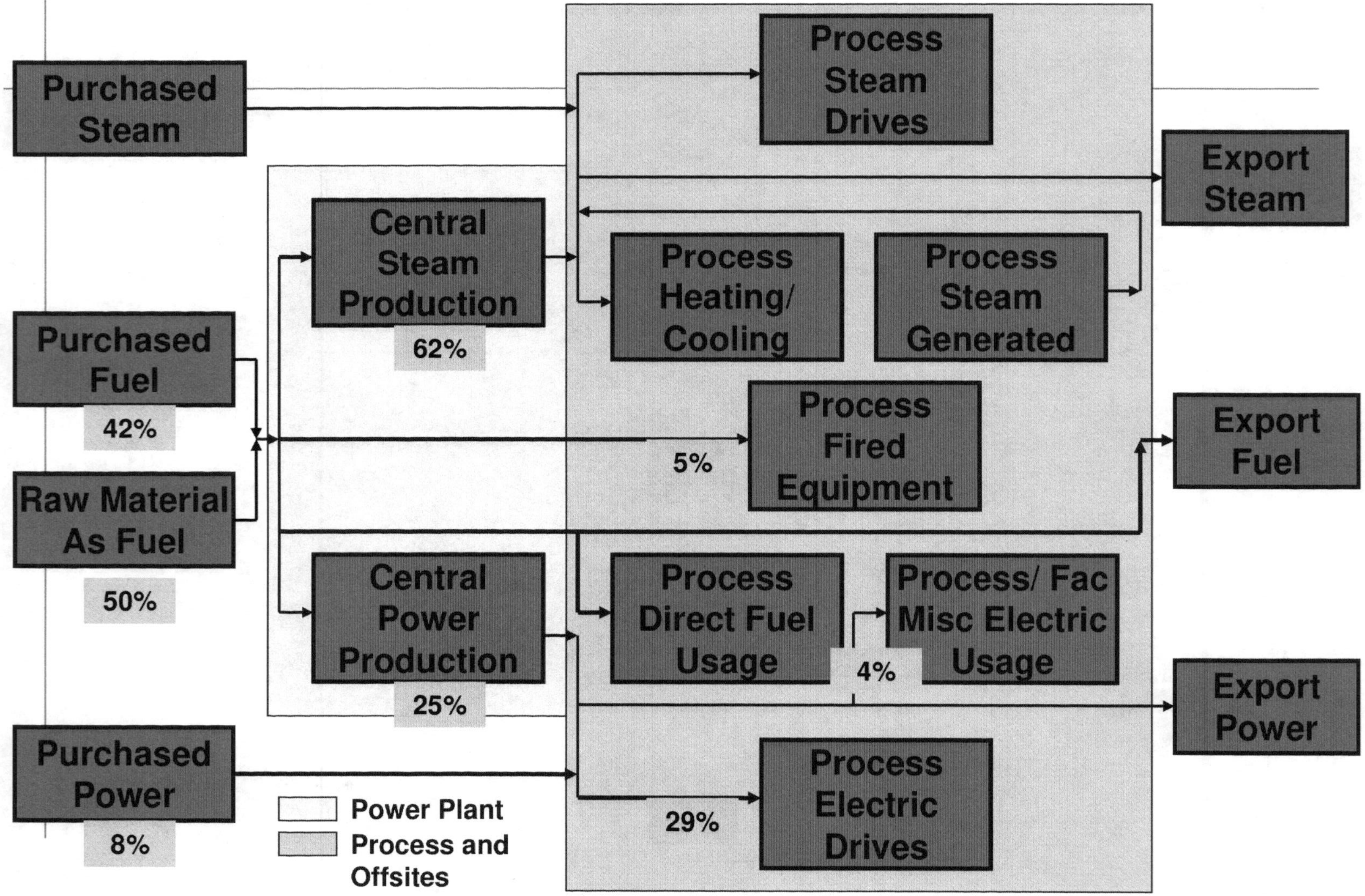

%-Equivalent BTU basis (Including losses) on total input

Source - DOE EIA

Oil Refining Energy Flow

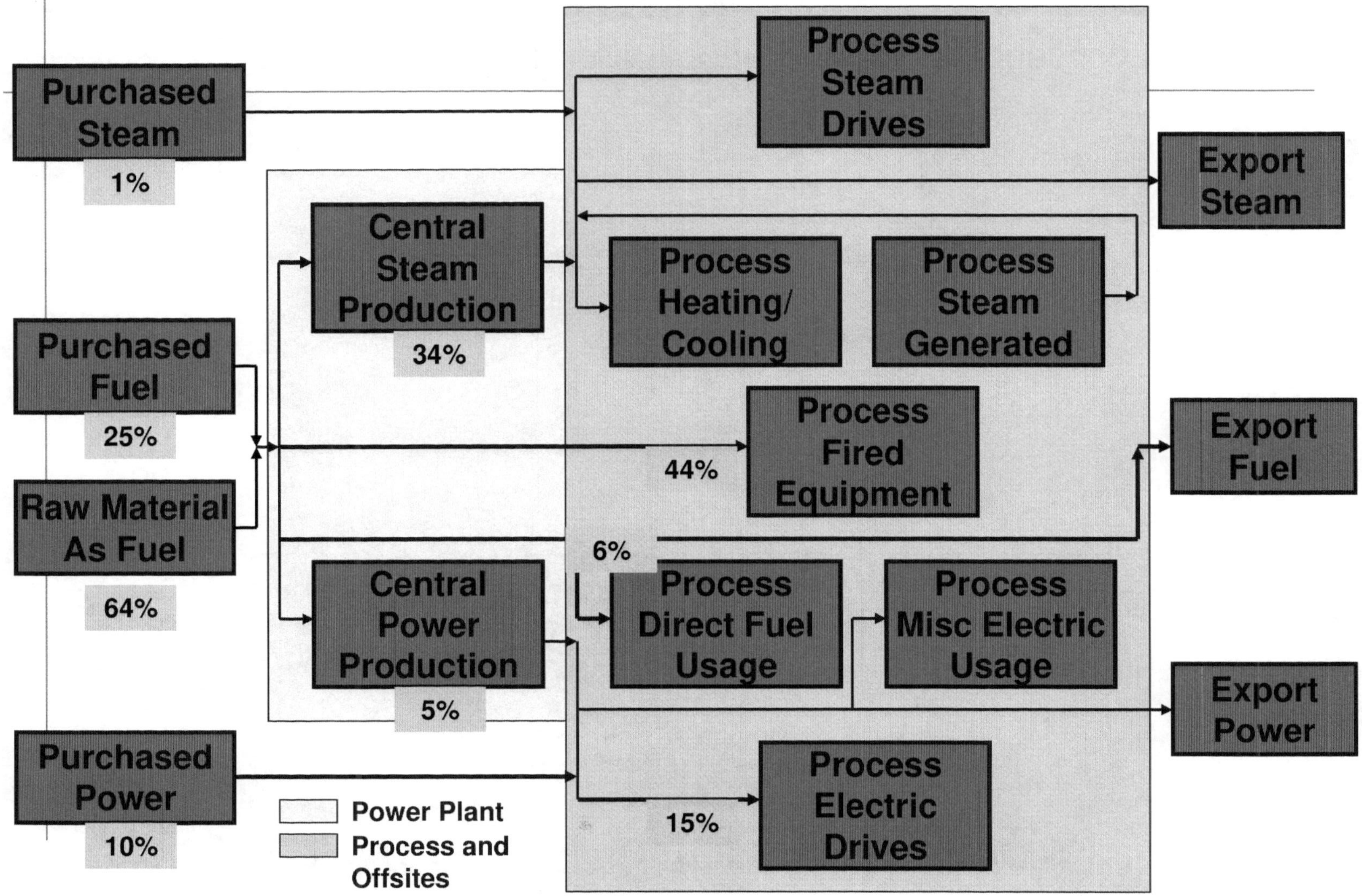

%-Equivalent BTU basis (Including losses) on total input

Source - DOE EIA

Opportunities For Energy Savings

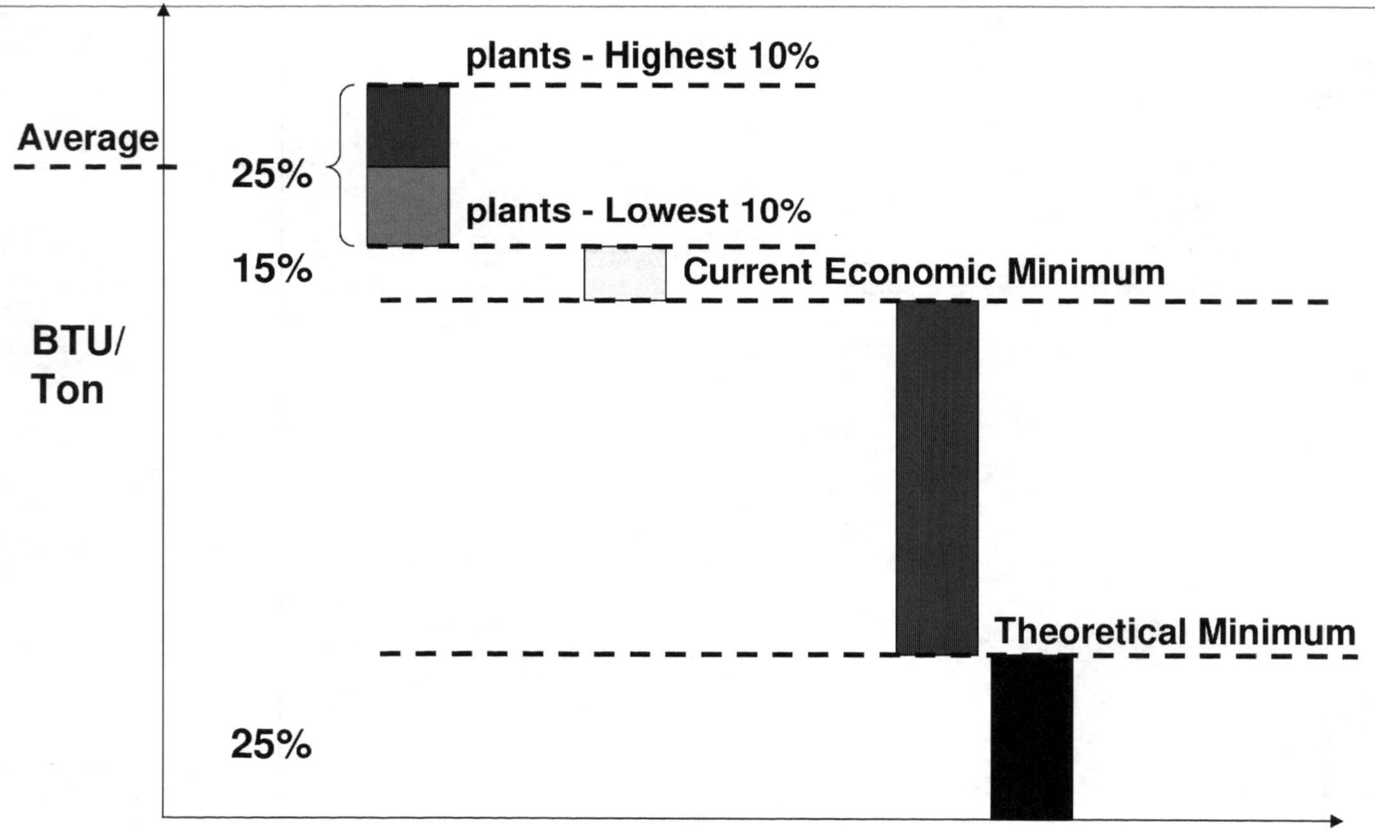

Corrected to Standard Conditions for Process Configuration, Product Grades, and Feed Quality

AIChE 2009 Spring National Meeting

Energy Saving Investment Options

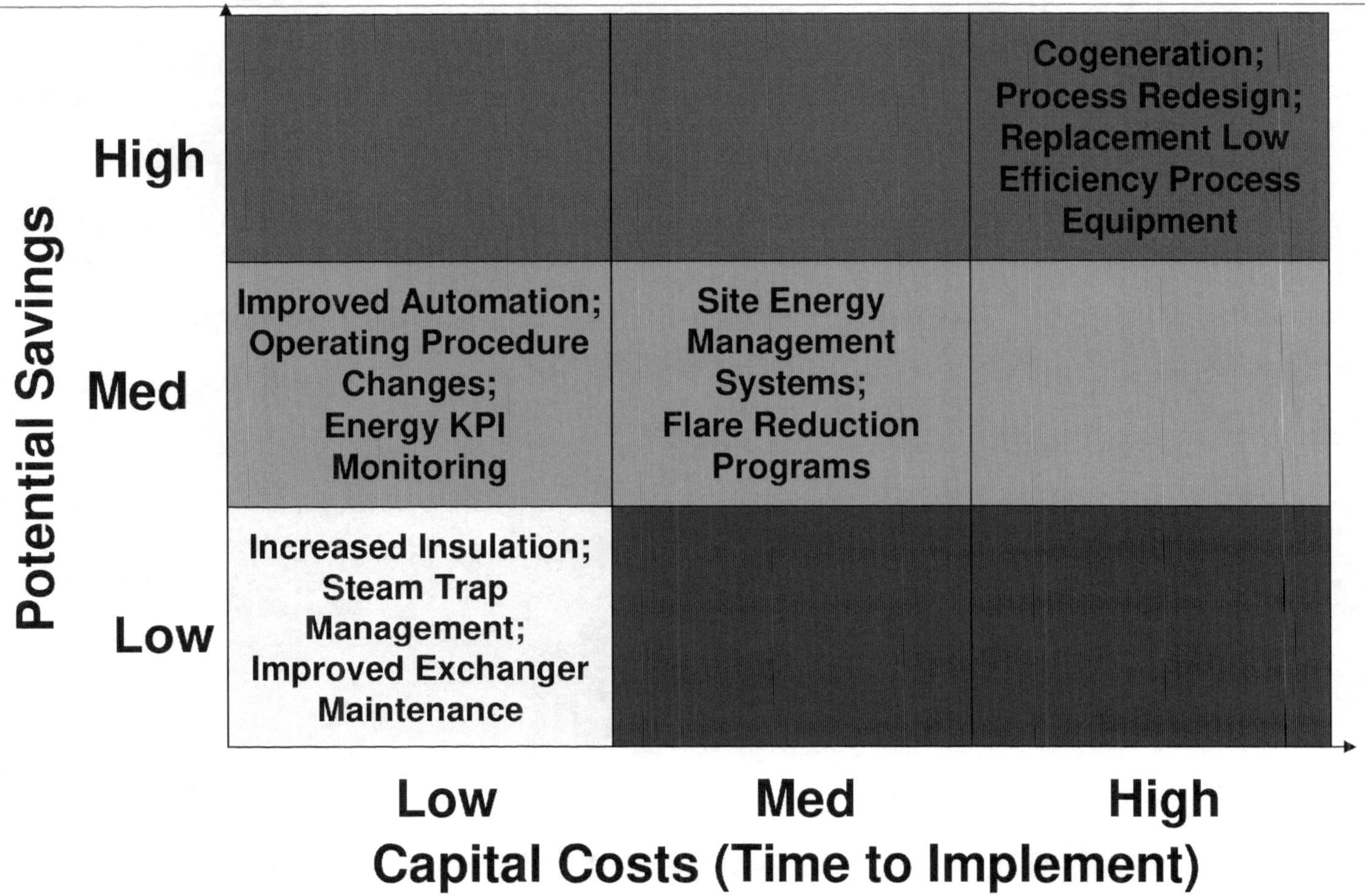

How Can Automation Reduce Energy Usage?

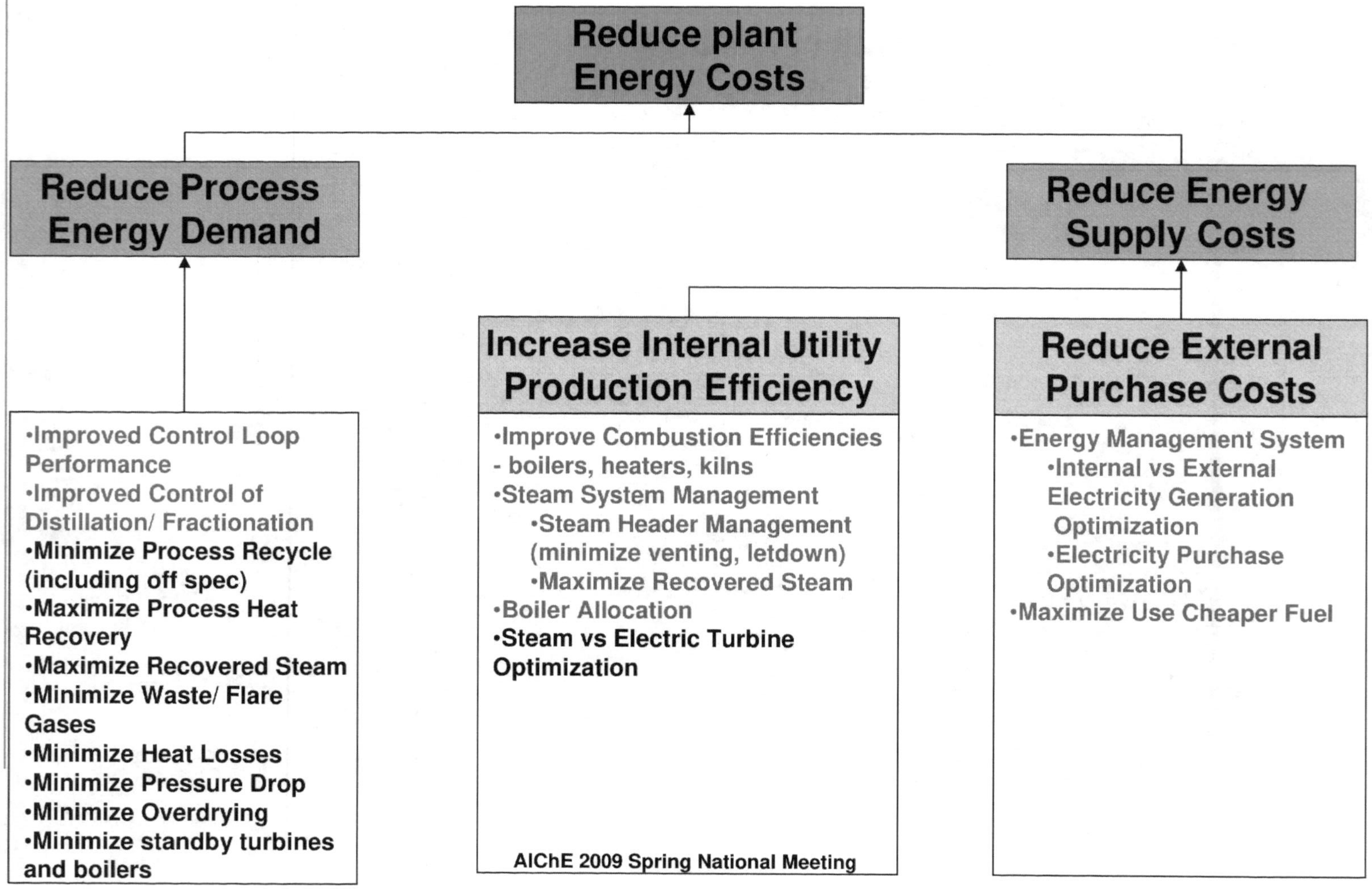

Energy Savings Through Automation – Target Areas

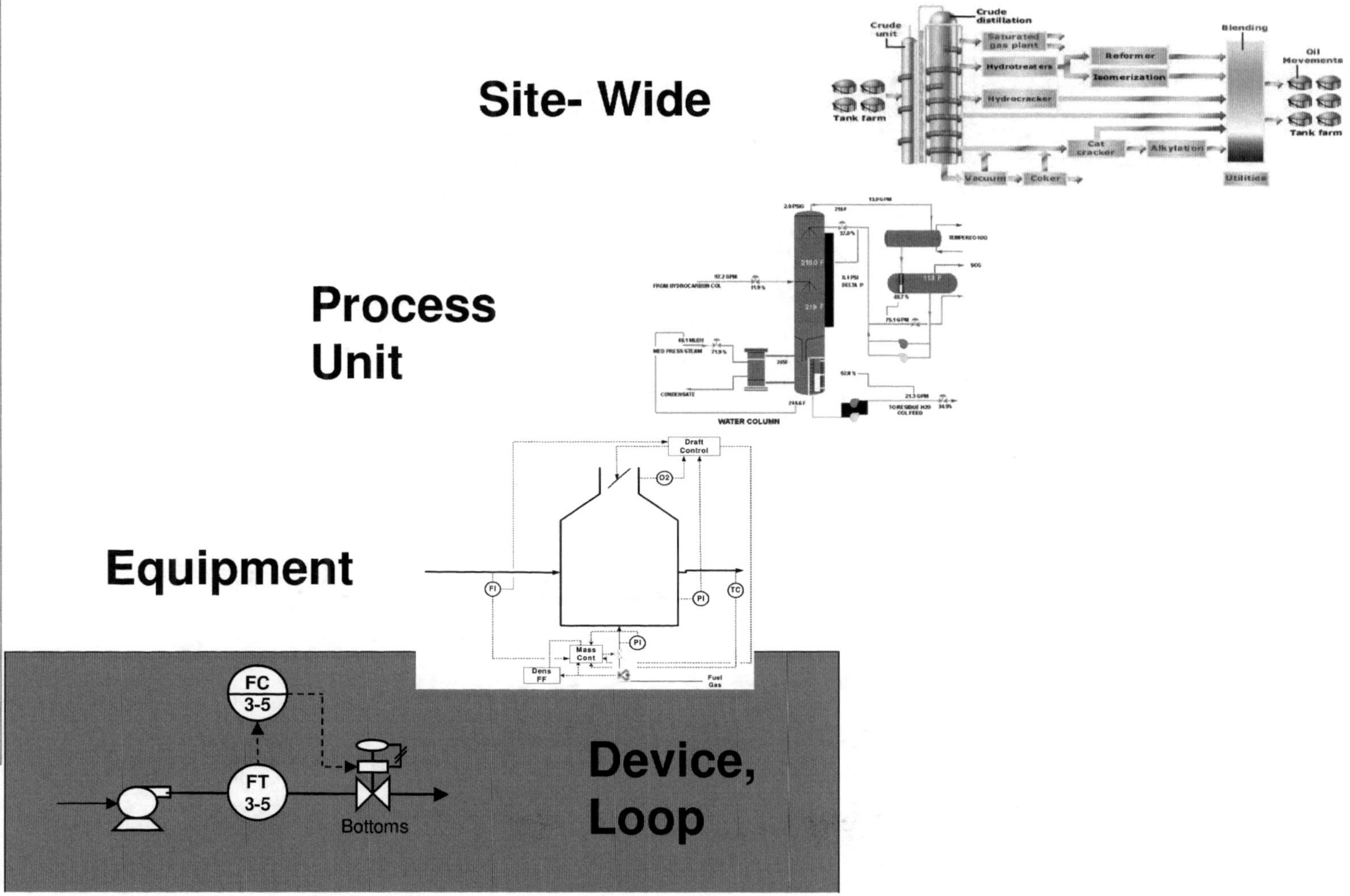

Saving Energy – Improving Loop Performance Examples

Improved Loop/ Multi-loop Control Performance

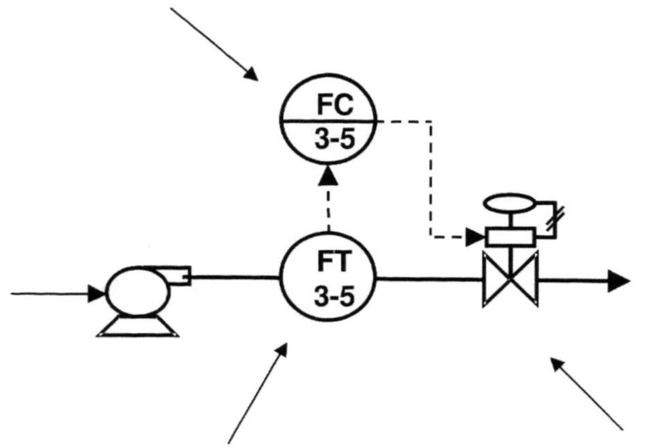

Improved Measurements

Better Control Valve Performance

Component Heating Values

Fuel Gas Component Heating Value		
Component	Heat of Combustion kcal/ NM3 (gross)	
Hydrogen	3020	
Methane	9520	
Ethane	16820	
Propane	24320	
Butane	32010	

Control Fuel Flows By Mass Instead of Volume

Energy Savings From Improved Measurements – Hydrogen Plant

Figure 1 Hydrogen Production in SMR Unit

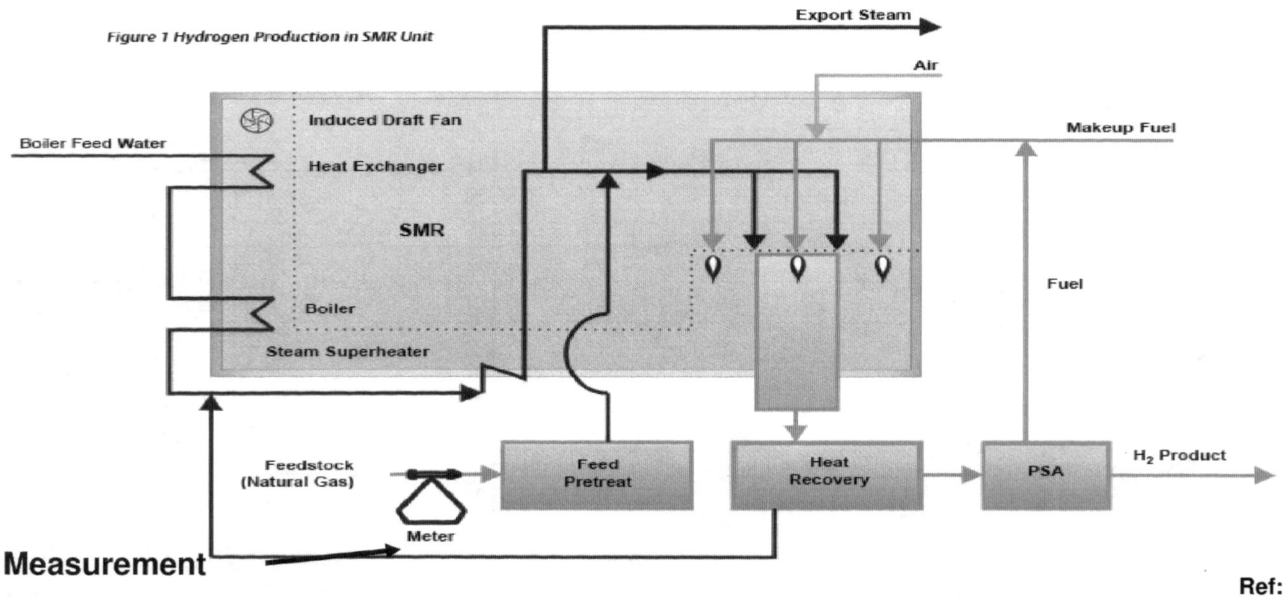

Ref: MMI WP_00724

Objective: Control S/C ratio as close to 3.2 as possible but avoid going below

Disturbance: Fuel gas C1 77 – 85%; C2 6.8 – 15; N2, CO also fluctuate

Test: Normal orifice plus GC – max error 0.2; MMI – max error – 0.02

Benefits: Moving 0.2 ratio closer to limit worth 8 BTU/SCF of H2;
80 MMSCFD plant; $12 MM BTU gas – *<u>$2.7 MM per year</u>*

AIChE 2009 Spring National Meeting

Control System - Energy Losses Through Bypassing

Bypass Option

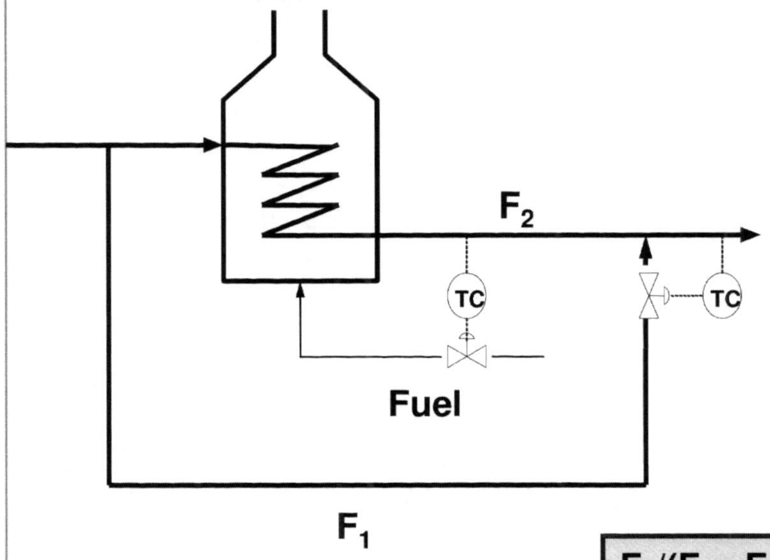

Direct Control Option

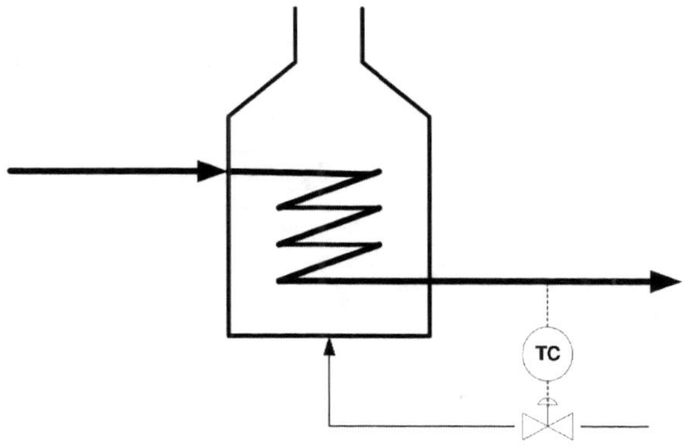

$F_1/(F_1+F_2)$	Heat Loss Increase - %
0.14	3.2
0.25	6.8
0.4	14.3

Reference: G. Shinskey; Energy Conservation Through Automation

Batch Reactor – Control Problem

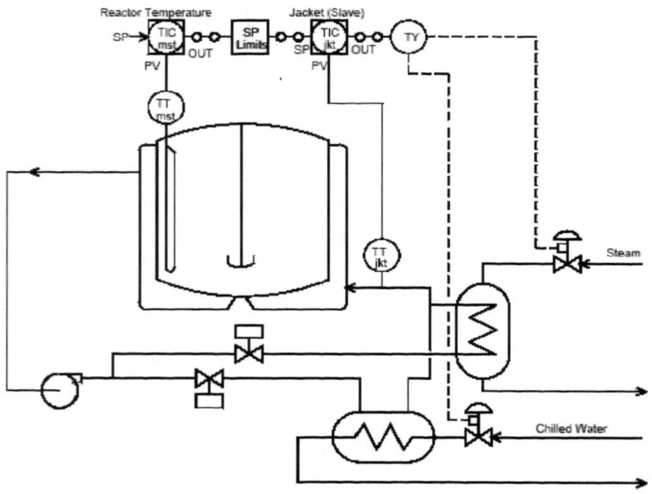

As-found reactor temperature response

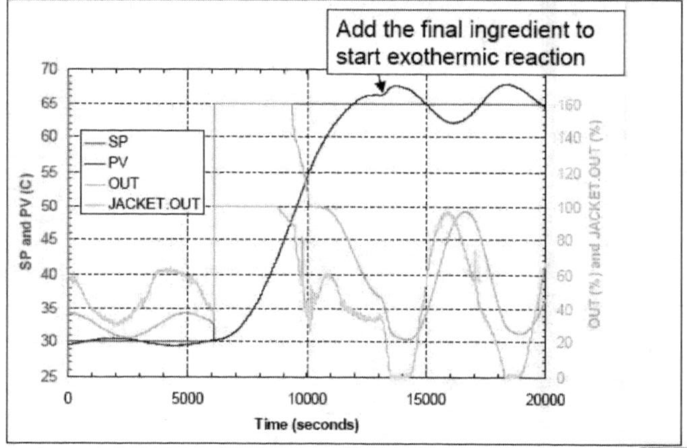

Ref: M. Coughran; "Temperature Control in Jacketed Reactors: New Case Studies;" Emerson Exchange, 2007

- All temperature loops cycled badly
- Wasted steam on up-cycle and chilled water on down cycle

Batch Reactor – Solution and Result

- **Solution**
  - Simulate reactor
  - Tune simulation
  - Implement Tuning in the field

Nonlinear simulation with controllers altered to reach 64 C quickly

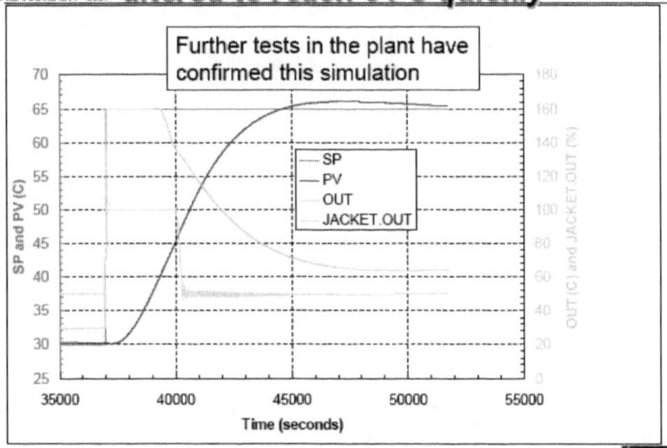

- **Results**
  - Oscillations Eliminated
  - Saved approximately 50% of the steam and 15% of the chilled water on each batch

Ref: M. Coughran; "Temperature Control in Jacketed Reactors: New Case Studies;" Emerson Exchange, 2007

Key Takeaways – Control System Improvements

- → Control System
  - – Identify Key Control Loops That Impact Energy Usage
    - Identify Cycling Loops
    - Look at loops always (mostly) run in manual – "They've always operated that way."
  - – Identify source of problem – external disturbances, field device issues, loop tuning
- → Measurements
  - – Consider if mass flow measurements will reduce variability and energy usage
  - – Evaluate changes in types of flow measurement reduce pressure drop and energy usage
- → Valves

Energy Savings Through Automation – Target Areas

Site- Wide

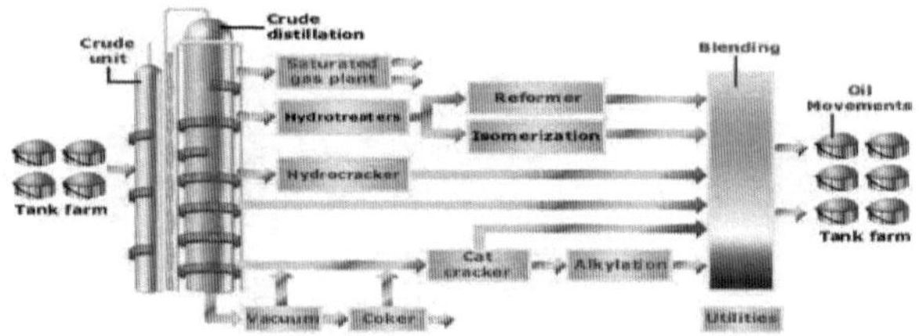

Process Unit

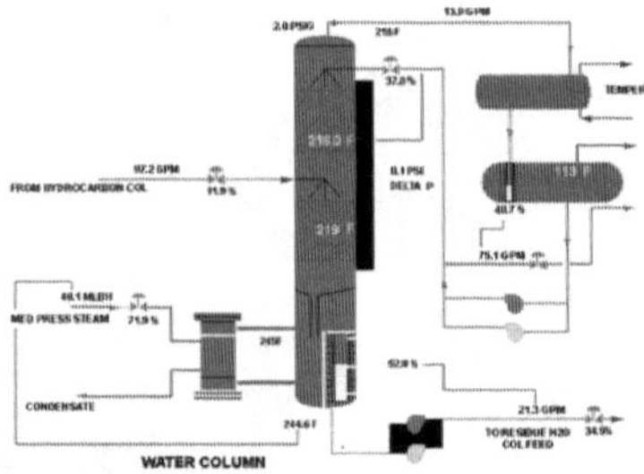

Equipment

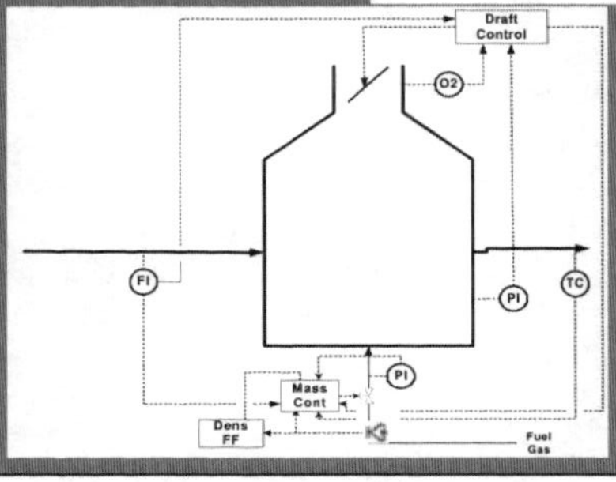

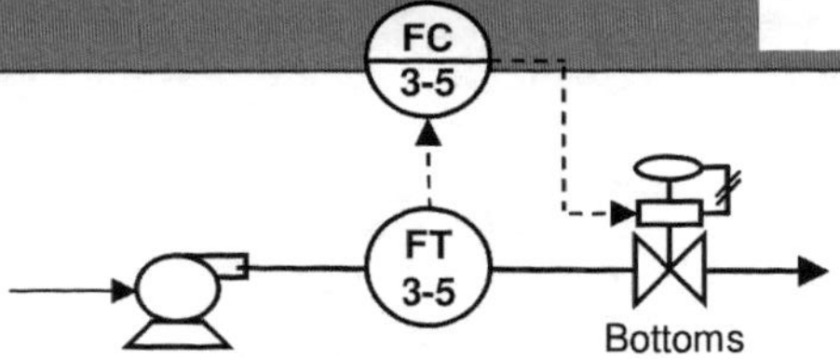

Device, Loop

AIChE 2009 Spring National Meeting

Equipment - Energy Saving Primary Targets

→ Fired Heaters/ Boilers

→ Distillation/ Fractionation

Fired Heater Controls

Heater Variability – Potential Energy Savings Example

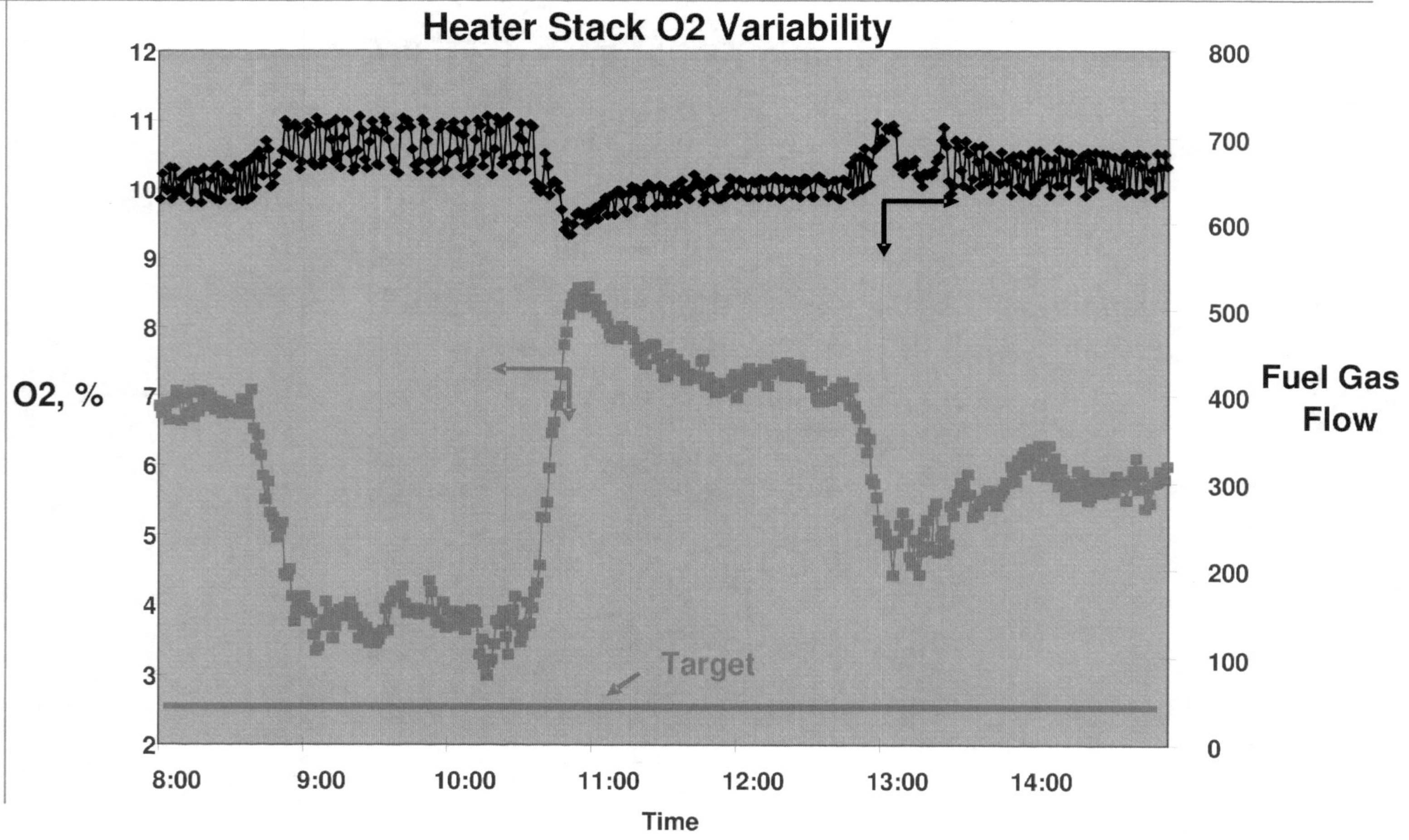

Excess Air Relationship

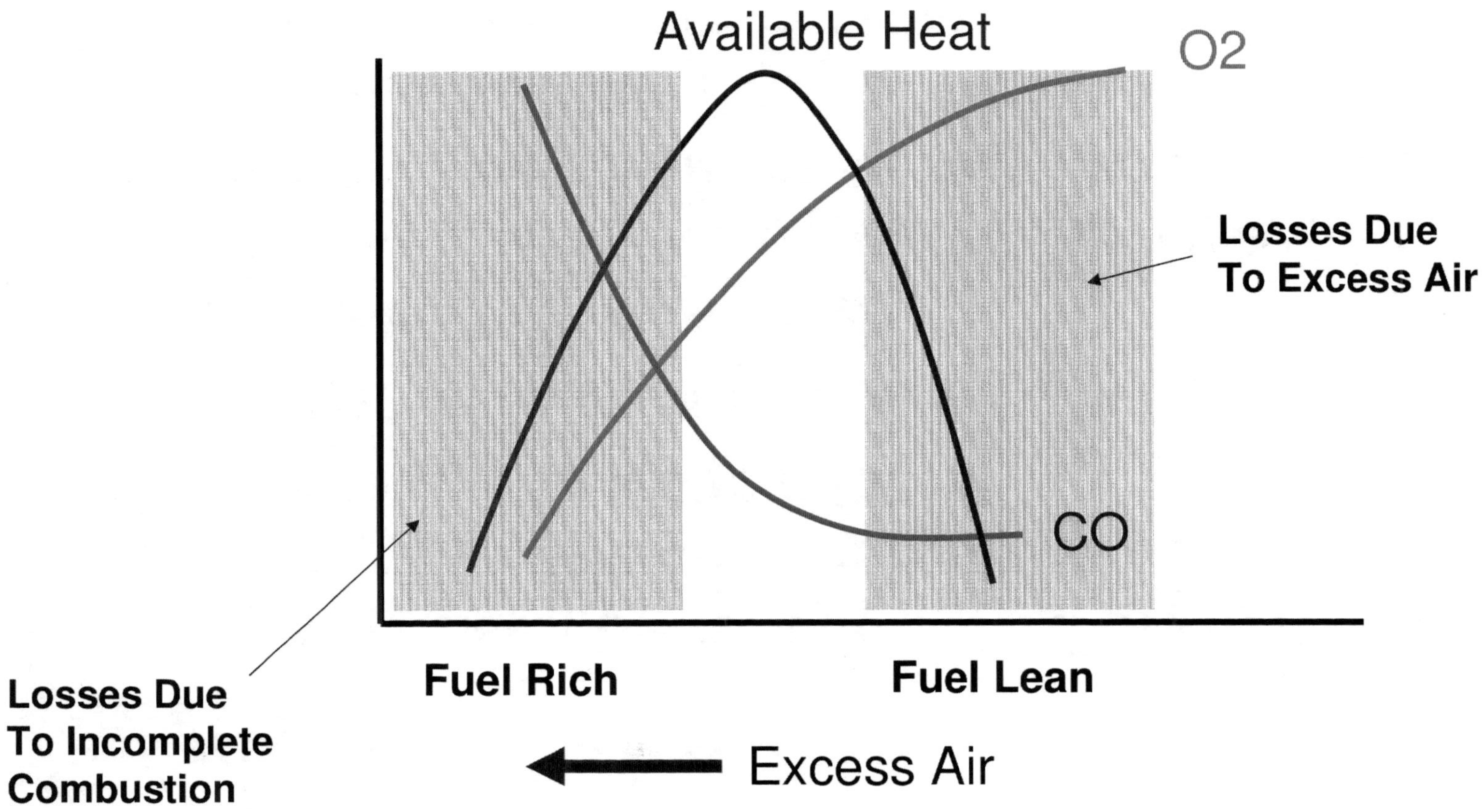

AIChE 2009 Spring National Meeting

What's The Problem?

- **Why do we need better control? Why can't we use a portable flue gas analyzer, set up the fuel/ air ratio and leave it alone?**
  - Ideal O2 and fuel/ air ratio varies with load
  - Fuel composition can vary
  - Changes in ambient temperature and relative humidity have a significant effect on combustion efficiency
  - Equipment performance changes

Fuel Types for Heaters/ Boilers

- Single Fuel (i.e. single gas stream)
  - Relatively constant fuel composition (i.e. purchased natural gas)
  - Varying Fuel Composition (i.e. plant fuel gas composed of a variety of streams)
- Multiple Fuels
  - Same phase (i.e. waste gas plus other gases fed individually)
  - Multiple phases (i.e. solid/ gas; liquid gas, etc.)

Automation requirements and opportunities vary with the different types of fuel

Varying Fuel Quality / Multiple Fuels - CO Analyzers

- Without CO Analyzers
  - New Heaters 3 – 5 % excess O2
  - Older Heaters – 3 – 10% excess O2
- With CO Analyzers
  - New Heaters – 0.2 – 2% excess O2
  - Older Heaters – 1 – 3% excess O2

Reduction of average O2 by 1% is worth at least $100,000 per year (100 MMBTU/ Hr Heater (released); $12 kscf fuel gas)

Best Practice – install CO analyzers on all major heaters and boilers

Typical Heater Instrumentation

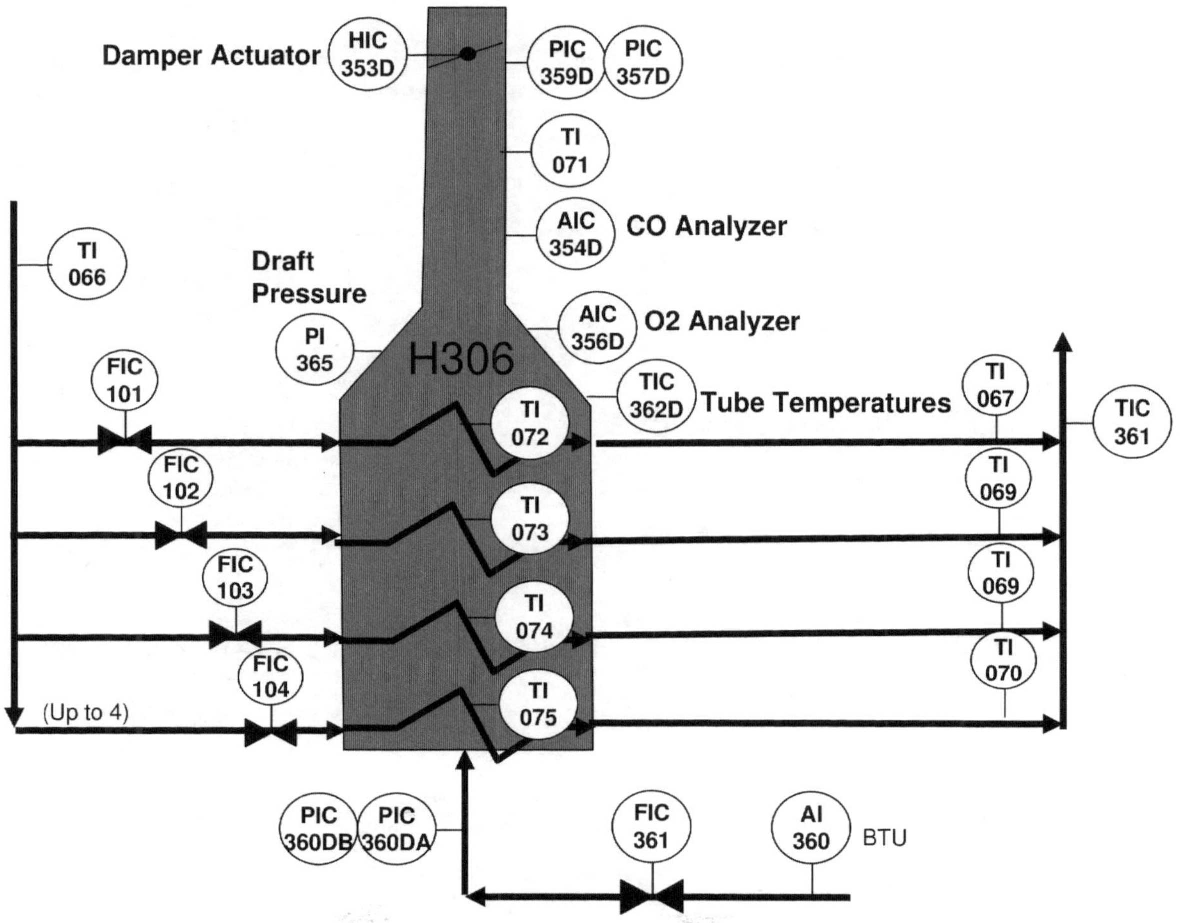

Typical Heater Advanced Automation Package

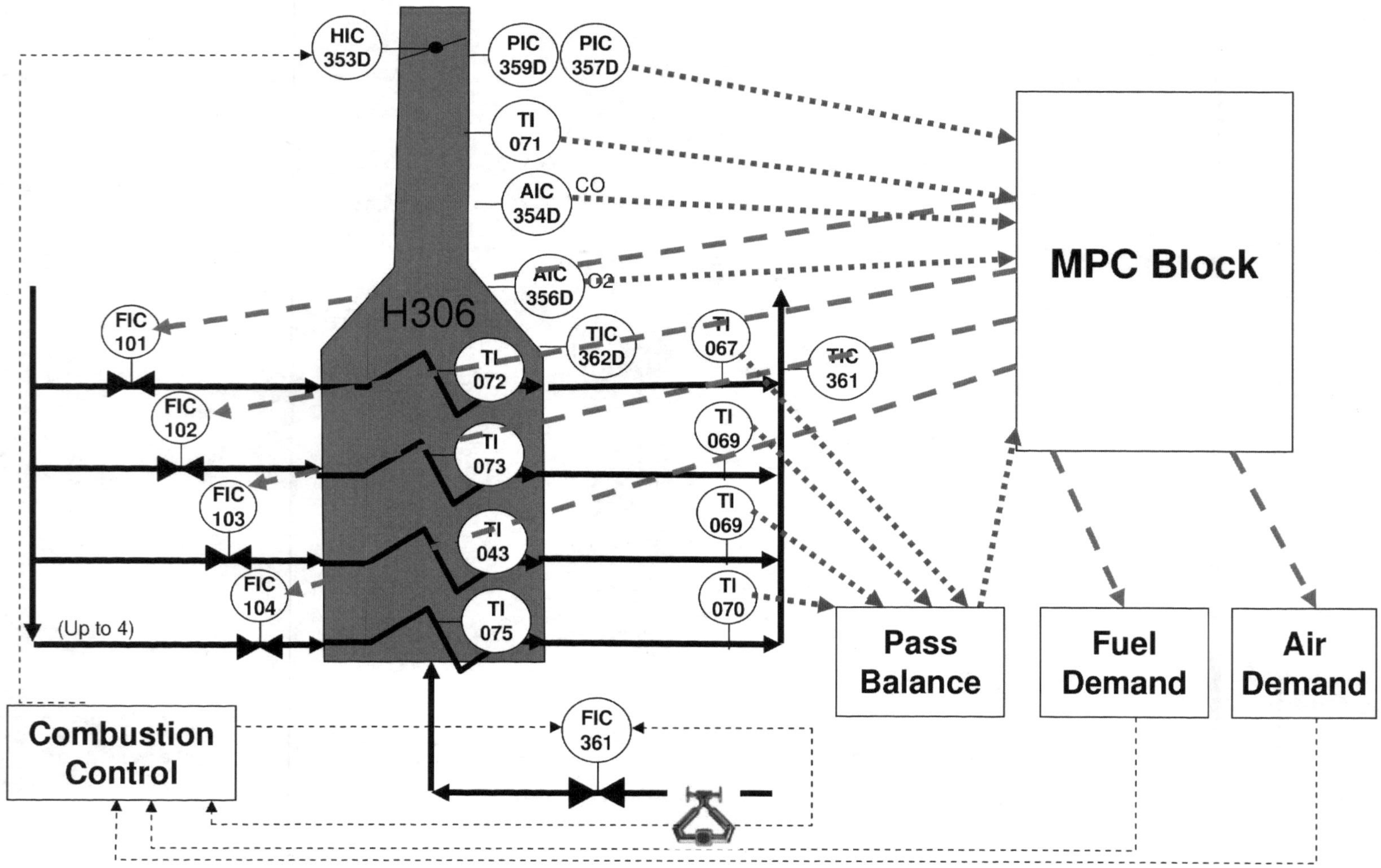

SmartProcess™ Heater Graphics

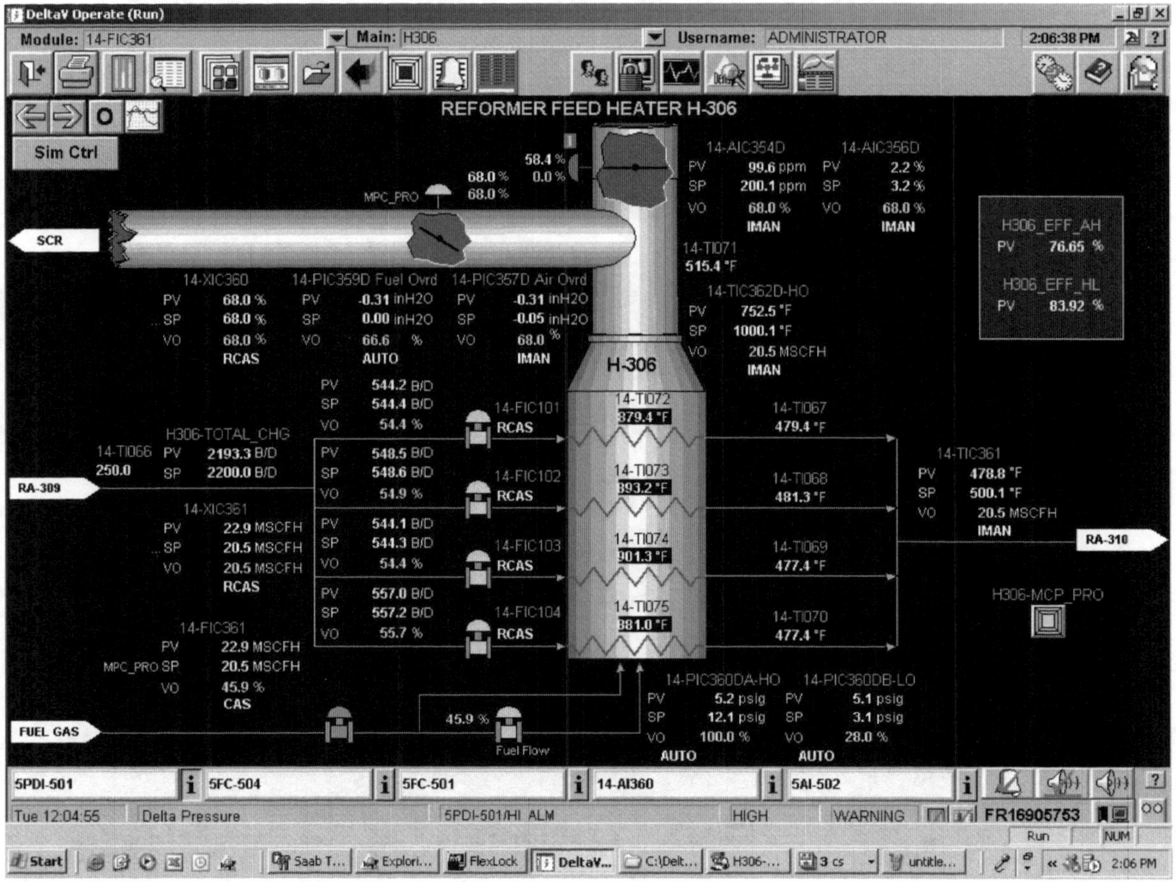

AIChE 2009 Spring National Meeting

Air Control in Fired Process Heaters

- Natural Draft
  - Outlet Damper
    - Controls furnace draft, and also affects excess O2
    - Needs to be automatically controlled
  - Burner Inlet Air Dampers
    - Usually manually controlled
    - May have Individual actuators or multiple burner inlets mechanically linked
- Forced Draft
  - Dampers/ Fans
    - Can have inlet and/ or outlet dampers and fan controls
  - Burners
    - Inlet vanes or outlet dampers used on large burners
    - Smaller burners connected via a common plenum

Automation requirements and opportunities vary with the different types of air control

Excuses For Not Improving Heater Controls

- **Damper/ Air controls are not reliable**
  - Answer: Add positioners to dampers, with feedback to control system; Analyze and fix controller problems

- **Don't have online analyzer/ can't maintain them**
  - Answer: Analyzers are cheaper and more reliable – particularly mass flow meters. With higher fuel costs, they are well justified.

Heater Controls – Key Takeaways

- For moderate to large size heater – evaluate adding CO analyzer if there is not one there already – implement updated combustion control
- Check load characterization curves – if they have not been checked and revised in the last year, they need to be updated
- Add actuator to permit damper to be manipulated by the control system
- For larger forced draft heaters/ boilers consider variable speed drives
- If fuel quality varies – try to reduce swings via control and compensate with mass measurements if possible

Distillation Columns

- → **Over 40000 distillation/ fractionation columns in the US alone**
- → **Consume 40% - 60% of the total energy used in chemical and refining industry**
- → **Consume 19% of the total energy used in manufacturing industry in the US**
- → **Consume 6% of total US**

Reference: Office of Industrial Technology:
Energy Efficiency and Renewable Energy;
US Department of Energy
Washington, DC
"Distillation Column Modeling Tools"

Distillation Column – Standard Controls

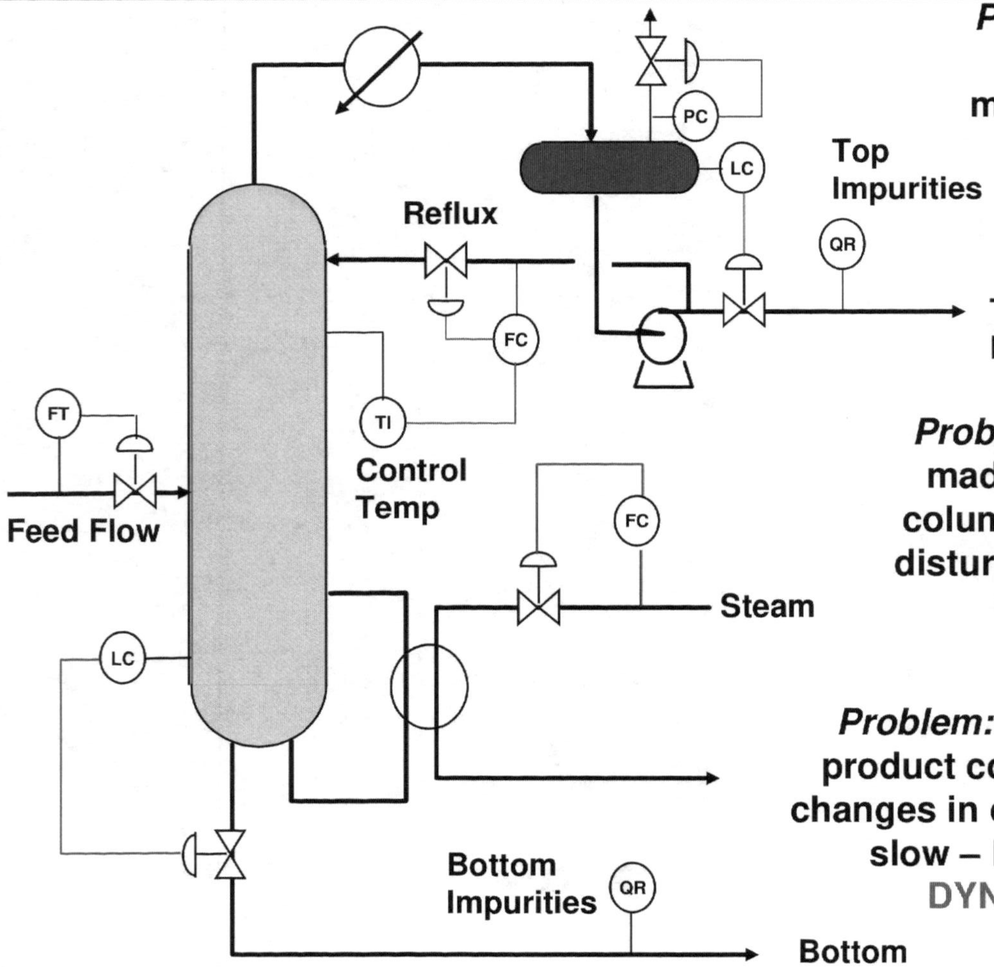

Problem: **There are many potential limits to maximum production- lots of CONSTRAINTS**

Problem: **If a change is made to the top of the column, the lower part is disturbed as well – lots of INTERACTION**

Problem: **Reaction of product composition to changes in controls is very slow – DIFFICULT DYNAMICS**

Distillation Column – Advanced Controls

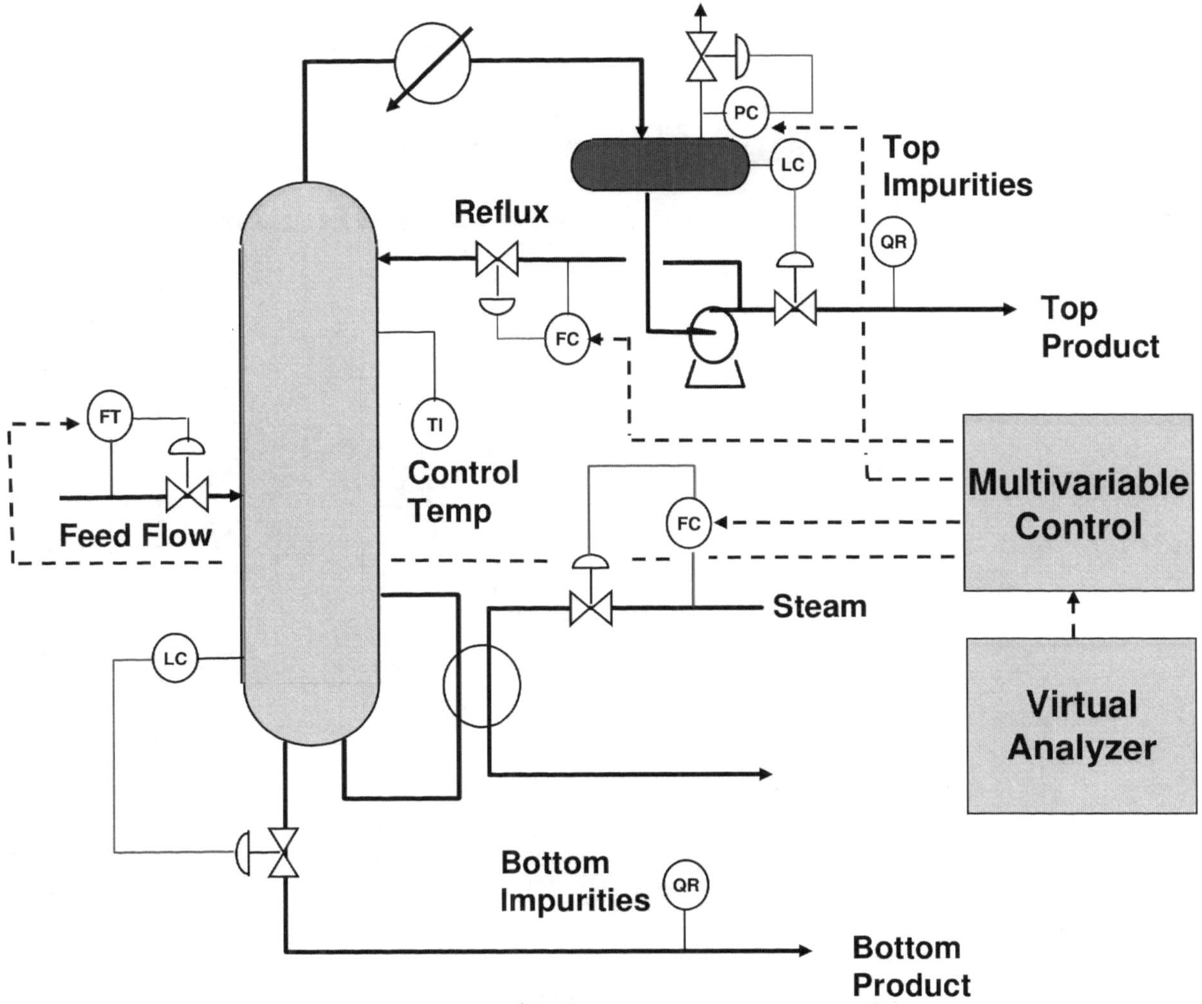

Effect of Reduced Variability

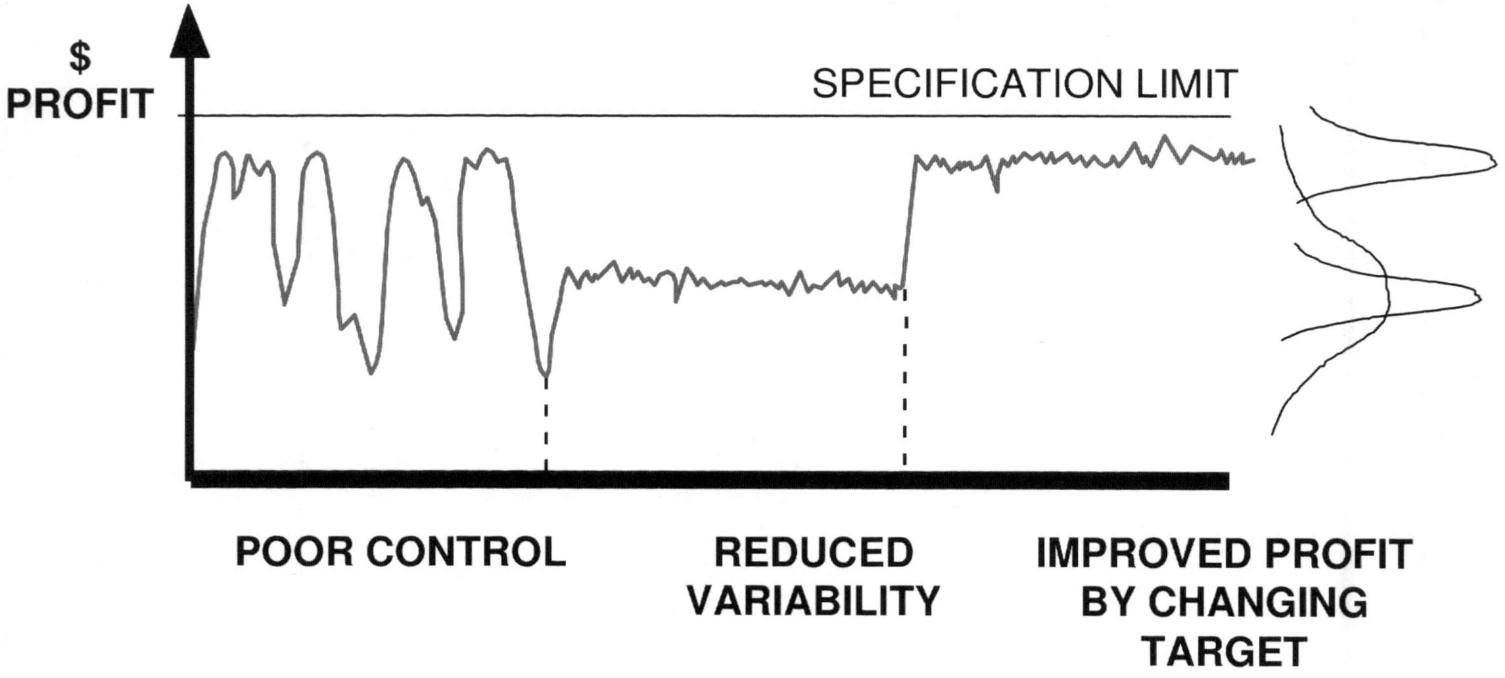

Product Value with Variability

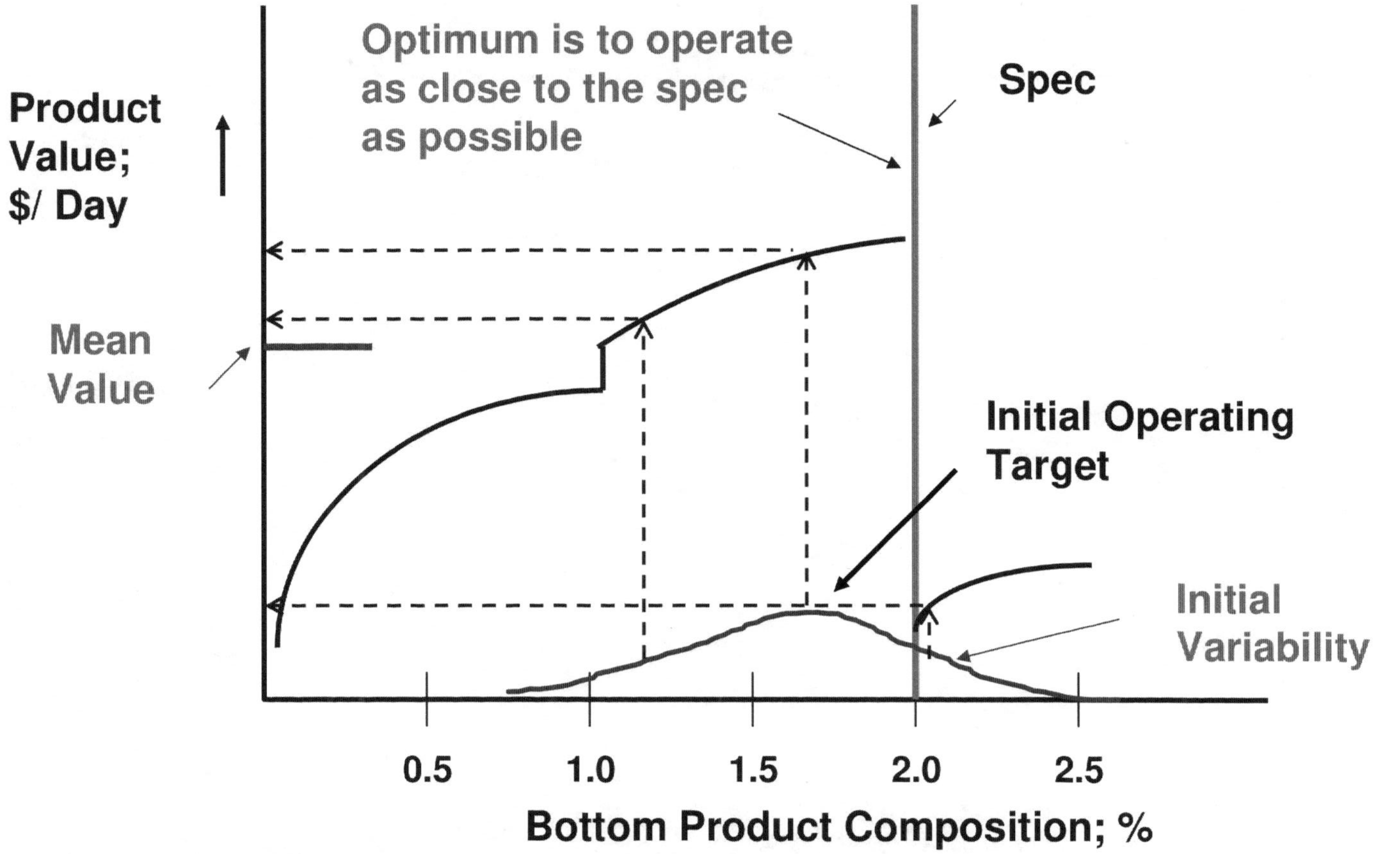

Product Value with Variability

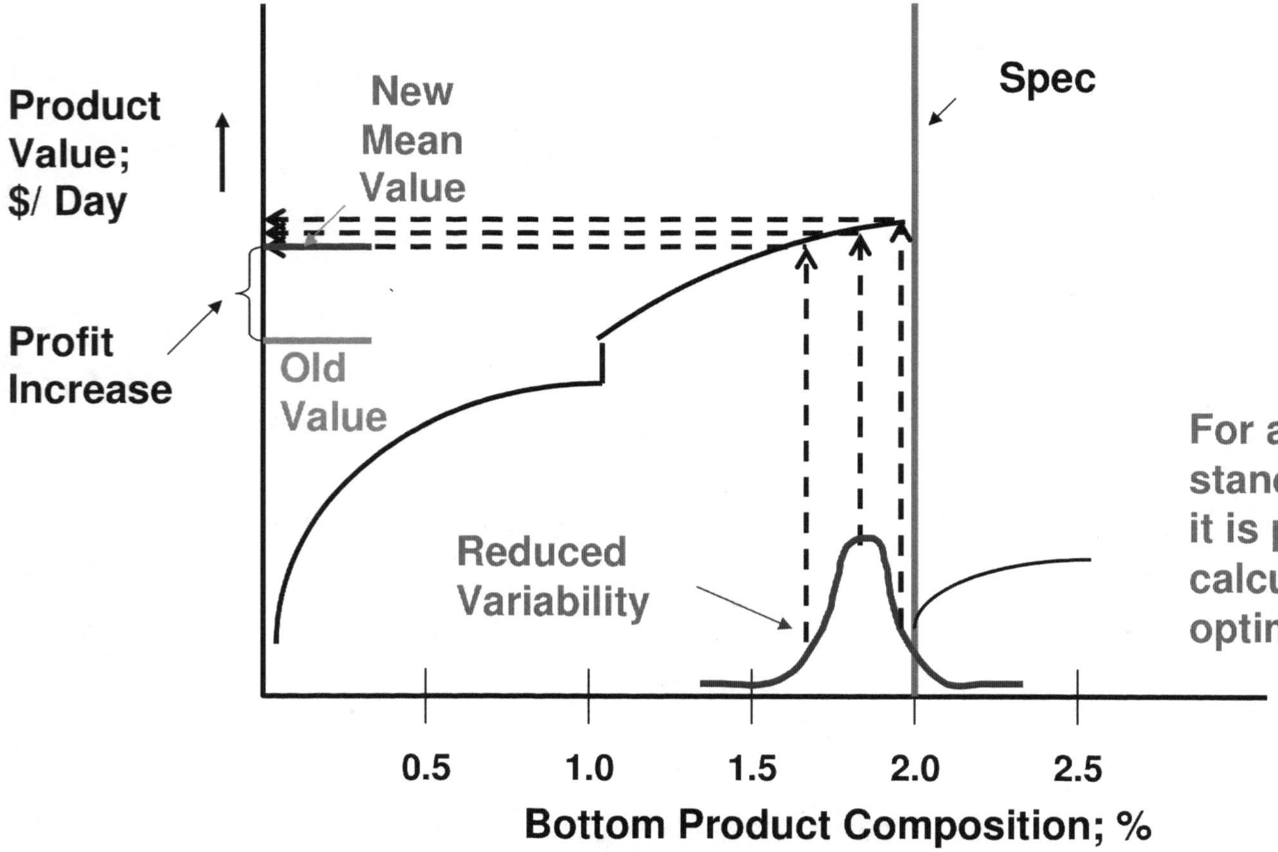

For any given standard deviation it is possible to calculate the optimum setpoint.

Example - Debutanizer

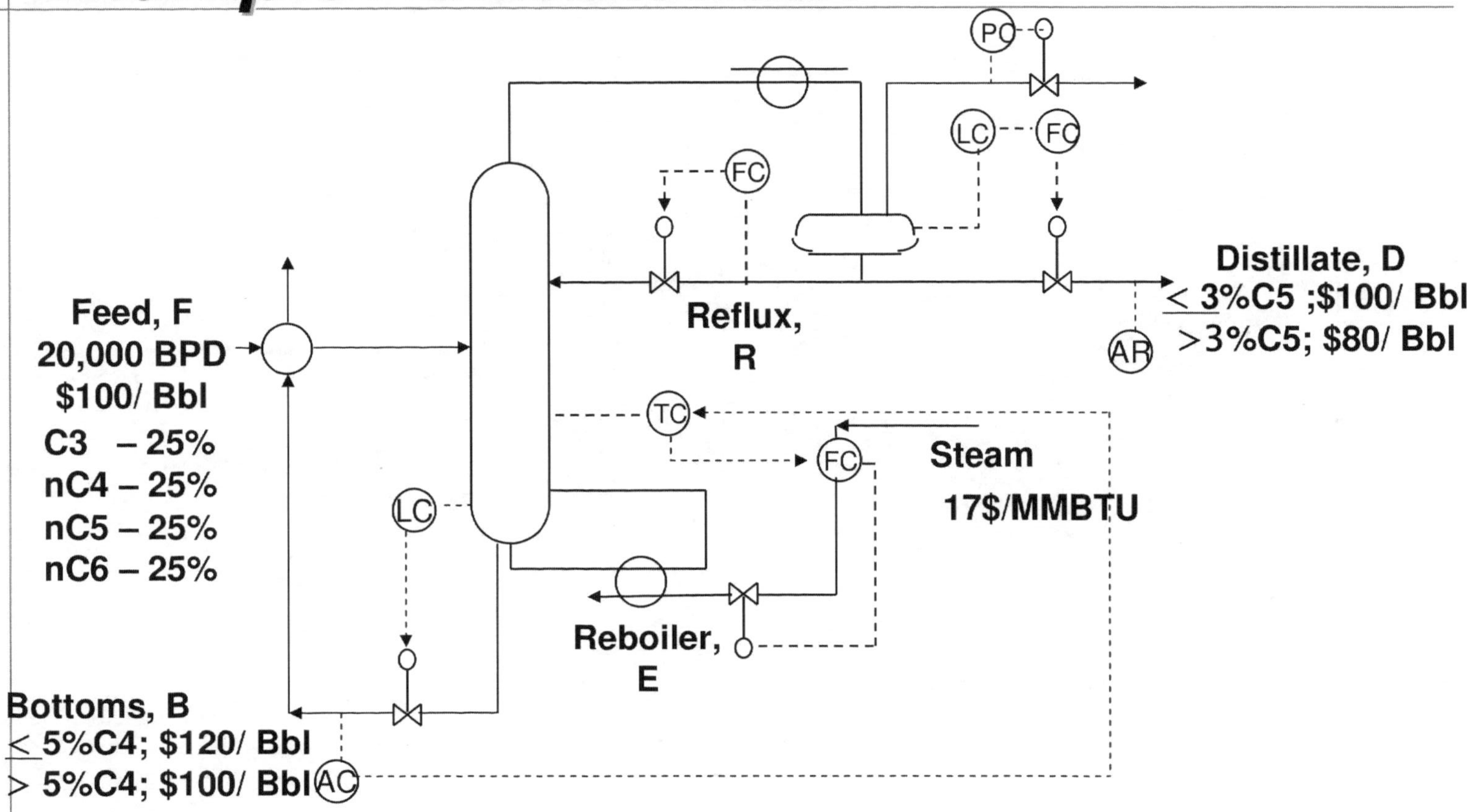

AIChE 2009 Spring National Meeting

Example - Debutanizer - Net Profit Curve

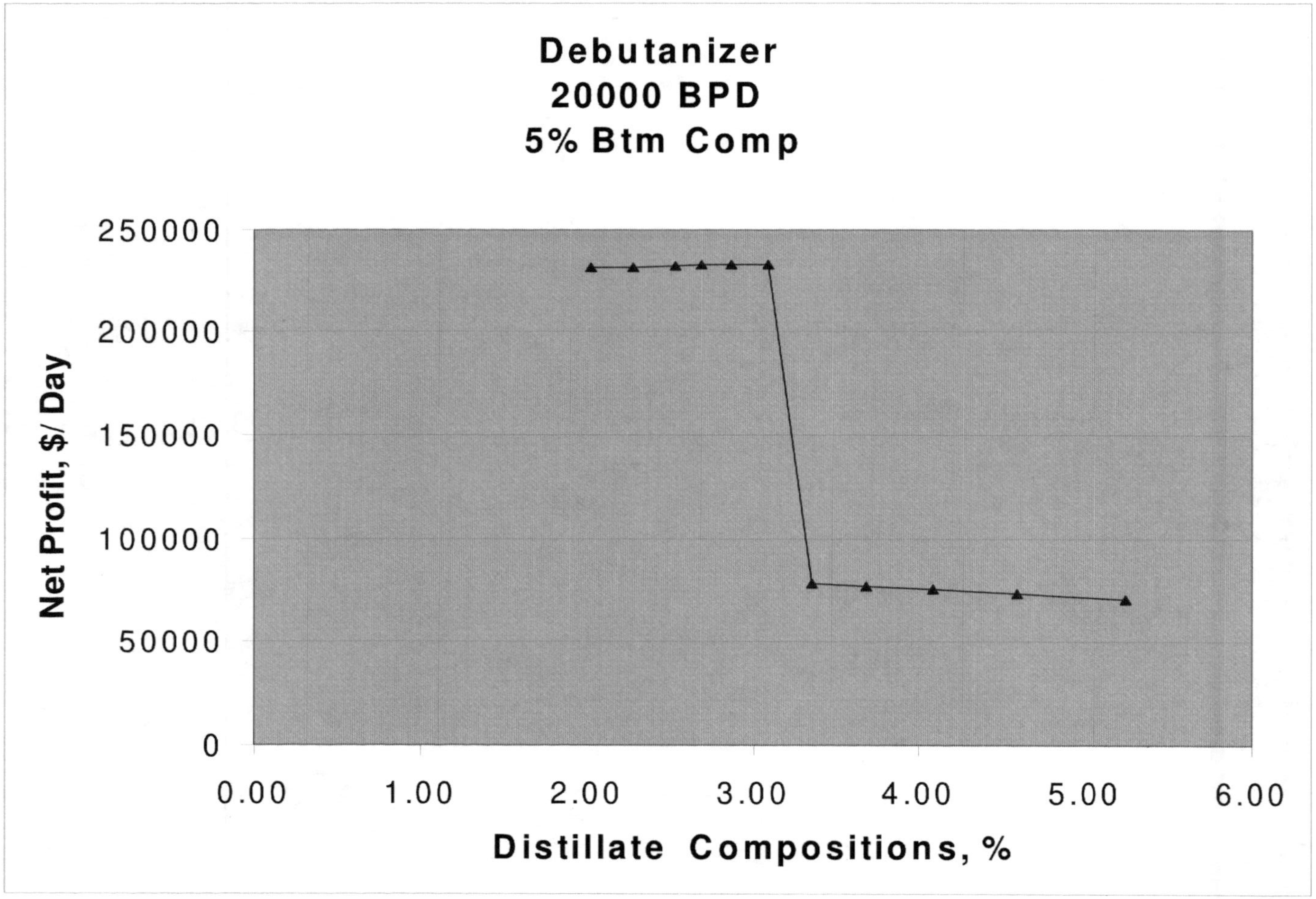

Debutanizer – Optimum Setpoint With Variability

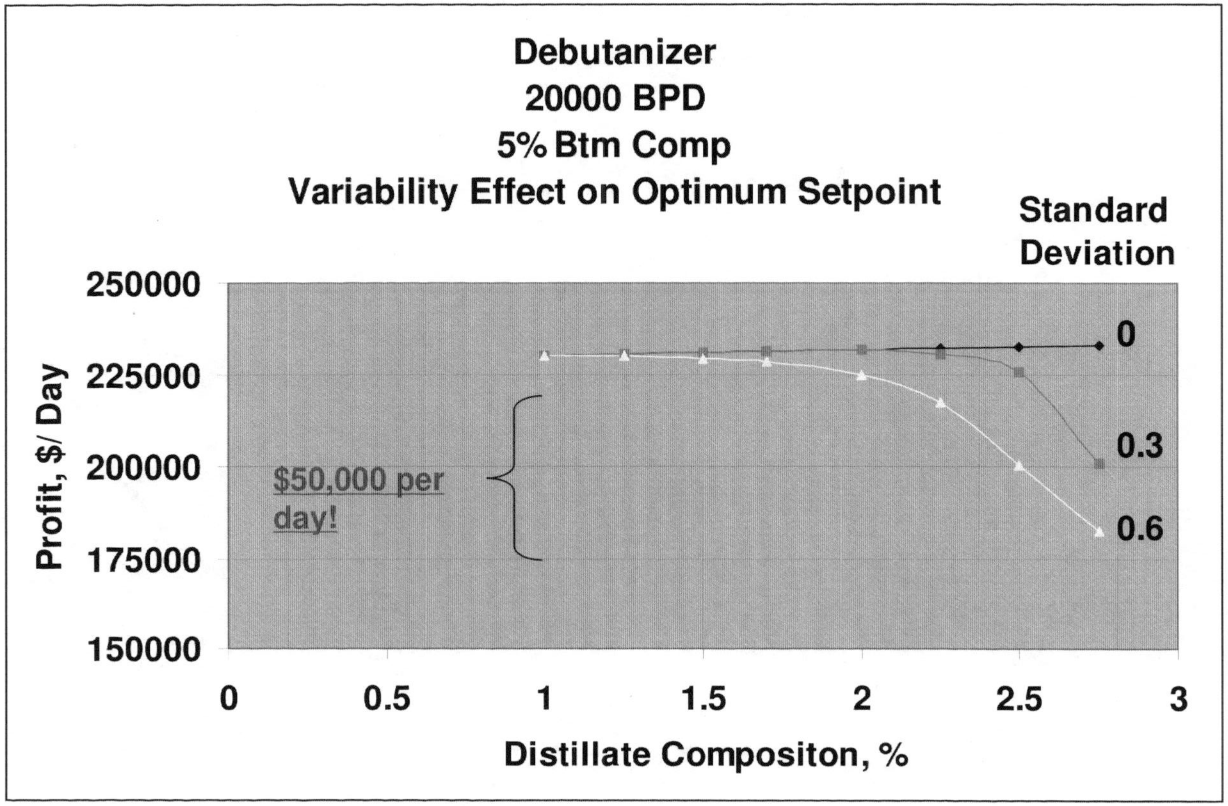

Optimum Setpoint Depends on Variability

AIChE 2009 Spring National Meeting

Profitability – Energy Use

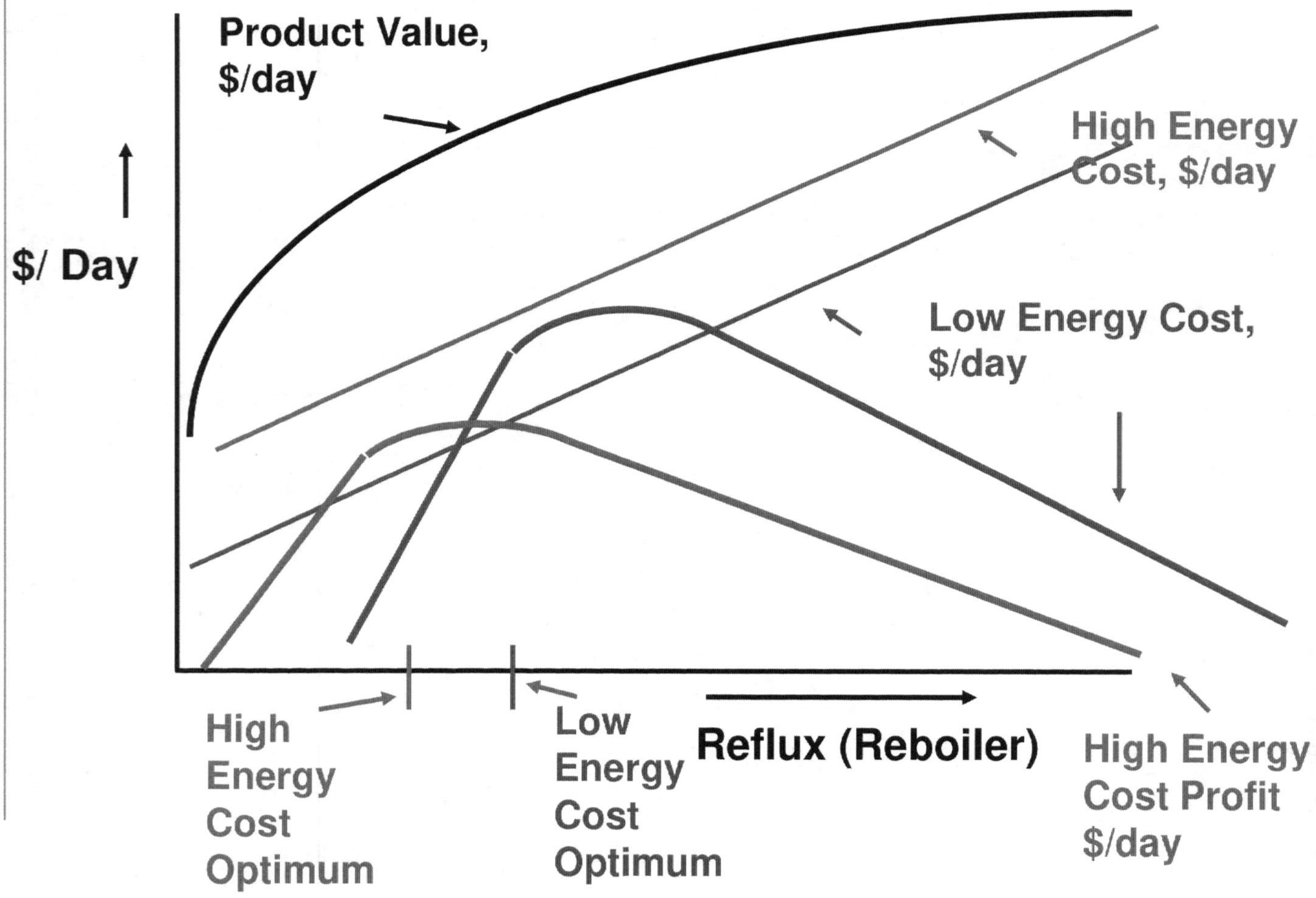

AIChE 2009 Spring National Meeting

Profitability – Energy Use - With Variability

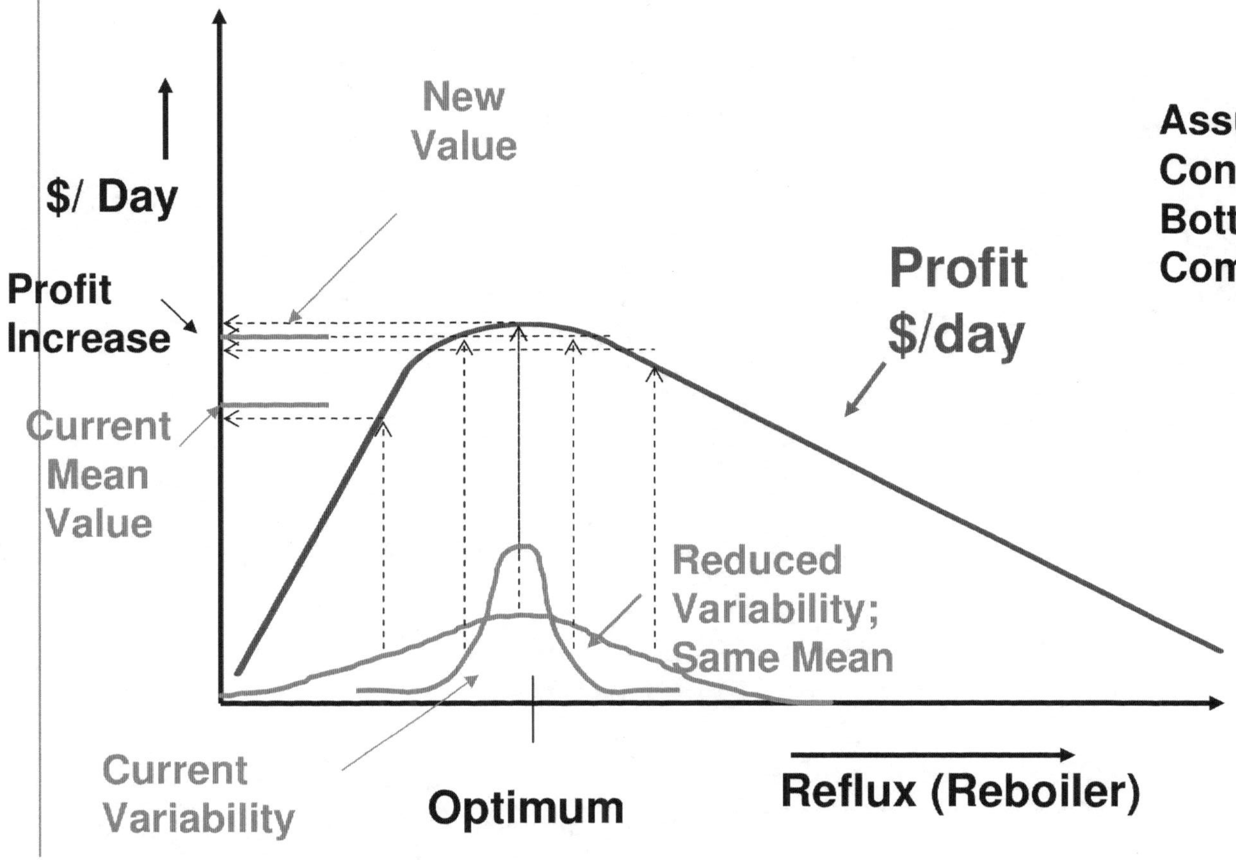

Assumptions:
Constant
Bottoms
Composition

Variability Vs Energy

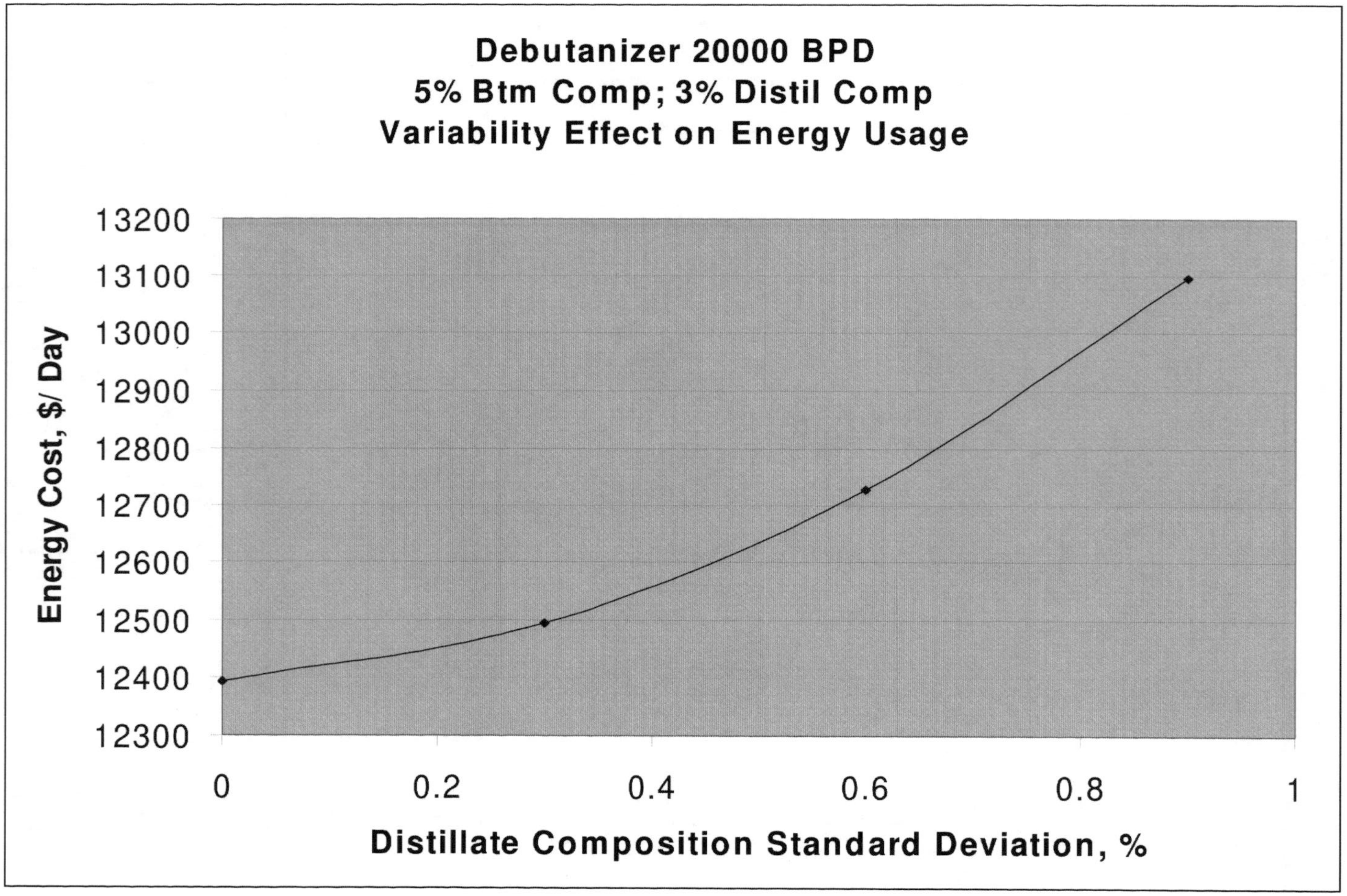

AIChE 2009 Spring National Meeting

Column Pressure Effect

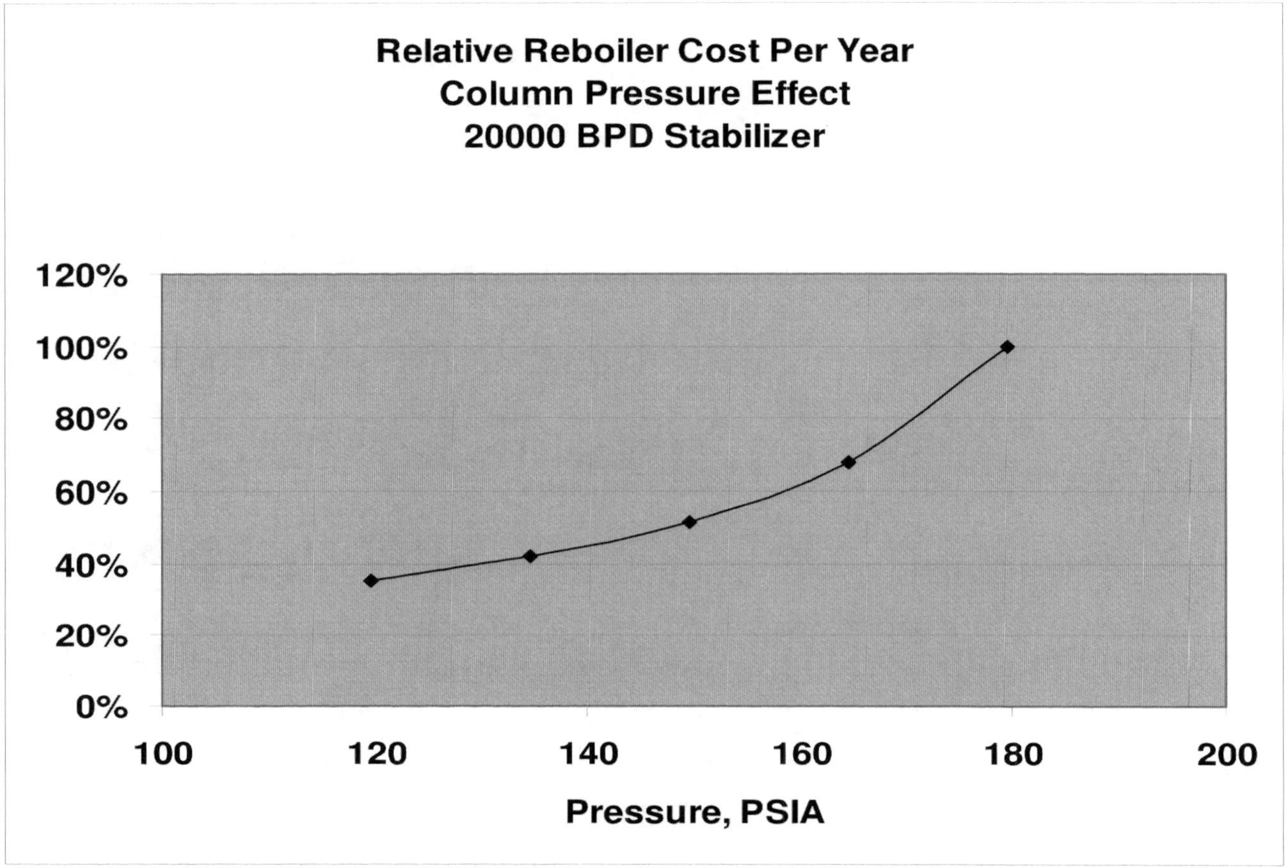

Basis: Constant Separation
Modeled With ChemSep
Peng Robinson Equation of State

AIChE 2009 Spring National Meeting

Key Takeaways - Saving Energy in Fractionation

→ Closer Control to Specifications

→ Optimize Energy Usage

→ Minimize Pressure

→ plant Crude/ Vacuum Columns

  – Minimize Stripping Steam

  – Minimize Overflash

Example

Utilities Example – Biomass Power Boiler

- Paper mill
- 160k PPH Fluidized-bed Boiler
- Fuels:
  - Sludge
  - Wood waste
  - Tires
  - Fuel gas
- Incentives:
  - Maximize use of cheap fuels (Tires & Wood)
  - Burn all the sludge to minimize land fill
  - Maximize steam production

Solid Fuel Composition Control

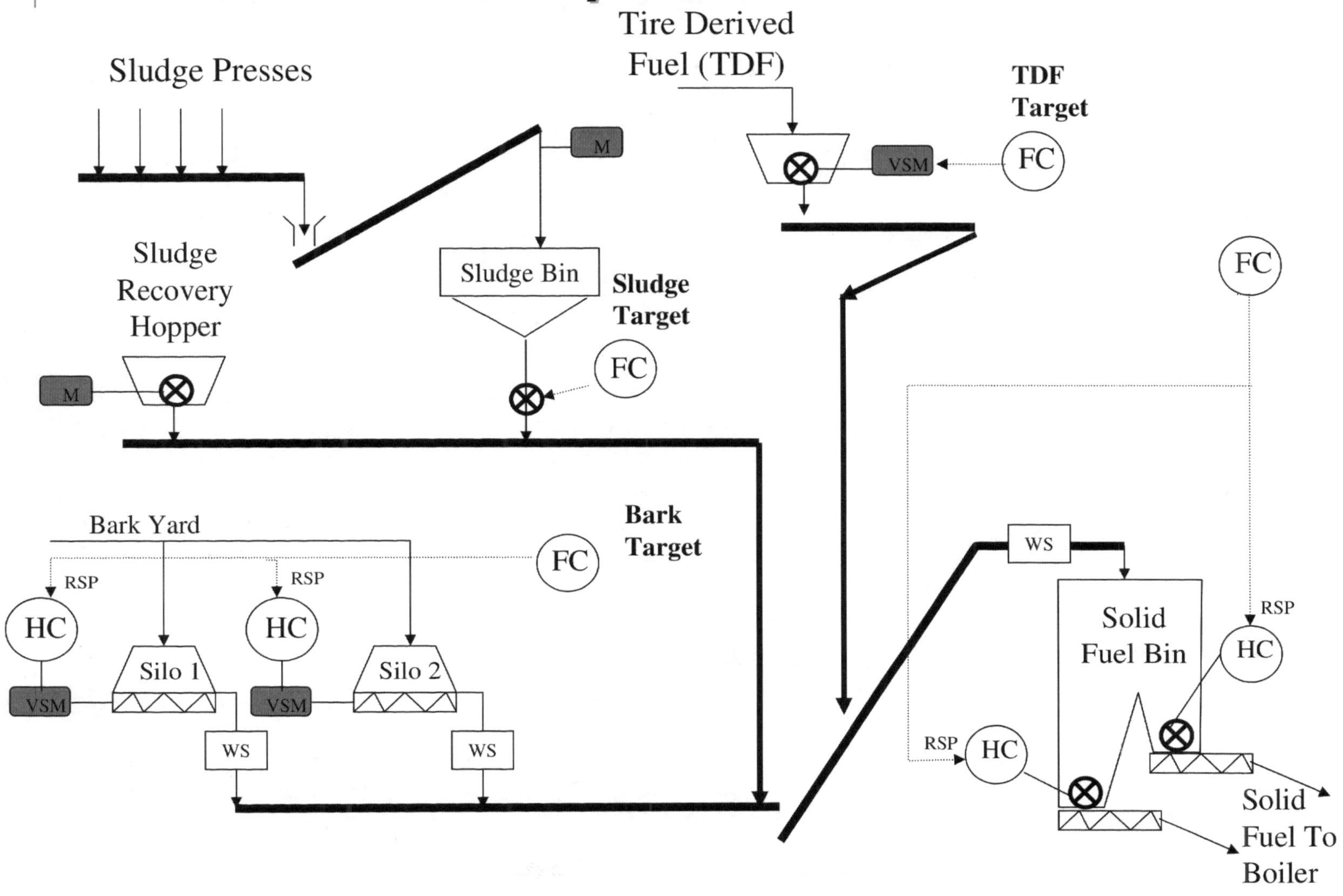

Boiler Control

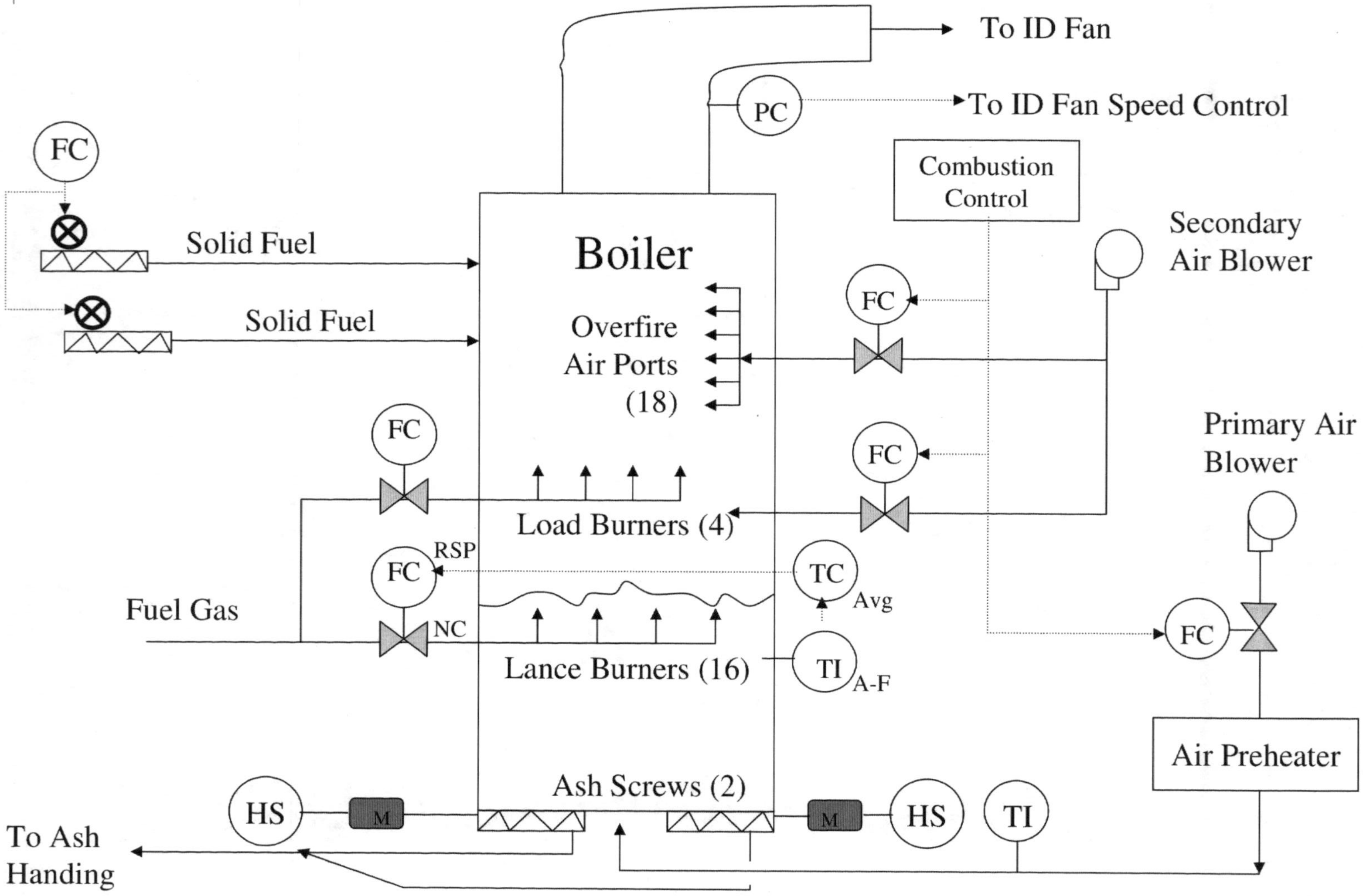

Boiler Process Control Issues

- → Varying water in sludge
- → Long delay & lag times (20 – 60 minutes) to change fuel composition
- → Fuel composition time constants are a function of fuel bin level
- → Solid fuel composition in fuel bin is unknown
- → Bed temperature constraints (max & min)
- → Multiple operators controlling same unit
- → Different operating philosophy used by each shift

Solid Fuel Composition Control

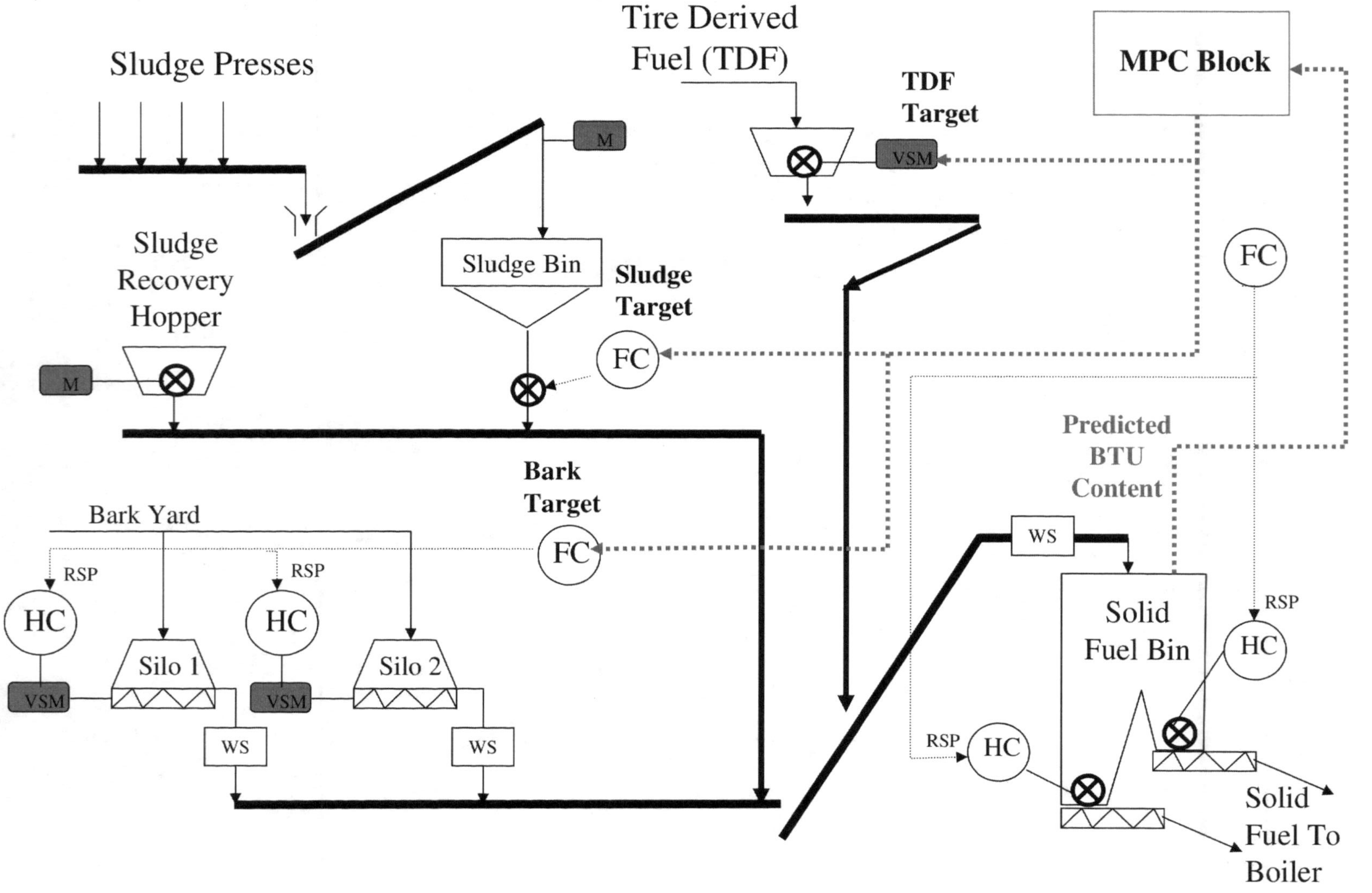

Boiler Control

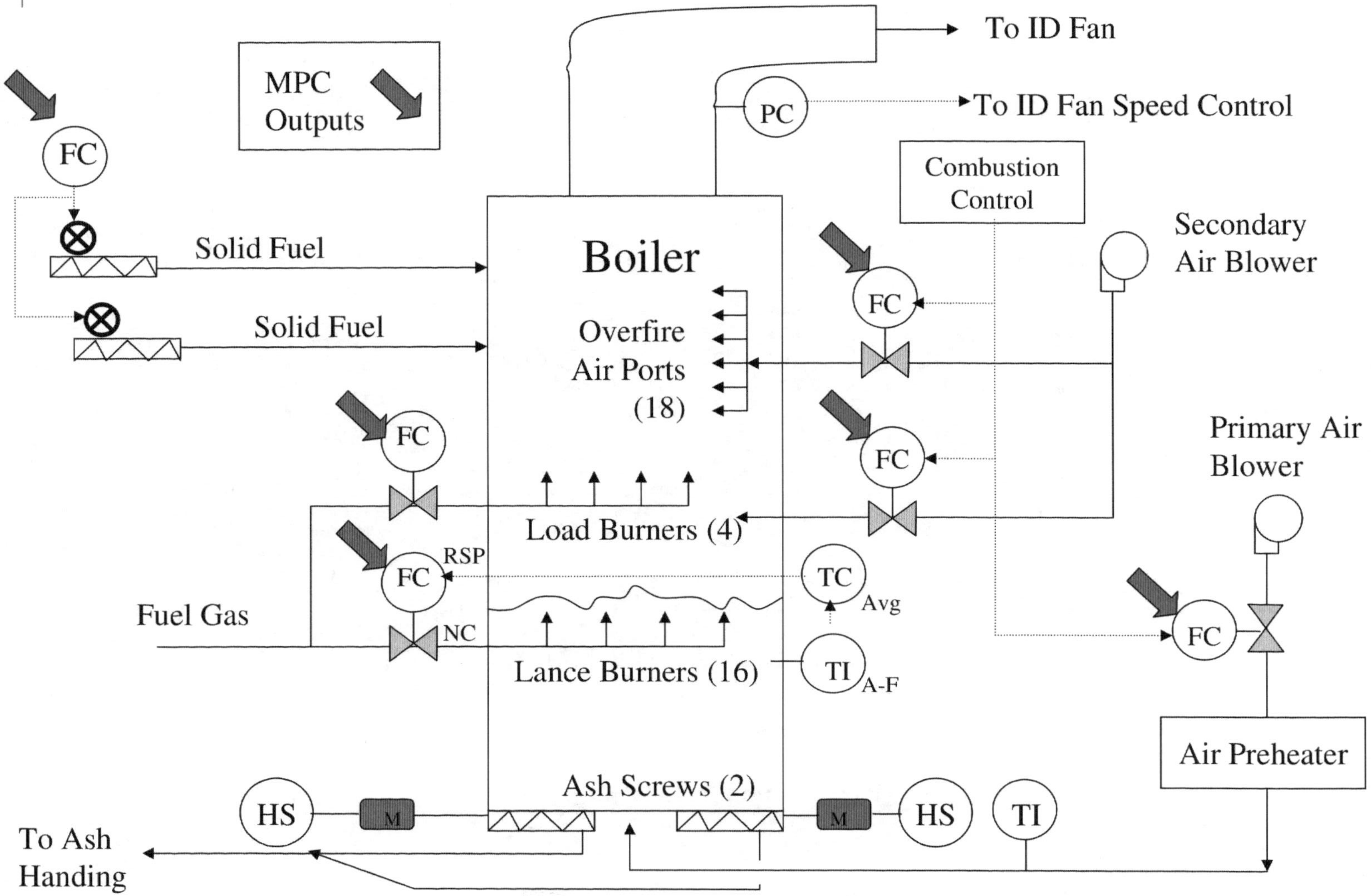

Power Boiler Sludge

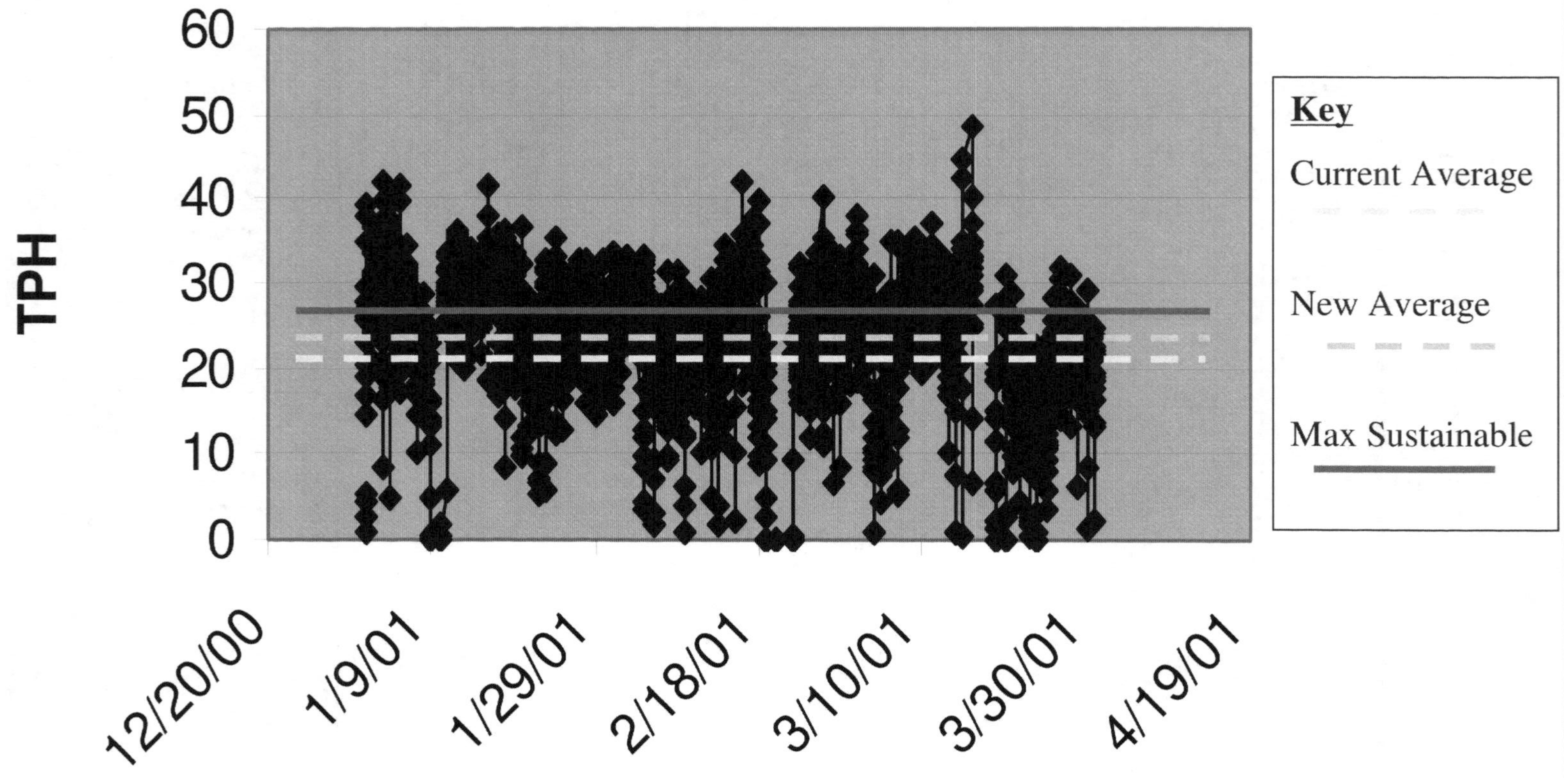

AIChE 2009 Spring National Meeting

Power Boiler Bark Fuels

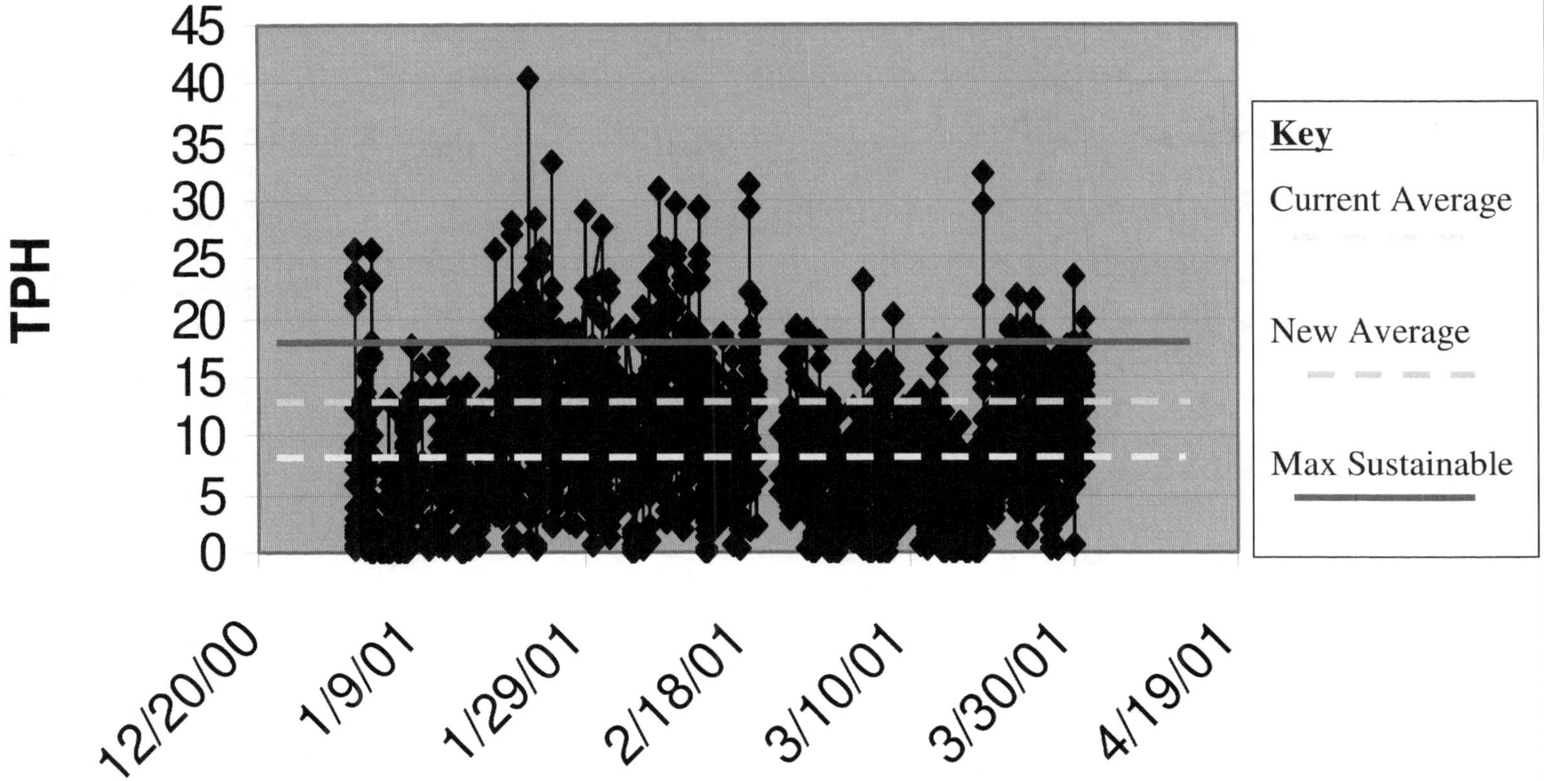

AIChE 2009 Spring National Meeting

Power Boiler APC Benefits

Power Boiler	Difference in Hourly Costs & (Savings)
Natural Gas	$2.22
Sludge Disposal	($23.04)
Sludge Ash Disposal	$9.18
TDF	$0.46
TDF Ash Disposal	$0.00
Waste Wood	$26.91
W Wood Ash Disposal	$0.77
Total	$16.50
Package Boilers	
Displaced Natural Gas	($98.42)
Net Savings, $/Hr	**($81.92)**

Total Savings

$56k / mo

$672k / yr

Project Justified:

- Replacement of required pneumatic instruments
- DCS Hardware / Software
- APC Tools
- Turnkey Engineering Services

Excuses for Doing Nothing

- → Not enough manpower - Too busy doing other things
- → Our plant is special – analysis based on other sites doesn't apply
- → We run our plant well already, there won't be any big savings found
- → We don't believe our measurements so your analysis is wrong
- → Ostrich - (If we find something obvious, management will ask why we didn't find it before)

Summary

- Energy is the largest controllable cost in process operation – it's efficient production and use are keys to plant profitability
- Automation and Advanced Automation are keys to effective use and management of energy in the plant
- Implementation of a program to save energy requires a disciplined approach to evaluation and analysis

Questions? Comments? Other References

1. More material on subject: http://www.emersonprocess.com/solutions/consulting/advancedautomation
2. Questions? – contact doug.white@emersonprocess.com

Process-Utility Interaction Analysis for a True Energy Saving Value Determination

Kamarul A. Amminudin[1], Mana M. Al-Owaidh[1], and Zaher S. Barri[2].
(1) Process Control and Systems Department, Saudi Aramco, Dhahran, Saudi Arabia, (2) Yanbu Gas Plant, Saudi Aramco, Yanbu, Saudi Arabia

Abstract

This paper shares the experience of Saudi Aramco's energy assessment team in finding a true energy saving potential from one of Aramco's oil and gas plants. It involves a comprehensive energy assessment of both the process and utility section of the plant simultaneously. In the case of process energy saving assessment, pinch analysis together with other process improvement tools (e.g., process simulation analysis), is normally used to identify the energy saving opportunities. During this analysis, no value of energy saving ($/year) is determined but rather the analysis only focuses on the amount of savings in both heating and cooling loads. During this stage too, the team highlights the pros and cons of such energy saving initiatives, which includes the proposed project cost estimates, operability aspects and safety considerations.

In a separate task, the utility team uses an in-house utility modeling tool to assess energy performance of the existing utility system. The advantage of having such a model is not only for its ability to optimize the existing utility performance but also the ability to "link" or interact with our energy saving findings from the process section. For example, the team can then see clearly the full impact of reducing the heating load in the process on the performance of the utility system itself. Hence, this allows the true energy saving value to be determined from the overall performance of the utility system.

Once the true energy saving potential has been determined for each of the identified opportunities, the proposed projects can then be prioritized according to the common decision criteria, like economics, operability and safety. A case study from one of Saudi Aramco's NGL fractionation facilities will be used to demonstrate the approach highlighted in this paper.

Introduction

Even though the energy price has gone down sharply over recent months, the continuous effort by Saudi Aramco on energy conservation initiatives throughout its operating facilities has been relentless. To date, many energy saving assessments have been carried out and based on the recommendations from these studies, there are many projects slotted for implementation, both in operational improvement ("low hanging fruits") and capital-cost related projects. This is indeed an achievement for Saudi Aramco and it is worth sharing.

The assessment phase is the most important aspect of any energy saving project. Its objective is simple and clear, i.e., to identify cost effective energy saving opportunities while considering operability and safety elements for implementation. To meet this noble objective requires a multidisciplinary technical support. A detailed deliberation of the work done is beyond the scope of this paper. The authors share their experience in finding the true value of energy saving potential from each of the identified opportunities in the project. This value is critically important to ensure that realistic energy savings can be realized.

Process-utility interface

Before the true value of energy saving can be estimated, it is required that the process-utility interface concept is highlighted. The process system is normally made up of the heat exchanger network, separation and/or reactor systems. Nonetheless, for the process to function effectively, it also requires, among others, heating and cooling, and most of the time this is satisfied by the plant site utility system. The utilities system typically consists of cogeneration units (gas turbines with waste heat recovery), boilers, steam turbines, letdown stations, de-super heaters, deaerator, condensate returns, furnaces and cooling utilities (cooling water system, fin fan coolers, and refrigeration system).

The utility system has one purpose, i.e., to provide the energy interface or link to the process. What is more important to us is that the utility system contributes significantly to the annual energy cost of the facility, and this energy cost is to be minimized in a cost effective manner. This is in stark contrast to a popular belief that process energy savings can be easily calculated from a "given" utility cost (parameter) rather than from the "marginal" cost. It is this marginal cost that takes into account the energy linkage between process and utility systems. Failure to establish this energy interface correctly leads to an inaccurate energy assessment and may pose a serious economic implication for the project.

The concept of marginal cost for the utility system is not new. It has been deliberated by several authors in the field of process integration (3,4). One of the potential reasons that this process utility interface is not given full attention is the failure of realizing the interdependent nature of the utility system itself. There exists the complex interaction among the individual utility hardware itself - let alone the interaction the hardware has with the process system. Thus, as in process system where a process simulation tool is used to assess various interactions within the process, we also need a simulation tool for the utility system.

Utility system models

Even though there are number of commercial software packages available for utility system, like Visual Mesa™ , AspenUtil™ and ProSteam™ (2), it is cost effective for us to develop the model in-house, especially to take advantage of the benefits from the MS Excel™ based spreadsheet. The model performs simultaneous mass, heat and power balances as direct access to the steam properties databank is now possible within the spreadsheet. Detailed models of the individual utility equipment such as steam turbines and boilers can be easily incorporated. The models also feature unit constraints such as minimum and maximum capacities of the equipment, and fixed heating and power demand, which are critical for realistic modeling. More importantly, the availability of an optimization solver within the Excel based spreadsheet allows the utility system optimization to be performed. Figure-1 shows the typical configuration of the utility system.

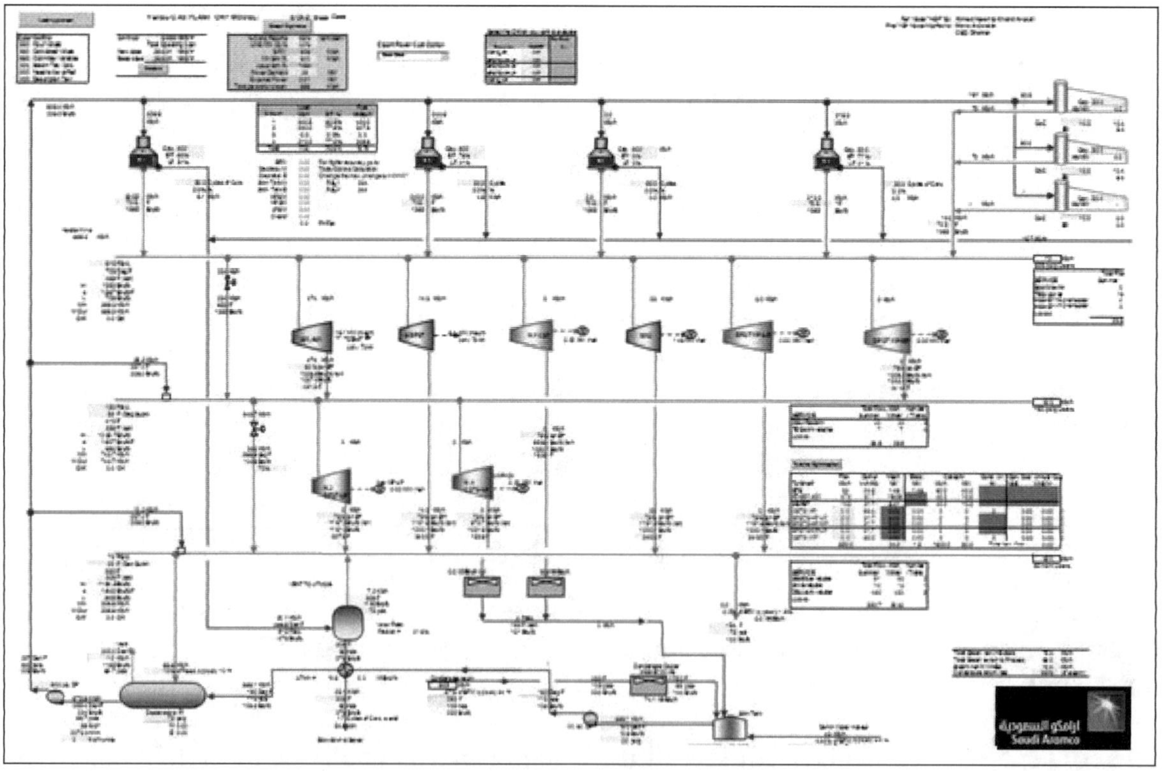

Figure 1: A typical snapshot of the in-house utility system model developed in an Excel based spreadsheet.

Approach

In a typical energy assessment, we first focus on the process area to find energy saving opportunities. This involves using a process simulation tool in conjunction with pinch analysis software. The amount of energy savings in each of the identified opportunities are then determined based on its energy value (MMBtu/hr) rather than on the cost. At the same time, these opportunities are deliberated for any operability concerns, should these opportunities be implemented.

In a separate assessment, a utility model is developed to reflect the existing performance of the utility system of the plant. Once this model is optimized to reflect the existing operating conditions of the utility system, the model can now be used to quantify the true value of energy saving opportunities from the process.

Based on the quantity of the energy savings determined from the process area (in MMBtu/hr), we then used this figure in our utility model as the amount of reduction in utility used, i.e. the amount of utility generated from the utility system. In our case, the amount of steam reduction from the process, as a result of energy efficiency improvement, should be reflected in the utility model by modifying the steam balance of the system.

There are two serious negative implications if we implement the energy saving initiative right away without using the utility model to account for this imbalance.

First, we may end up of having this saved or "unused" steam to be vented off unnecessarily and obviously the *true value* of energy saving is zero! Or we simply lose the opportunity to take advantage of the excess steam to generate more power from our steam turbines, provided that we have additional capacity to do so. This in essence reduces our demand for external power requirements and hence, a true value of energy saving could have been realized.

Second, one may reduce the fuel consumption in the boiler to reflect the reduction in steam demand. Even though intuitively this is the logical way to go, in many cases, this approach has an adverse impact on the steam turbines to sustain their power generation as it simply reduces the power generation. The worst part is that we have to resort to external power, which is relatively costly, to make up the loss of power generation from these turbines. This obviously defeats the purpose of the energy saving initiative to start with.

Hence, the advantage of having this utility model is obvious as it allows us to maximize the true value of energy saving in monetary form. The approach we adopt here is somewhat similar to the one used to determine the marginal utility cost (1,4). We instead use the utility model to link "directly" to the potential energy savings from the process, as we determined earlier in terms of either MMBtu/hr or ton of steam. Since the savings from the process (in terms of MMBtu/hr) reflect the utility imbalance throughout the utility system, the utility model is then used to correct the imbalance by optimizing the system taking into account all physical and capacity constraints of the utility system.

Case study

This case study relates our experience in applying the concept mentioned in this paper to a process improvement of a refrigeration system to condense C2 product in a deethanizer. The compressor is a steam turbine driven machine and hence, any improvement on the refrigeration system or a reduction in deethanizer condensing requirement should have a direct impact on the overall steam system performance. Figure 2 shows the simplified configuration involved, which highlighted this interface.

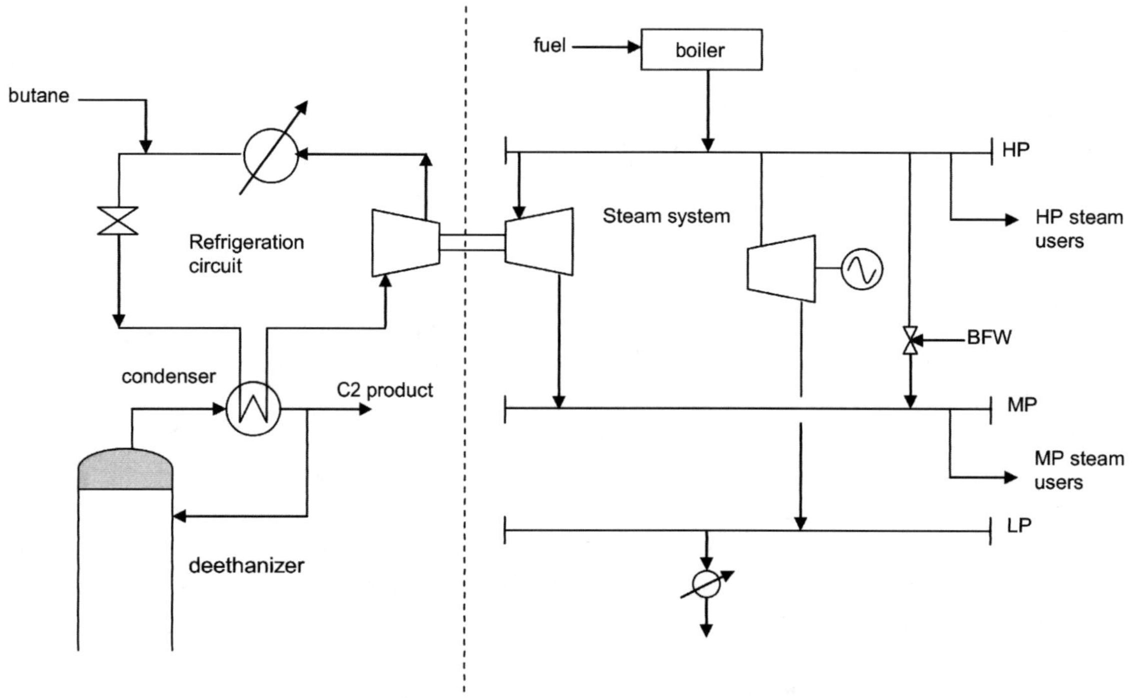

Figure 2: Introducing butane to the propane refrigeration circuit reduces compression requirement and results in a lower power needed from the steam turbine.

This initiative is to introduce a mixed refrigerant instead of pure propane. As a result of this new mixture, the refrigeration compressors power will be reduced by about 350 hp, hence the consumption of steam for the compressor steam turbines will be reduced by 10,000 lb/hr.

With the model, the true value of the potential energy savings can be determined realistically by exploring all possibilities of maximizing the savings value within the utility system. For example, the model can investigate whether the steam saved from the reduction in compression requirements can be used to generate extra power from other steam turbine generators, or to minimize fuel consumption in the boiler.

The total annual potential saving out of this initiative is about $30 K/yr.

Conclusions

In this paper, the process-utility interface concept is used to determine the true value of energy savings from an energy assessment project. It involves the use of process simulation and the utility system model extensively. Such utility models should provide accurate assessment for all energy saving opportunities as the utility model features realistic representation of the system. Various utility constraints, such as equipment capacity can be incorporated and the model can then be optimized. This has been demonstrated in the case study where the process improvement of the refrigeration system creates steam imbalance in the system and the model corrects this balance through optimization.

References

1. Kumana, J.D. "Use Spreadsheet-Based CHP models to Identify and Evaluate Energy Cost reduction Opportunities in Industrial Plants", 23rd IETC, Houston, May 2001
2. Eastwood, A. "ProSteam – A Structured Approach to Steam System Improvement," Steam Digest 2002.
3. Kemp, I.C. "Pinch Analysis and Process Integration – A User Guide on Process Integration for the Efficient Use of Energy," 2nd Edition, IChemE, Rugby, UK, 2007.
4. Smith, R. "Chemical Process Design and Integration," John Wiley & Sons Ltd, Chichester, UK, 2005.

The U.S. Department of Energy's (DOE) Industrial Technologies Program is run by DOE's Office of Energy Efficiency and Renewable Energy (EERE).

Save Energy Now is a national initiative of the Industrial Technologies Program (ITP) to drive a 25% reduction in industrial energy intensity in 10 years. Companies nationwide can participate in no-cost energy assessments and utilize ITP resources to reduce energy use while increasing profits.

Organizations can partner with Save Energy Now and use ITP's proven technical resources to enhance the value of customer and member communications.

Energy Assessments
Save Energy Now energy assessments have helped U.S. manufacturing facilities save an average of $2 million, or 8% of their total energy costs. Plants across the country can begin the process of reducing energy intensity by applying for a no-cost energy assessment performed by a DOE Energy Expert. Assessments help companies identify key opportunities for savings by focusing on energy-intensive systems such as process heating, steam, pumps, fans, and compressed air. View current results, read case studies, and see a list of participating plants to see how companies are saving.

Partnerships
To help industry meet energy reduction goals, ITP works with trade associations, state and local agencies, utilities, non-profit organizations, universities, and other groups that support improved energy management throughout supply chains. Your organization can partner with Save Energy Now and use ITP's proven technical resources to enhance the value of your customer and member communications. Learn more.

Resources
Save Energy Now offers a portfolio of resources to help companies save money and energy. Use DOE's system-specific software tools to analyze your plant's energy use. Attend training to enhance your energy management expertise and learn how to take full advantage of the software tools. Stay informed about best practices and improvements in industrial energy efficiency technologies through ITP publications such as tip sheets and sourcebooks.

ITP has performed over 1900 energy assessments for US industrial plants over the past three years. The largest 500 facilities that received assessments found over $850 million in energy savings with $125 million of the saving having already been implemented to date. The assessments also serve as hands-on demonstration training for plant staff on using DOE's software tools to identify energy projects. Many more energy projects are being planned based on plant staff doing their own analysis using the free DOE software tools. DOE also has an on-going training program that provides one & two-day training classes on the software tools. All these resources are provided at little or no cost to companies.

As of early March, 2009 when this abstract was submitted, DOE continues to accept applications from companies to apply for their energy assessments. You apply on-line at http://www1.eere.energy.gov/industry/saveenergynow/

Also available at this website are all the DOE energy analysis software tools as well as information about scheduled training sessions on the tools.

General questions about DOE Industrial Technologies Program can be addressed to the Information Center which is funded by DOE's Office on Energy Efficiency and Renewable Energy. The phone number is 877-337-3463 or you may submit question via email through their website at http://www1.eere.energy.gov/informationcenter/

70a Measuring and Optimising the Emulsion Layer in Desalter Vessels

Darren Steers, Shell Global Solutions, 3333 Highway 6 South, Houston, TX 77082

Maintaining the correct crude oil/water level and emulsion thickness in a desalter vessel is essential for successful desalter operation. Mechanical displacers, capacitance probes, and Agar probes are the three types of interface level transmitters that are most commonly used in desalter vessels. Each provides an indirect measurement of interface level based on material properties of the liquids, and each has its own unique limitations.

This paper presents the results of a refinery field trial from a first stage desalter vessel of a novel interface level transmitter based on ultrasonic (sonar) technology. It is both a transmitter and receiver that emits sonic pulses through a liquid and then measures the reflected sounds or echoes. The sonar transmitter takes advantage of the property that sound traveling through a liquid medium will be reflected when it encounters step changes in a fluid medium's sonic impedance. Thus, the device can detect an interface between two liquids or between a liquid and a solid.

The results have shown that the sonar probes have successfully measured the top and bottom of the emulsion layer in a desalter. In addition to liquid/liquid level measurement and control, the technology has enabled other properties that influence the desalter process to be investigated and potentially optimised. The measured reflected sonar signal quality has shown a link to chemical emulsion breaker usage, the properties and qualities of the crude diet, and the mix valve setting. The probe has also shown that it can indicate several hours in advance when the desalter is approaching a possible upset condition, enabling pre-emptive action to be taken to ensure smooth, upset free operation.

70b Optimizing Separator Operation with Liquid Measurements

Dave Joseph, Emerson Process Management, 2400 Barranca Parkway, Irvine, CA 92606

Petroleum refineries process large amounts of crude oil to make gasoline, diesel, heating oil, and other products. The processing steps produce some water in oil emulsions that are difficult to treat and dispose of. The water produced by a refinery tends to extract the water soluble components of the crude, including heavy metals, sulfides, and salts that can require substantial wastewater treatment prior to discharge. Various separator units are used to allow collection of the water phase separately from the organic products and, later on, to remove solid contaminants from the water phase prior to discharge.

Selecting appropriate liquid analytical measurement instrumentation for these vessels optimizes removal of waste products and minimizes product loss. However, in-situ measurement in this environment requires resistance to the physical and chemical effects of both the hydrocarbon and the aqueous phases, which can include substantial amounts of heavy metals and sulfur compounds. We will discuss some of the latest electrochemical solutions for meeting these challenges and how online diagnostics can be used to prevent downtime and predict maintenance requirements.

Maximize Existing Assets Through Innovative Revamps
CCR Platforming™ Process Unit

Refiners generally strive to maximize the utilization of existing assets regardless of the economic climate. During periods of economic expansion, there is a strong incentive to increase throughput in order to capture favorable margins. During economic downturns, potential scenarios include consolidation of assets, maintaining flexibility to capture niche market opportunities and increasing the production of a key unit by-product, such as hydrogen from a reforming unit. One of the biggest challenges Refiners are facing is how to meet these objectives when capital budgets are limited. Revamp options that require significant turnaround implementation time are becoming less attractive as unit down time becomes increasingly expensive. Revamp options with limited turnaround scope are therefore preferred.

This presentation will highlight a number of innovative revamp options that expand the processing capabilities of an existing CCR Platforming unit at a fraction of the cost of traditional options. These revamp options include:

- Alternative catalyst solutions to meet revamp objectives without equipment modifications
- Reactor bypasses as a cost effective means to debottleneck key reactor circuit constraints
- Auxiliary reactor section heaters connected in parallel to existing heaters and constructed primarily before the implementation turnaround
- Radiant followed by convective heater design to avoid costly turnaround scope modifications to an existing reactor section heater

A case study showing how a refiner was able to revamp an existing CCR Platforming unit to achieve a 90% capacity increase above name plate utilizing revamp innovations will also be described.

Topical session: "Refinery Processing; Expansion, Revamps, and Debottlenecking"

INCORPORATING FUTURE AUTOMATION TECHNOLOGIES AND CAPABILITIES WITH INSTALLED LEGACY DCS INFRASTRUCTURE

Chris King, Emerson Process Management, Austin, TX

Biography

Chris King joined Emerson Process Management in 2002 as a Migration/Modernization Consultant after a 23 year career as a TDC2000/TDC3000 expert with Honeywell Process Solutions in Phoenix.

He worked on a large variety of Honeywell systems, controllers and devices, and was an early adopter of Fieldbus technology developments.

Chris provides consulting and technical support to customers, as well as for Emerson global sales and Emerson's industry centers, which provide new solutions for our industries.

He conducts System Modernization studies and walkthroughs to evaluate existing Honeywell DCS, recommend modernization and/or migration approaches, and suggests performance improvements which might be achieved.

Legacy automation investments have delivered significant value over the years, but with today's increasing global competition and rapid technology advancements older technology may no longer provide the required safe operation and competitive edge. As the age of our legacy systems approaches the danger level, we are all cognizant of the need to modernize to continue operating safely and remain competitive.

Many engineering & operations people feel the only low-risk solution is to stay with the original vendor of their legacy system. But automation technology has changed significantly in recent years, becoming more open, such that even new systems from the original vendor will be different with more features to optimize the process and now may proactively alert operators of potential reliability issues. Therefore, some training will be required no matter what system is selected.

Refiners need a checklist of evaluation items when considering a system migration. The solution must address short-term needs, including maintaining safe operation, but should also match long-term goals related to operational efficiency and increased reliability. Many automation manufacturers have developed low-risk solutions that eliminate the now perceived "pain" of upgrading your system to a modern one. This paper will describe the preferred method for such a migration project, show the major steps, and provide an example of a successful implementation.

Legacy Systems are a Huge Issue

An example of the challenges facing companies is one of the largest US refining companies, which has grand plans for expansion, but faces a potential obstacle: not enough *qualified* people to get the work done.

Their CEO has stated that the lack of talent is one of the most dangerous threats to his company's long-term health. "People are a big concern," he said. Half of their employees are eligible for retirement within five years. Unless these workers can be replaced, the refiners expansion could be costlier and slower than planned.

What is Required for Migration and Modernization

In an ideal world, migration from an old control system to a new one would be straightforward, simple, sure and fast; so the vendor chosen to implement the work should have the following characteristics and qualifications:

- *Flexible Approach* — The vendor's migration solutions and capabilities should allow the company to evolve system components that will have the greatest impact, while preserving those components with remaining useful life.
- *Control System Platform Expertise* — The vendor should have platform and subject matter experts on control systems (both existing and new) to ensure successful modernization.
- *Migration Experience* — The vendor should have extensive experience planning and implementing migration projects with dedicated migration/conversion project experts ensures successful projects.
- *Superior Risk Mitigation* — The vendor should provide conversion services using automated tools to convert displays, logic and database configuration to the new control system to reduce the risks of migration.
- *Justifying Modernization* — The vendor should assess site-specific needs for system modernization, outlining best-in-class migration options, and help project justification by building the business case for modernization.
- *Total Migration Solutions* — The vendor should combine world-class migration products with best system expertise and key consulting with superior project services to provide the best total migration solution.

The ideal migration solution combines consulting and project services, domain expertise and system transition specialists, and a full set of transition products and services.

The consulting and project services should include a preliminary site investigation, modernization studies, plant walk downs, design services (architecture, I & E and buses, for example) and such site work as installation and hot cutover procedures.

The vendor should have extensive domain expertise and system transition specialists familiar with all legacy products.

The set of transition products and services would include both the target platform and the tools required. Essential parts of the platform may include a single-cable solution for connecting the existing termination panels and auto-marshalling the signals to the new control system's I/O or controllers, plus solutions that make it possible to add workstations as needed without disturbing existing controllers, I/O, and wiring. Automated Conversion Tools should be available for database and graphics, with methodological services to fill in any gaps for logic migration.

Transition Philosophy

An important factor to consider when evaluating vendors to help a plant make the transition to a more modern control system is to examine its transition philosophy. This philosophy should include several elements:

- *Flexible Approach* — The vendor should be able to work with all legacy systems, using the same technology and methodology, with only such variations as are really necessary. It should also be possible to implement the transition in phases, to minimize downtime. If the change can be made piecemeal it may be possible to pay for a good portion of it out of MRO funds rather than as a capital expense.
- *Leverage Up-To-Date Technology* — The proposed solution should be built on open and standards-based technology that works with field buses; it should have full asset management capability built in or available for integration; similarly, it should have embedded advanced process control functionality available.
- Minimize Switching Costs and Risk — The proposed system should built around integrated consoles; it should provide interoperability and be capable of simple integration with legacy systems. Part of the latter includes the ability to make use of existing panels and wiring.
- Strong Domain Expertise in Legacy Products — The vendor should be intimately familiar with all existing systems in the plant.

Transition Methods

An example of a typical transition might include replacing PLCs with new DCS controllers while maintaining the existing I/O cards, routing existing field signals to the new DCS without re-wiring using a single cable method, and installing an integrated operator interface that will make it possible to operate both systems from the new DCS using supplied connection software. The new system may use a common data historian to create a joint data repository for PI, IP-21, PHD, etc. Data exchange and integration would be via OPC/Serial/Industrial Ethernet Data Exchange. The next step might include implementing advanced process control in the new DCS.

Concerns: Existing Systems

There is always something to worry about, and any currently existing system will supply enough to maintain a healthy level of concern. Will there be enough legacy system support? With some older systems local support may be becoming difficult to obtain, and experienced support people may have retired.

Replacement equipment may be unavailable; maintenance costs may be very high, and it may be expensive to update an older system to its current version. On top of that, it may be difficult to get a straight answer from representatives of the original vendor or supplier as to how long support will continue to be available. On top of that, the existing system may not have (or will take too long to have) the future technology for the site to remain competitive.

Reducing Modernization Costs and Risks

There are several things to consider which reduce the costs and risks on a modernization project. Pay close attention to all aspects of interoperability and integration of the new system with the existing one. Make sure that it's possible to utilize as much of the existing hardware as possible; while operator interfaces and controllers will be replaced, most of the original field instruments and wiring will likely remain in place. Make sure this is so, and make sure that the investment in engineering will be protected. Be sure there is a formal method stated in the contract for managing risk, and make sure there is a clear understanding of the total costs of modernization.

Typical Component Lifecycle

Different parts of a control system age at different rates; while some parts may become obsolescent or even obsolete in just a few years, others can last for decades, as shown in Fig. 1. Consoles, for example, typically last for three to seven years before breaking down, and for systems using PC-based consoles perhaps just two to four years. The lifetimes of displays and controls vary greatly. Controllers typically last five to 15 years, while I/O and termination panels typically last 15 to 20 years. Longest-lasting is wiring, which typically lasts 30 to 40+ years. Clearly, a transition to a new control system should not replace perfectly good (and often expensive) equipment that will work as well with the new system as it did with the old.

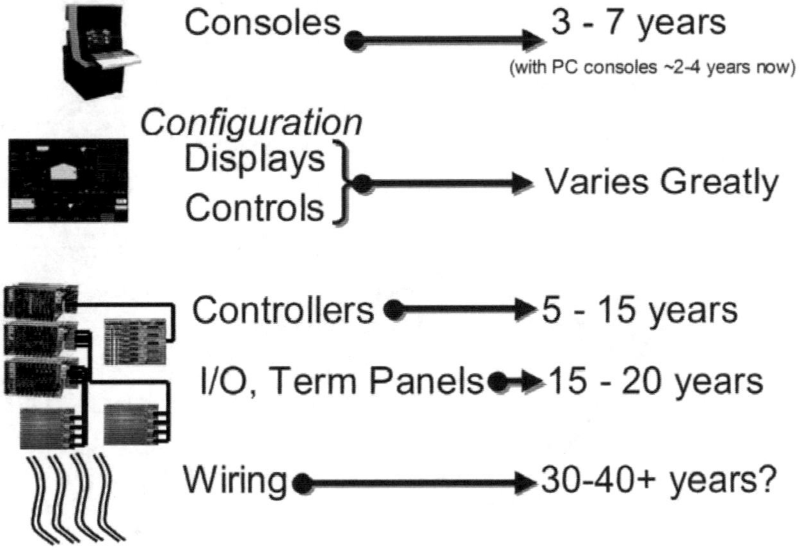

Fig. 1: Typical Component Lifecycle

Implementing the Transition Philosophy

When it comes time to implement the transition philosophy previously discussed, make sure that the new system takes advantage of today's technologies. It should use modern digital buses for field communication; it should have advanced control capabilities embedded; it should take full advantage of self-diagnosing instrumentation; it should be scalable on the fly, without having to do a major shutdown for reconfiguration. And finally the included software tools should allow for plug-and-play business integration

For obvious reasons, the new design must reduce engineering, installation and startup costs, and it should provide ongoing operating and maintenance savings.

Example: Transition from LCN-Based System to DeltaV

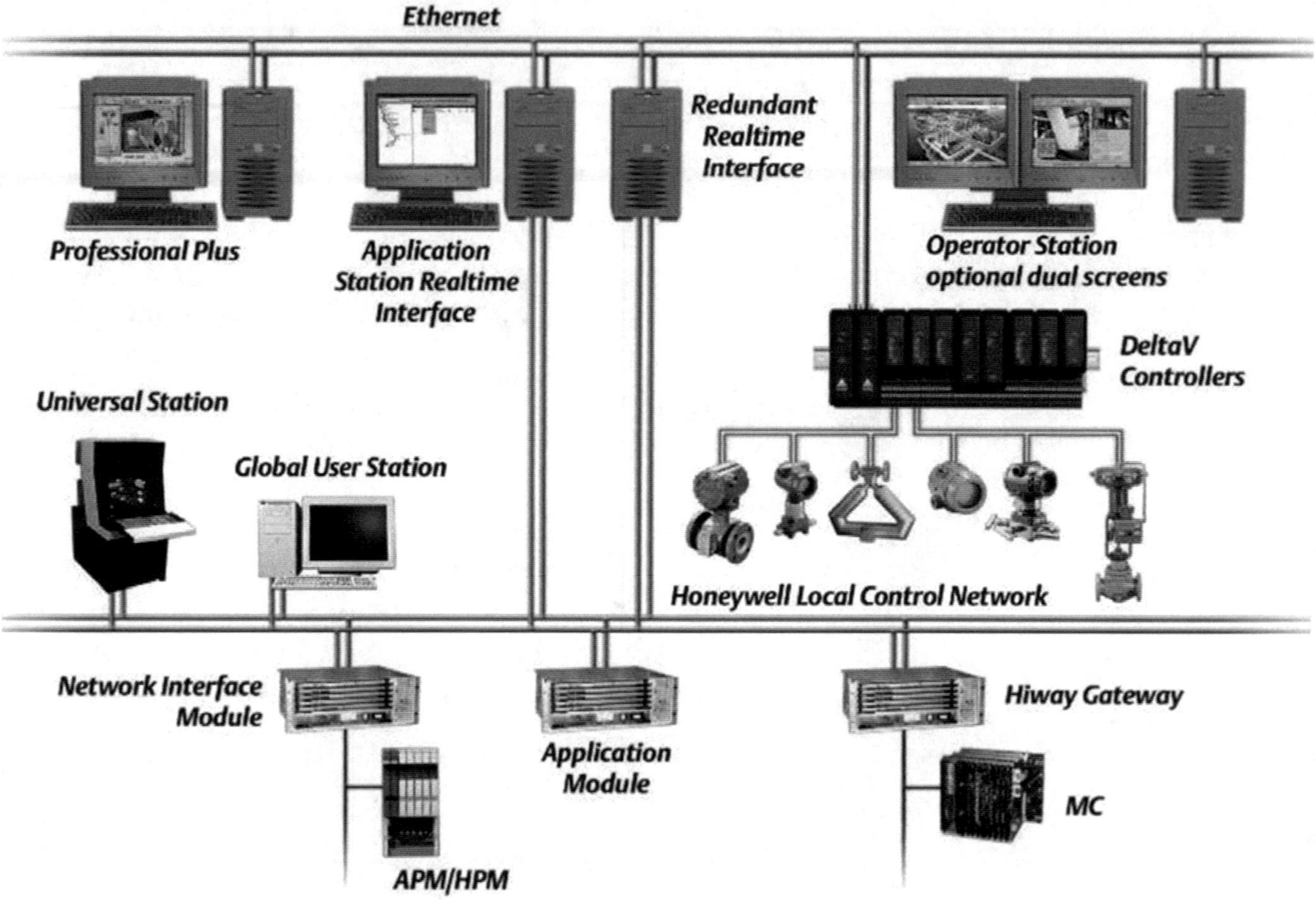

Fig. 2: Connecting New Control System to Existing System's Local Network

Figure 2 shows an example of the integration of an existing local control network (in this case, Honeywell TDC3000) with a new control system (Emerson's DeltaV) using redundant real-time interfaces. The interface does not appear as an LCN node, no Honeywell configuration is required, and there is no need for a shutdown to make the transition.

In this particular application, after weighing its advantages and disadvantages, it was decided not to use OPC for data exchange. OPC is a widely used industry standard communications technology that uses standard system functionality, and is typically satisfactory for low to medium throughput applications. On the other hand, it may cause bandwidth issues in high-throughput applications. In addition, it requires a fair amount of effort to implement and maintain, and introduces considerable complexity in managing such functions as alarming. It is also not likely to be a fully redundant solution. And, since it is not a "product" approach, integration is left up to the user, as a "special".

In summary, OPC is appropriate for sharing limited process data or implementing supervisory control; it is useful in support of a broader migration strategy, but was not used in this case. Instead, Emerson's DeltaV Connect™ Solution for Honeywell® Systems was used.

Cable Solutions: Basic Architecture

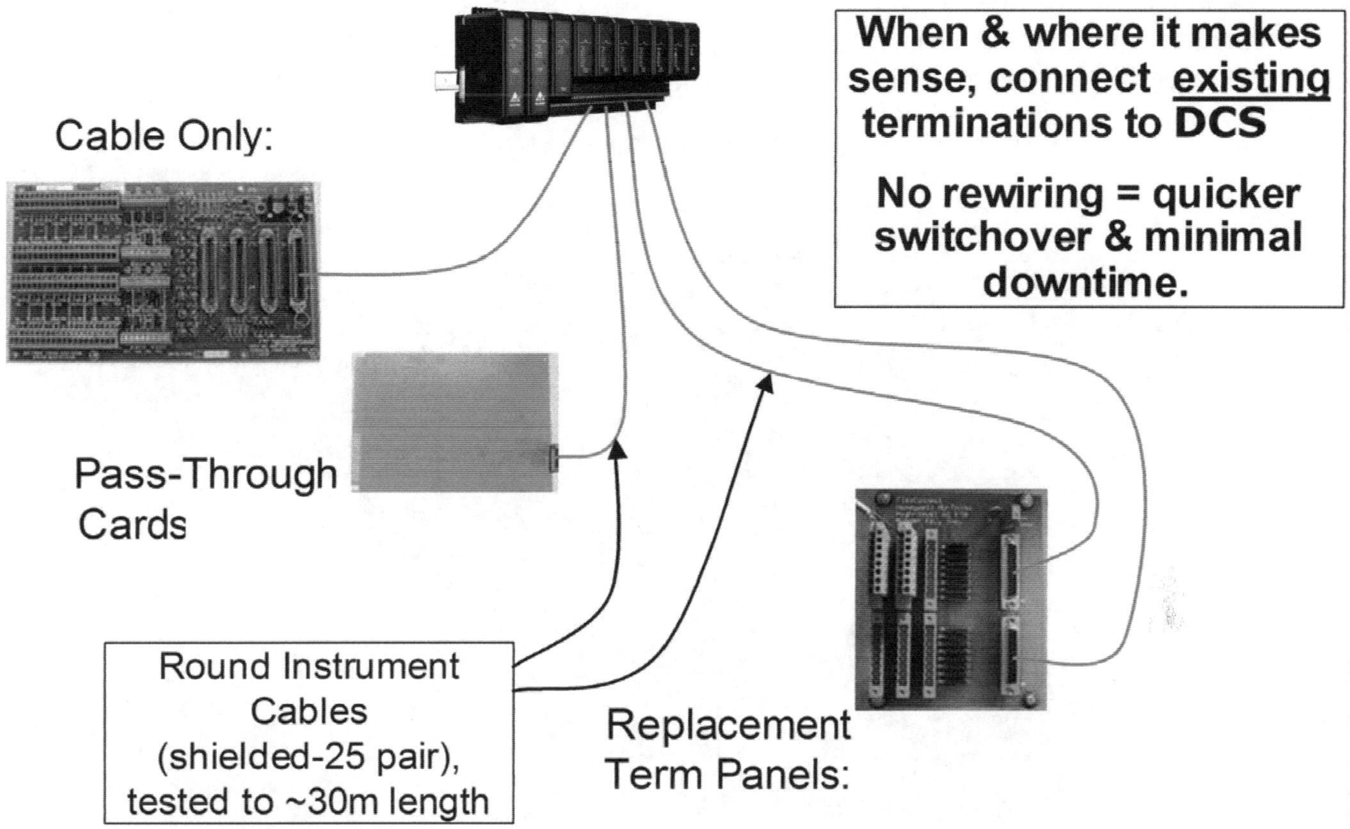

Fig. 3: One Cable Solution: Basic Architecture

Cable solutions like FlexConnect (Fig. 3) provide a method of connecting pre-engineered cables and/or boards to existing termination panels — marshalling the field signals to the new control system using cables or a combination of interface panels and cables. This pre-engineered marshalling solution saves time by avoiding the need to rewire existing field wiring, minimizing process downtime when upgrading legacy systems to new technology. It is appropriate when there is some time available for a process shutdown but not enough time for rewiring. It not only saves time but greatly reduces costs and the amount of rewiring required.

Automated Conversions via Tools

Transition projects are considerably improved by using tools for automated conversion services; examples include tool assisted services for graphics conversion, the appropriate interface modules, as well as control data conversion (via bulk edit). The goal is to be able to save the configuration, connect up the appropriate hardware, and have the software automatically produce the new configuration. In most cases, however, some hands-on intervention will be needed. For example, you must re-address the new controller hardware.

The tools should also (with, perhaps, some manual input regarding individual preferences) change the operator display screens from the old style — with a black background and many bright colors — to the current operator-friendly and efficient style de facto style that typically features gray backgrounds and colors reserved for abnormal conditions and alarms. Figure 4 shows the difference.

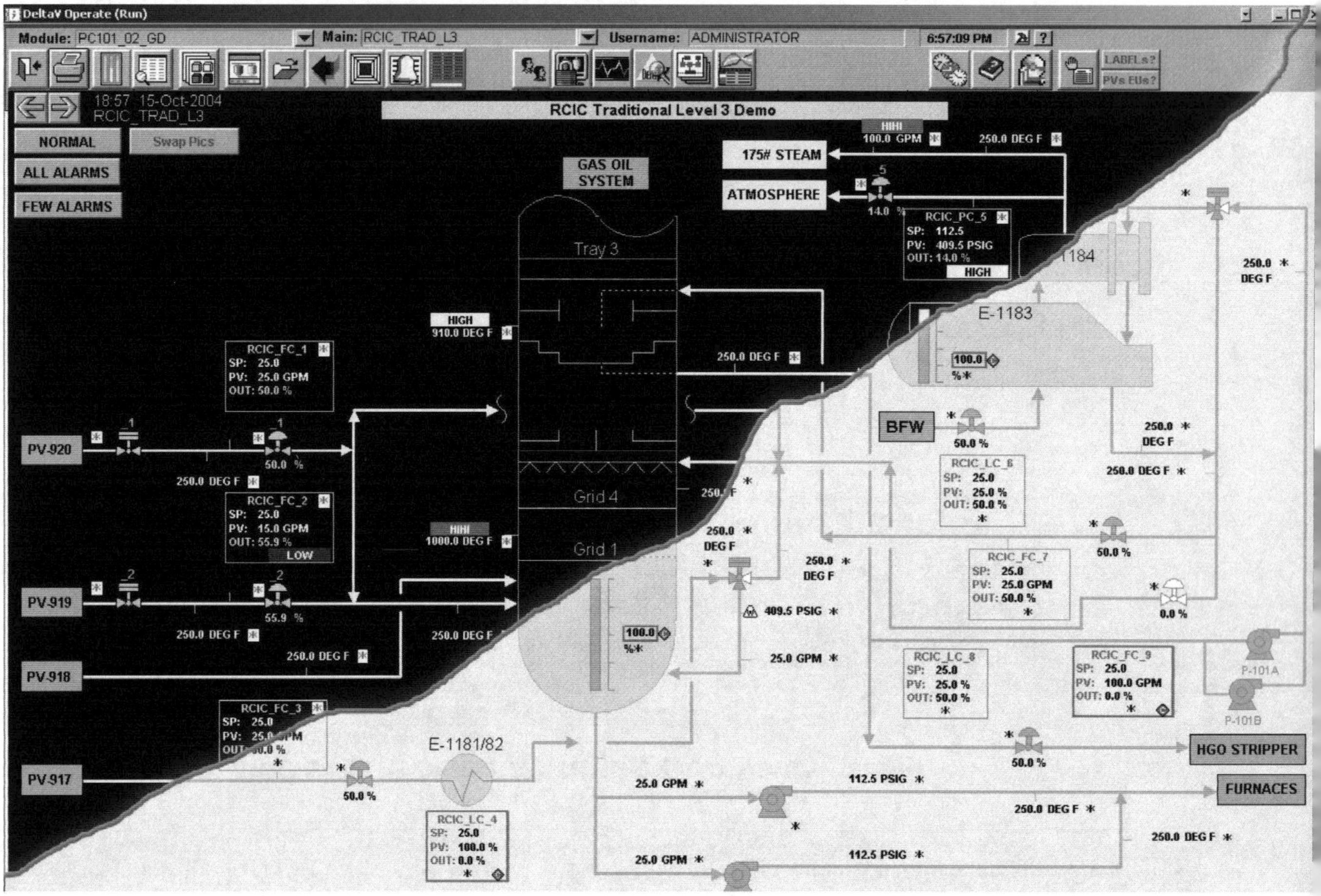

Fig. 4: Old vs. New Operator Displays

Modernization Study Reports

In general, the more tasks the vendor chosen to do the transition can take on the better, so it is important to make sure that it has both the necessary expertise and the practical, real-world experience. One part of this is to have the vendor create a modernization consulting study. Such a study defines sources and magnitudes of measurable plant improvements that can be expected, and provides a detailed roadmap and flexible implementation plans. It should also provide a sound business case and financial return-on-investment analysis for project justification and prioritization.

A typical study might begin with an executive overview and a discussion of risk mitigation. It would then define the scope of work in migrating from the existing control architecture to the new architecture and define the migration plan, which would include the operational philosophy, the modernization philosophy, the maintenance philosophy, the factors influencing the choice of hot cutovers, and the training plan. It should also include a description of the migration execution process, an application overview, a connectivity overview, a list of benefits, financials, required documentation, and schedules.

Transition to Advanced Control and Vertical Industry Applications

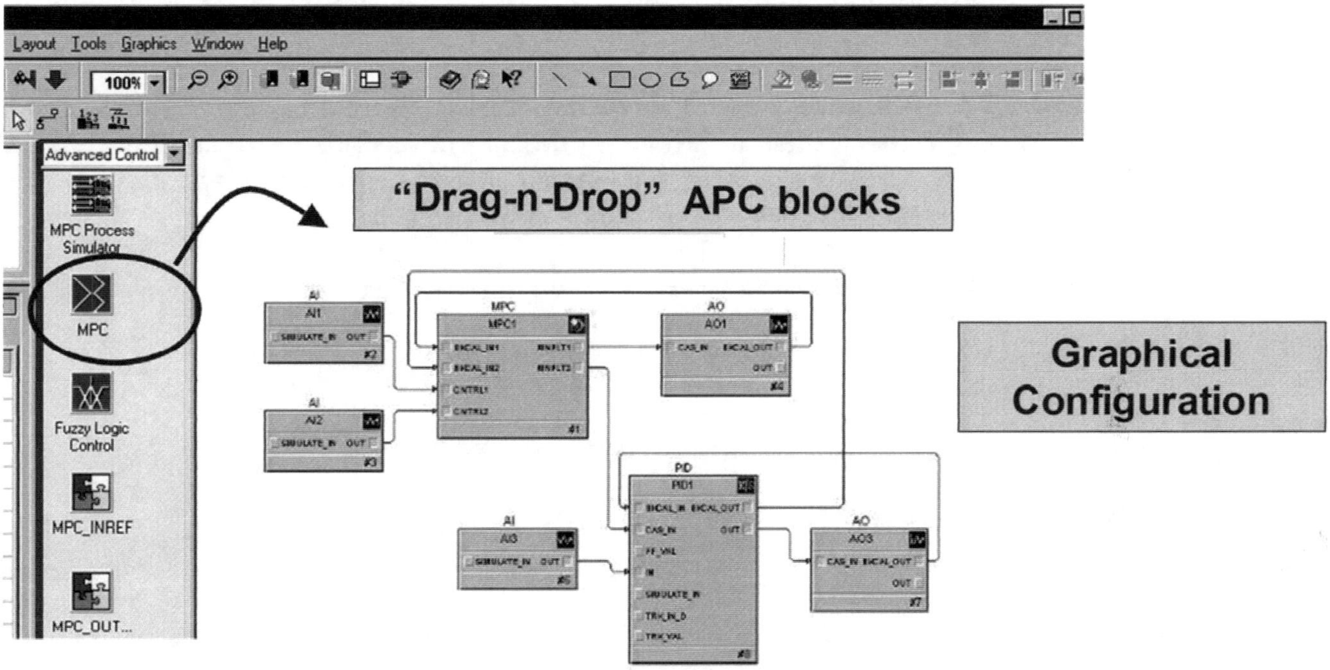

Fig. 5: New Approach to APC Configuration

A new control system should not simply replace an existing obsolescent one, it should provide a way to add more capabilities — and in a form that may be implemented by anyone. For example it should include an integrated method for implementing advanced control, with graphical tools for doing the integration (Fig. 5).

It should also provide access to vertical industry applications designed for specific processes. These should include re-usable functions, calculations and graphic libraries that are pre-engineered and used multiple times. These often consist of bundled combinations of:

- Composite blocks, modules & templates;
- APC tools (e.g. Predict, Neural);
- Application code (e.g. VB, .Net);
- Graphics;
- Example configuration;
- Documentation;
- Services (in some cases)

Also useful are templates to be configured for a particular site

Case History

One example of the use of the methods explained here was the conversion of part of a BASF plant in Hannibal, MO from Honeywell TDC2000/3000 to multiple Emerson DeltaV systems using FlexConnect. The original conversion took just eight hours for the plant's critical path systems, including shutdown of Multifunction controllers (process), disconnecting the cables between the MCs and terminal panels, pulling out the MCs and installing DeltaV controllers in the same rack, plugging in the FlexConnect cables and the unit startup.

Results

The use of FlexConnect and Profibus saved the company $500,000; the operators quickly came to prefer the DeltaV interface to the old one, and the maintenance people are well pleased with the DeltaV diagnostics. Plans call for further conversion work from the present 14,000 I/O to the remainder of the plant in the near future.

70e NHT Reactor High Pressure Drop Alleviation with Catalyst Grading

Abdulrahman H. Harbi and Rabea Saggaf. Yanbu Refinery, Saudi Aramco, Yanbu, Yanbu, Saudi Arabia

Saudi Aramco Yanbu' Refinery experienced over years high pressure drop in the Naphtha Hydrotreater unit reactor which caused several interruptions to the NHT and downstream Platformer unit leading to loss of gasoline production. The NHT unit usually undergo shutdown every 18 months to open and inspect the NHT reactor and to skim or dump the reactor catalyst and start up again. This activity takes around a week, interrupting the downstream Platformer unit and other H2 dependent units.

With the removal of the NHT reactor trash basket and installing a new configuration of catalyst grading materials, the reactor pressure drop maintained at low level and the cycle improved to five years.

The paper will share the refinery experience with original design of NHT reactor trash basket, high pressure drop across the reactor, new configuration of catalyst grading materials that can be implemented in other refineries to mitigate the high pressure drop problem

70f Building a Grassroots Refinery in the 21st Century: What Are the Effects of the Financial Crisis?

Ahmed Khogeer, New Business Development- New Yanbu Export Refinery Project, Saudi Aramco, PO Box 5836, Dhahran, 31311, Saudi Arabia and Mohammed A. Balamesh, Process and Control Systems Department, Saudi Aramco, PO Box 5836, Saudi Aramco, Dhahran, 31311, Saudi Arabia.

In this paper I will explain the positives and negatives of the financial crisis on large mega projects and new refinery projects

86a Biofuel and Biochemical Productions from Mixed Sugars Derived from Lignocellulosic Biomass by the RITE Bioprocess

Hideaki Yukawa, Molecular Microbiology and Biotechnology Group, RITE, 9-2 Kizugawadai, Kyoto, 619-0292, Japan

In order to combat global warming and find alternatives to the increasing scarce fossil fuel resources, the world's attention is currently focused on biofuel and commodity chemicals from renewable biomass resources.

In the U.S., developing new technologies based on the biorefinery concept is a strongly promoted national policy which has resulted in a significant increase in production of bio-ethanol, a representative product of the biorefinery concept. Having said that, most feedstocks for current ethanol production processes are food crops, and bio-ethanol production has lately been blamed for the current high food prices. There is, therefore, a growing need to use "soft-biomass" obtained from non-food crops as well as inedible parts of food crops, as the demand for the ethanol production increases.

We developed the "RITE-bioprocess" to produce ethanol from soft-biomass in collaboration with Honda R&D. The RITE-bioprocess utilizes *Corynebacterium glutamicum*, which is widely used in industrial production of amino acids. The process is based on a new concept that fundamentally resolved the technological problems of the existing bioprocesses. We constructed ethanologenic *C. glutamicum* strains to demonstrate ethanol production and using resultant strains, various substances have been produced at high volumetric productivity with cells packed in a reactor to high density under growth-arrested conditions. In addition, we have improved the ethanol production process from mixed sugars containing hexose and pentose sugars derived from lignocellulosic biomass. In this presentation, trends in the technological developments made in the biofuel production process and our current research into biofuel production for industrialization will be discussed.

86b One Step Cellulosic Ethanol Production

Seunghyun Ryu, Department of Chemical Engineering, Texas Tech University, MS 43121, Lubbock, TX 79409-3121 and M. Nazmul Karim, Chemical Engineering Department, Texas Tech University, Lubbock, TX 79409-3121.

A whole-cell biocatalyst system was developed in our research to directly produce ethanol from amorphous cellulose (Phosphoric acid swollen cellulose, PASC) in a single step. This whole cell biocatalyst was constructed with LY01, which is one of the most developed ethanologenic Escherichia coli strains, as a host cell. The cellulase genes, celCCA, celCCE and beta-glucosidase, were prepared from the mesophilic strain Clostridium cellulolyticum. To enhance the stability and activity of the cellulosic enzymes, these enzymes were co-displayed with the anchor protein PgsA on the surface of the host cell. For inducing the synergism of enzymes, this recombinant cellulolytic microorganism co-expressed endoglucanase, cellobiohydrolase, and beta-glucosidase simultaneously. The saccharification product, monosaccharides, can be uptaken immediately by the host cell and produce ethanol so that the inhibition of the catalytic activity of enzymes can be efficiently prohibited. In this research, we also applied the whole-cell biocatalyst system in bioethanol production process with the lignocellulosic biomass as a substrate.

A Systematic Framework to Determine Optimal Economic Performance of Biorefineries

Norman Sammons Jr.,[a] Wei Yuan,[a] Susilpa Bommareddy,[a] Mario Eden,[a] Burak Aksoy,[b] Harry Cullinan[b]

[a]*Auburn University, Department of Chemical Engineering, 132 Ross Hall, Auburn University, AL 36849-5127, USA, edenmar@eng.auburn.edu*
[b]*Auburn University, Alabama Center for Paper and Bioresource Engineering, 242 Ross Hall, Auburn University, AL 36849-5127, USA, culliht@auburn.edu*

Abstract

The integrated biorefinery has the potential to provide a strong, self-dependent alternative to the use of fossil fuels for the production of chemicals and energy, but difficulties arise in measuring the potential economic and environmental benefit of the biorefinery. A myriad of products and production pathways are possible in this growing field of biorefining, and the production path with maximum value and minimum environmental impact cannot be determined on heuristics alone. A framework is needed to determine the most optimal route based on measures of economic and environmental performance. Gross profit and net present value are used as economic metrics in the short term and long term respectively, and environmental impact is measured using the WAR algorithm developed by Young and Cabezas [1999]. Top candidates in economic and environmental performance are then subject to process integration techniques in order to minimize mass and energy usage, and these integrated biorefineries are once again analyzed for optimal performance.

Keywords: Environmental impacts, Economic, Biorefinery, Product allocation, Optimization

1. Introduction

The integrated biorefinery provides a unique opportunity for reinvigorating an entire manufacturing sector by creating new product streams from a renewable resource [Bridgwater, 2003]. Economic and environmental sustainability are achieved through the optimal use of renewable feedstocks, and a need exists for a process systems engineering (PSE) approach to ensure maximum economic and societal benefit through minimizing the usage of raw material and energy resources as well as the cost involved in supply chain operations intrinsic to biorefining. To maximize the applicability of such systematic methods and to integrate experimental and modeling work, a unique partnership has been established consisting of researchers, government entities, equipment vendors and industry stakeholders to procure the wide range of information necessary such as data needed for process models, capacity constraints, financial data, and optimization techniques. The overall goal of this work is to develop a system that will enable decision makers to evaluate production pathways in biorefining in order to maximize value while measuring and minimizing environmental impact.

2. Development, Optimization, and Evaluation of Process Designs

In biorefining, the large number of possible process configurations and products results in a highly complex problem that cannot be solved using simple heuristics or rules of thumb. Thus, it is necessary to develop a framework which includes environmental metrics, profitability measures, and other techno-economic metrics. Such a framework should enable policy and business decision makers to answer a number of important questions like:

- For a given set of product prices, what should the process configuration be, i.e. what products should be produced in what amounts?
- For a given product portfolio, how can process integration methods be utilized to optimize the production routes leading to the lowest environmental impact?
- What are the discrete product prices that result in switching between different production schemes, i.e. what market developments or legislative strategies are required to make a certain product attractive?
- What are the ramifications of changes in supply chain conditions on the optimal process configuration?

The introduction of PSE methods into biorefining research provides a systematic framework capable of seamlessly interfacing results generated in simulation studies as well as experimental work. Such a framework is imperative when attempting to combine knowledge and information from a variety of research areas and disciplines. The objective of this approach is to create a library of rigorous simulation models for the processing routes along with a database of corresponding performance metrics.

Figure 1 shows a schematic representation of the strategy employed for identification of characteristic performance metrics of the individual subprocesses. An initial superstructure lists feasible technologies for a given feedstock, and basic simulation models are constructed for those corresponding processes. CAMD and property clustering techniques are employed to identify alternative solvents that minimize environmental and safety concerns [Eden et al., 2004; Harper and Gani, 2000].

Process integration techniques are then used to optimize the simulation models. This is an integral step in the model development as it ensures optimal utilization of biomass and energy resources. The optimized models are used to generate data for the economic as well as environmental performance metrics. The estimation of environmental performance is achieved through the use of the US-EPA Waste Reduction (WAR) algorithm [Young and Cabezas, 1999]. The end result of this strategy is a superstructure of biorefining simulation models and a database of corresponding economic and environmental performance metrics.

As stated in previous work, this superstructure and database is then exported into a mathematical optimization program whose objective is to identify candidate solutions based on maximum profitability while measuring and ranking the environmental impact of these solutions [Sammons et al., 2007]. If a candidate solution satisfies both economic and environmental objectives, then the optimal production pathways have been identified. However, if none of the candidate solutions satisfy the environmental constraints, then the framework is then modified by relaxing economic performance constraints until one or more solutions will satisfy both criteria. Economic and environmental performance are decoupled in order to avoid the pitfall of identifying the zero production facility that will identify minimum environmental impact [Sammons et al., 2007].

It should be noted, that only the economic and environmental performance metrics are incorporated in the solution framework described below, thus decoupling the complex models from the decision making process. This approach allows for continuously updating the models as new data becomes available without having to change the selection methodology. Similarly, if new processes are to be included for evaluation, an additional set of metrics are simply added to the solution framework, thus making it robust and flexible.

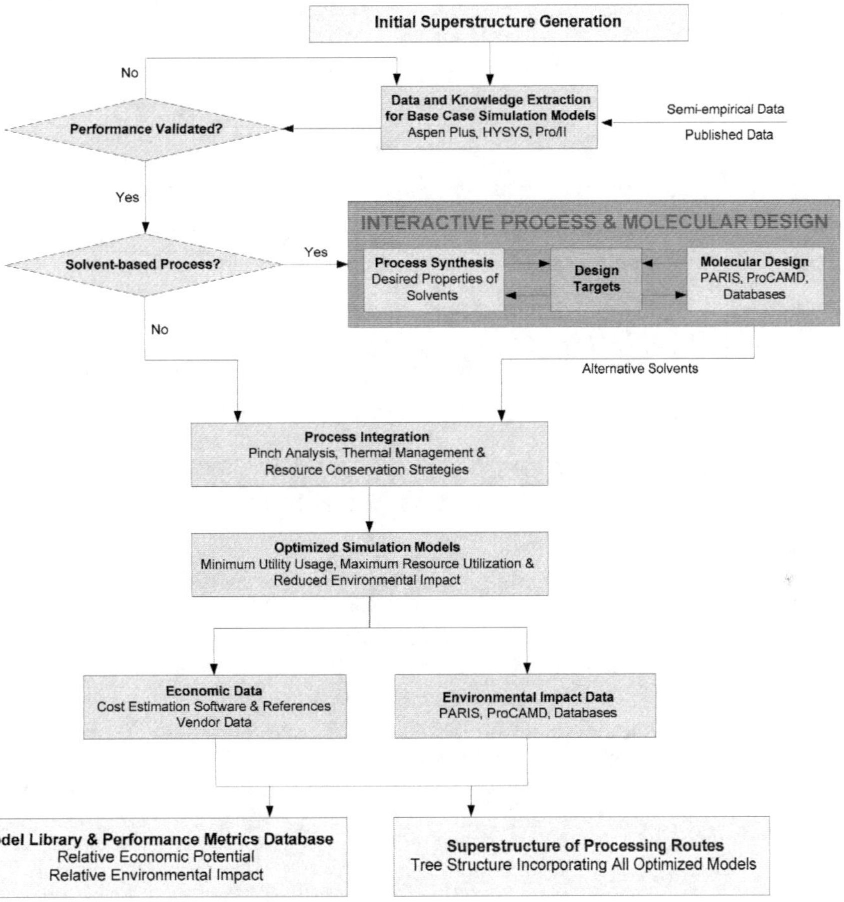

Figure 1 – Strategy for identification of performance metrics.

3. Generalized Biorefinery Model and Economic Optimization Strategy

A generalized biorefinery model given in Figure 2 has been used to develop the structure of the optimization framework. The model structure was formulated to include a variety of basic complexities encountered in the decision making process, e.g. whether a certain product should be sold or processed further, or which processing route to pursue if multiple pathways exist for a given product. The objective function maximizing the overall gross profit of the biorefinery is illustrated in Equation 1, but this may be replaced with an objective function maximizing net present value as shown in Equation 2.

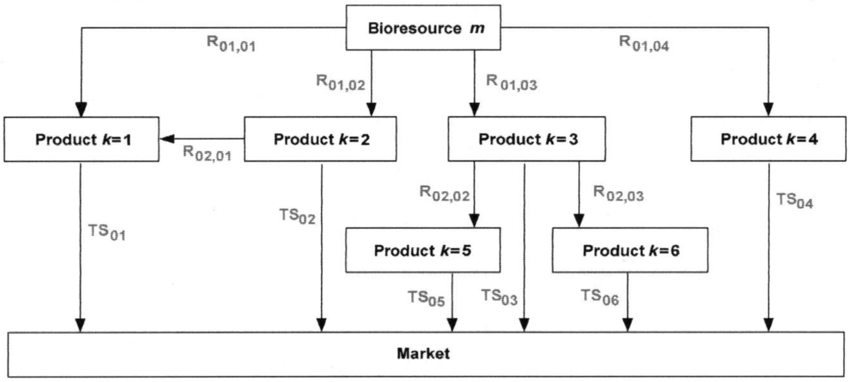

Figure 2 – Generalized model to illustrate possibilities in decision making tree.

$$GP_t = \sum_m \left(\sum_k TS_{mkt} C^s_{kt} - \sum_i \sum_j R_{mijt} C^P_{mijt} - C^{BM}_{mt} \sum_j R_{m1jt} \right) \qquad (1)$$

$$\max NPV = \sum_t \left[\frac{GP_t(1 - Tax_t) + Dep_t Tax_t - Hedge_t + Gov_t}{(1+R)^t} \right] \qquad (2)$$

Using this nomenclature, the first set of terms in Eq. (1) represents the sales revenue from the products made from each bioresource m. TS_{mkt} is a variable that denotes the production rate of product k from bioresource m that is sold to the market. C_{kt}^{s} is the sales price of product k which is a scalar and is determined through a survey of published prices and vendor quotes. The second set of terms represents the total processing cost incurred by the pathways pursued in production. R_{mijt} is a variable that represents the processing rate of route ij while C_{mijt}^{P} is a scalar that represents the cost of processing bioresource m through route ij and is determined through simulation models and process economics. The third set of terms represents the total cost of the biomass resource m, and this is broken down into the scalar purchase price of bioresource m in C_{mt}^{BM} and the combined rate of biomass processed by the plant in R_{m1jt}. Although both TS_{mkt} and R_{mijt} are variables in the optimization program, they are not independent since the variables are related via mass balance constraints around the product points.

This gross profit term may be maximized by itself or incorporated into an objective function that takes many more issues into account. Tax_t represents the marginal tax rate, $Hedge_t$ represents the expenses associated with hedging against catastrophic market actions, Gov_t is the net benefit realized through government incentives or penalties, and R is the expected rate of return, or cost of capital involved with the time value of money. The gross profit version of the model may be utilized for short-term decision making with existing equipment and market arrangements, while the net present value version is better suited for long-term planning that will involve extensive capital expenditure, anticipated government action, and supply chain considerations.

Without including any constraints on capacity of the processing steps, the solution is a single-product configuration in which all available biomass is converted into the most profitable product. However, if constraints are imposed on the most profitable route, the framework identifies the additional products and processing routes required to maximize the overall profit, thus leading to a polygeneration facility [Sahinidis et al., 1989]. Approximate capacity constraints are based on a variety of sources, e.g. existing equipment, vendor data and qualitative process information provided by academic and industrial collaborators. In order to effectively address the strategic planning objectives of business decision makers, it is necessary to incorporate the total capital investment as a constraint in the formulation. Inclusion of capital cost constraints is crucial for practical application of the results, i.e. enabling evaluation of the potential benefits to be obtained for a given maximum investment through retrofit or new construction.

4. Case Study

A case study was performed on a potential biorefinery involving the conversion of chicken litter to syngas, hydrogen, and electricity. Actual data on conversion rates were obtained from experimental work performed by the university and affiliated agencies as well as simulations constructed in ASPEN. Figure 3a shows the possible pathways for production and sale of these chemicals on the commodity market. The execution of the optimization code verified the results obtained from manual calculation; producing syngas from chicken litter and selling it on the market would maximize profit due to the high costs involved in converting the syngas to hydrogen or electricity. Figure 3b illustrates the active pathway chosen by the optimization program. This simple case study will be expanded to include a much wider range of products and feedstocks in order to become a crucial decision support tool in the emerging field of biorefining.

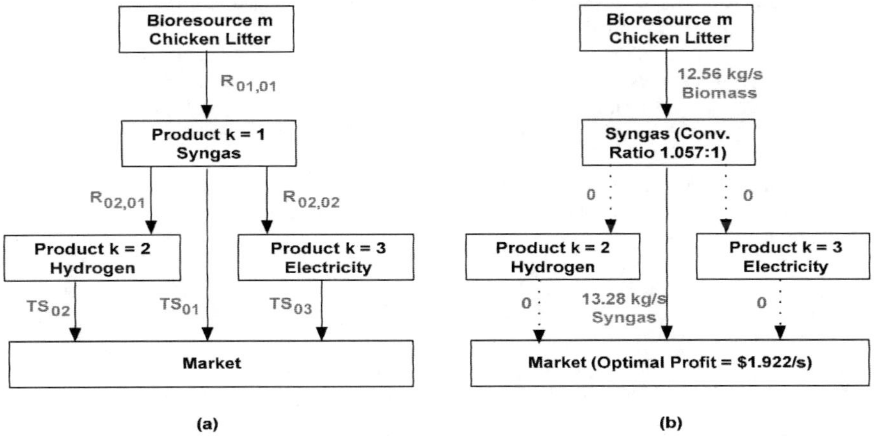

Figure 3 – (a) Unsolved decision making tree with variable designations. (b) Solved decision making tree with flowrate values and objective function.

5. Conclusions and Future Work

A general systematic framework for optimizing product portfolio and process configuration in integrated biorefineries has been presented. Decoupling the process models from the decision-making framework reduces problem complexity and increases robustness. The next phase of this work involves development of additional process models for the generation of performance metrics, specifically information on conversion, yield, and production cost for economic metrics. These additional process models will focus on a greater number of product streams (e.g. electricity, process steam, dimethyl ether, Fischer-Tropsch fuels, and mixed higher alcohols) as well as feedstocks (e.g. black liquor, forestry residues, crop and crop residues, animal byproducts).

The EPA WAR algorithm will be incorporated into the mathematical optimization software, and data from the additional process models will be used to generate numerous time-based and mass-based measures of environmental impact. From there, process integration will be utilized to optimize these process models by reducing energy usage, material consumption, and waste streams.

The framework will also become a stronger financial tool through the incorporation of various economic ideas and analyses. The further development of qualitative predictive models for capital investment and inclusion of capital amortization into the objective function will also increase the strength of the framework. Incorporation of options theory into the framework will allow management to develop financial strategies in response to events in the market or legislative environment. Optimization under uncertainty will be studied to quantify the effects on process configuration resulting from minute changes in product prices [Banerjee and Ierapetritou, 2003]. This, in combination with implementing superstructure generation techniques, will lead to increased robustness of the methodology and thus better recommendations [Chakraborty and Linninger, 2003].

6. References

A. V. Bridgwater. (2003). Renewable fuels and chemicals by thermal processing of biomass. Chemical Engineering Journal, 91, 87-102.

A. Chakraborty and A.A. Linninger. (2003). Plant-wide waste management 2: Decision making under uncertainty. Industrial and Engineering Chemical Research, 42, 357-369.

M. R. Eden, S. B. Jørgensen, R. Gani and M. M. El-Halwagi. (2004). A novel framework for simultaneous separation process and product design. Chemical Engineering and Processing, 43, 595-608.

P. M. Harper and R. Gani. (2000). A multi-step and multi-level approach for computer aided molecular design. Computers and Chemical Engineering, 24, 677-683.

I. Banerjee and M. G. Ierapetritou. (2003). Parametric process synthesis for general nonlinear models. Computers and Chemical Engineering, 27, 1499-1512.

N. V. Sahinidis, I. E. Grossmann, R. E. Fornari and M. Chathrathi. (1989). Optimization model for long range planning in the chemical industry. Computers and Chemical Engineering, 13, 1049-1063.

N. E. Sammons, W. Yuan, M. R. Eden, H. T. Cullinan, B. Aksoy. (2007). A flexible framework for optimal biorefinery product allocation. Journal of Environmental Progress, 26(4), 349-354.

D. M. Young and H. Cabezas. (1999). Designing sustainable processes with simulation: the waste reduction (WAR) algorithm. Computers and Chemical Engineering, 23, 1477-1491.

86d A Framework to Suggest Enzyme Candidates for Desired Bio-Based Products

Ayoun Cho[1], Hongseok Yun[2], Sang Yup Lee[2], and Sunwon Park[1]. (1) Department of Chemical and Biomolecular Engineering, Korea Advanced Institute of Science and Technology, 373-1 Guseong-dong, Yuseong-gu, Daejeon 305-701, Korea, Daejeon, South Korea, (2) Department of Chemical and Biomolecular Engineering, BioProcess Engineering Research Center, Bioinformatics Research Center,, Korea Advanced Institute of Science and Technology, 373-1 Guseong-dong, Yuseong-gu, Daejeon 305-701, Korea, Daejeon, South Korea

At 2004, the department of energy of USA announced high value-added building blocks from biomass. All the reported top value-added building blocks were carbon monomers while the synthetic pathways in the metabolic network need to be developed or found for commercial purposes. To synthesize those carbon monomers with exiting sources, it is necessary to predict the promising enzymes to catalyze novel reactions producing desired products. We designed a system framework which predicts promising enzyme candidates for desired carbon monomers. Since a biological reaction occurs when substrates are attached to an enzyme, the similarity of binding site and reaction rule can qualify whether an enzyme would catalyzes the reaction or not. Through the qualification step, numerous enzyme candidates are identified. All the identified candidates are structurally meaningful; however, more promising enzymes should be noticed for the experimentations. Thus, the quantifying methodology is consequently used to evaluate qualified enzyme candidates. By five factors, such as chemical similarity, binding site covalence, thermodynamic favorability, pathway distance, and organism specificity, qualified enzymes are ordered by their quantified values. If an enzyme gets higher quantified value than others, the enzyme is called as the most promising enzyme candidate. In the system framework, synthetic pathways for any desired carbon monomers can be suggested. It is expected to contribute to speeding up the research cycle of high value-added chemicals from biomass.

87a FCC Operations Overview

Pravin Sheth, Process Design, Sunoco Refining and Supply, 10 INDUSTRIAL HIGHWAY, Lester, PA 19029

This presentation will discuss a Fluidized Catalyst Cracking Unit (FCCU) operation in a Refinery. Emphasis will be given in Operation Monitoring, Asset Support and Troubleshooting operating problems. Discussion will include Unit Heat and Pressure balances, catalyst circulation, steam stripper and cyclone efficiencies.

87b Determination of Diesel Properties through Comprehensive Multi-Dimensional GC×GC and SFC-TwinGC×GC

Cyril Dartiguelongue[1], Benoit Celse[2], Frederick Adam[1], Marion Courtiade[1], and Didier Espinat[1]. (1) Physics and Analysis Division, IFP, BP3, Vernaison, 69390, France, (2) Technology, Computer Science, and Applied Mathematics Division, IFP, BP3, Vernaison, 69390, France

Refining processes produce varieties of petroleum streams which have to be characterized in order to evaluate their quality, to monitor the process or to ensure that the products meet the required specifications. For middle distillates, global properties (density, refractive index) are thus routinely measured in refining products.

The cetane number (CN), which characterizes the combustion behaviour of middle distillates in diesel engines, is one of the most stringent specifications for the European market (EN 590). The reference method (ASTM D613), which consists in the measurement of the fuel auto-ignition delay, is time consuming and requires a large amount of sample (ca. 1L).

Several predictive models for cetane number have been developed based on correlations with fuel global properties (density, simulated distillation). The main drawback of these global models is the lack of explicit link between the molecular composition of fuels and the cetane index which can be explained by the poor separation power of conventional analytical techniques.

Comprehensive two-dimensional gas chromatography (GC×GC) enables within one single injection the characterisation of diesel components by families (paraffins, naphthenes, mono-, di- and tri-aromatics) and within each class of components by groups of isomers. The integration of 2D-Chromatograms by means of a dedicated software (2DChromTM, developed by IFP and commercialised by Thermo Scientific) yielded approximately 150 groups of compounds. An average cetane number was then attributed to each group, based on experimental cetane numbers of model compounds obtained from literature. A linear regression was carried out in order to optimize this cetane number for each group of compounds taking into account chemical constraints (cetane number must be positive, normal paraffins have greater cetane numbers than iso-paraffins...) The cetane number of each diesel could thus be obtained from the molecular composition provided by GC×GC analysis, using a simple linear correlation. For most of the diesels investigated, excellent agreement was reached between the cetane numbers calculated from GC×GC and the reference engine values, which validates the strategy followed.

However, significant differences were observed for diesels enriched in polynaphthenes (3- and 4-ring naphthenes) and olefins. These differences can be attributed to an incomplete separation in GC×GC, due to co-elutions between polynaphthenes/mono-aromatics on one hand, and olefins/naphthenes on the other hand. To solve this problem, the analysis of such diesels was performed using Supercritical Fluid Chromatography (SFC) hyphenated on-line to GC×GC. Using this powerful technique, saturated and unsaturated compounds were separated and recovered in two different fractions, each of which could be analysed on-line by GC×GC. Therefore, polynaphthenes and olefins could be quantified by families (3- and 4-ring naphthenes, n-, iso and cyclo-olefins) and within each class of components by groups of isomers. A cetane number was attributed to each group following the same strategy as described for GC×GC. The cetane index obtained from SFC-twinGC×GC analysis of these diesels was found to be in excellent agreement with the engine ASTM method. Therefore, SFC hyphenated on-line to GC×GC enables to predict the cetane number of olefinic and naphthenic diesels, unlike single GC×GC.

In conclusion, a simple composition-based model was successfully developed for predicting the cetane number of diesel fuels. Due to the explicit link between the cetane number and the detailed composition obtained from GC×GC or SFC-twinGC×GC, cetane numbers can also be estimated for narrow cuts of diesel samples. This is very useful for blending studies.

To our knowledge, this is the first cetane number predictive model based on the molecular composition of middle distillates obtained through a single analysis. This strategy is currently being applied to the prediction of various diesel fuel properties (aromatic carbon content, cold flow properties...).

Analysis of Operating Cases for a Complex Refinery Hydrogen System

AIChE 2009 Spring National Meeting
April, 2009
Tampa, Florida

Authors:
Chip Allen, Virginia Brown, and Jaime Robles
Fluor Corporation, Sugar Land, Texas
Greg Willing
BP, Whiting, Indiana

Introduction

As part of a major revamp project, BP's Whiting, Indiana Refinery had a need for an integrated model of its hydrogen system to assess changes due to the project. The refinery hydrogen system supplies hydrogen to a variety of hydrotreating units throughout the complex. Some of these units use hydrogen in a once-through operation, while other units contain a recycle compressor. In addition, hydrogen-rich effluent gas from several hydrotreaters cascade to other hydrotreating units for further use. This system has evolved as units were added through the years to maximize the efficient use of hydrogen in the refinery.

The refinery team desired a simulation model which could easily be used by engineers without significant proficiency in running simulation software. They also wanted a model which could be easily changed to reflect actual operating scenarios within the refinery. Using the standard commercial simulation software program, HYSYS, the process team built a model which met the refinery's needs and gave the operating engineers a tool for better prediction and optimization of the system.

BP Whiting

BP's Whiting Refinery is the fourth-largest of all US refineries, processing more than 400,000 barrels of crude oil daily into gasoline, diesel, asphalt and many other products people rely on every day. It is the second largest refinery in BP's worldwide refining system. Construction began on the refinery in 1889, and it has undergone continuous expansion and innovation since.

The refinery is currently in the midst of a major revamp project to increase crude flexibility. The Whiting Refinery Modernization (OCC) Project includes installing the following major process units and upgrades to existing equipment:

- A world-scale coking unit, a revamped crude distillation unit and petroleum coke handling facilities that increase resid conversion capacity
- A hydrogen production plant
- A gas oil hydrotreater and revamped existing hydrotreaters to maintain compliance with sulfur specifications for fuels

- Sulfur recovery facilities
- Refinery infrastructure facilities to support the additional heavy crude oil processing

Refinery Hydrogen System – Consumers and Producers

As with any refinery hydrogen system, its primary purpose is to supply hydrogen to hydroprocessing units and miscellaneous users throughout the refinery. Prior to the OCC project, Whiting's system supplied hydrogen to the following units:

- Distillate Desulfurization Unit (DDU) – hydrotreater for distillate-range material
- Cat Feed Unit (CFU) – FCC feed hydrotreater
- Blending Oil Unit (BOU) – hydrotreater capable of processing either gas oil or distillate-range material
- Diesel Hydrotreater (DHT) – hydrotreater for production of Ultra Low Sulfur Diesel
- Catalytic Reforming Unit (CRU) – former reforming unit revamped to a hydrotreater capable of processing naphtha or kerosene
- Isomerate Unit (Isom) – Isomerate production unit with front-end naphtha hydrotreating

Pre-OCC hydrogen production consisted of a small onsite high purity hydrogen plant (steam-methane reformer with pressure swing adsorption purification) as well as generation from a reforming unit. High purity hydrogen was also purchased from a neighboring facility.

The system also contained a membrane system which was used to purify a portion of the CFU/BOU effluent hydrogen, returning the purified hydrogen to the system and rejecting the heavier components to the hydrogen plant feed.

The OCC project will add new consumers to the system; a new gas oil hydrotreating unit (GOHT) and a coker naphtha diolefin saturator. Additionally, hydrogen consumption in the existing units will increase due to the higher sulfur content of the heavier crude slate.

The additional hydrogen needed post-OCC will be imported from a neighboring facility where a new high-purity hydrogen plant is being constructed to meet the needs of the refinery.

Pressure and Purity Levels

The Whiting Refinery hydrogen system has grown with the ever-changing refinery over many years. As units were added, attempts were made to minimize changes to the hydrogen system as well as to maximize the efficient use of hydrogen. As a result, the system consists of multiple pressure levels designed to minimize the compression necessary to supply hydrogen to high pressure units. Additionally, the system uses cascade flows of hydrogen from one unit to another. Hydrogen-rich effluent from units which require higher purity hydrogen can be used as make-up gas in units which can tolerate lower purity hydrogen.

As a result of this progression, the system consists of several hydrogen headers operating at various pressures and hydrogen purities. In general, these may be described as follows:

- High purity hydrogen headers, operating at approximately hydrogen plant outlet pressure, containing only SMR-produced hydrogen
- Reformer hydrogen ("A") header, operating at low pressure, containing hydrogen produced by the reformer plus effluent hydrogen which has cascaded through all of the higher pressure headers
- 1200 psi header, containing effluent hydrogen from the CFU
- 900 psi header, containing effluent hydrogen from the BOU and CFU
- 450 psi ("B") header, containing effluent hydrogen from several units
- 130 psi ("C") header, containing effluent hydrogen from several units. Feeds back into the reformer hydrogen header.

New units due to the OCC project were integrated into this existing structure, adding a new layer of evolution and additional level of complexity to the system.

Model Construction

In order to ensure that the new system could supply hydrogen adequately to all users, BP sought a process model capable of predicting the hydrogen purity levels and flow rates at all points in the system. Such a tool would also be useful later on in operation to optimize the system, and predict system changes necessary to handle unit shutdown / turndown scenarios. The project challenged the process team to develop a model of the system which could accomplish these objectives. Additionally, it was desired that the model be as simple as possible to manipulate, preferably by engineers with limited process simulation experience. The capability to simulate all hydrotreater units at any time in their cycle from Start-of-Run to End-of-Run was also requested.

The commercial process simulation package HYSYS was used to develop the hydrogen system model. Both BP and Fluor use this simulator as a reference system, and the process team was more familiar with HYSYS than any other commercial simulator.

Individual Unit Simulation

After a kick-off meeting to clarify the objectives of the modeling effort, development began with the simulation of the individual hydrotreater units. Since this model was not intended to be a detailed simulation of the hydrotreaters, certain shortcuts and assumptions could be used. In general, only the high-pressure reactor loop within each unit was simulated; peripheral operations such as product strippers were not included. The heart of the model for each reactor loop is a mass balance operation which calculates the reactor effluent flow and composition based on unit feed and predicted yield structures. The reactor effluent stream is then set to flow to hot and cold separators as appropriate. Amine contacting is not rigorously simulated, but is approximated by component splitters which remove hydrogen sulfide and ammonia. Heat exchange is also not modeled rigorously; instead, streams are set to known conditions. The unit pressure profile is set based on separate hydraulic calculations. Makeup gas compressors are included in the model for reference. As in the actual refinery, these are also used to control certain flows.

Each hydrotreater unit was constructed as a subflowsheet within the main model. Although each unit has its own individual configuration, for example, some units use once-through hydrogen and do not have a recycle compressor; in general, most of the units modeled resemble the schematic shown in Figure 1 below.

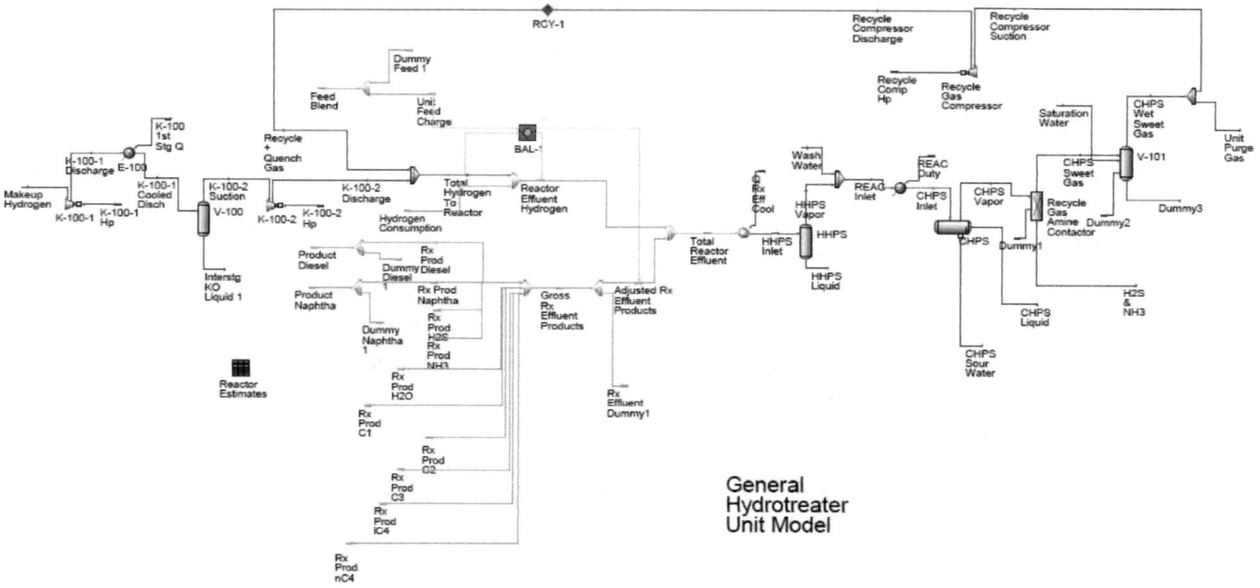

Figure 1 – General Hydrotreater Unit Model

Reactor yields are modeled using a spreadsheet containing yield data predicted by the catalyst vendor. This is illustrated in Figure 2 following.

Although the process team followed a simplified approach to model each unit, the process could be made more detailed if desired. Rigorous heat exchange models, pressure profiles, amine contacting, and other specific features could be added to produce a more complete model if the developer has need of them. This process team did not incorporate such detailed features since the principle objective for this model, to predict the hydrogen purity and flow rate at any point in the system, did not require more detail.

Oil Charge Flow, bpd	40000						
Oil Charge Flow, lb/hr	545703.3						
Fraction EOR	0						
Reactor Yields	**SOR**	**EOR**	**At Fraction EOR**			**Reactor Product Mass Flows**	
H2O	9.00E-02	9.00E-02	9.00E-02	Wt% of Fresh Feed		491.13	lb/hr H2O
H2S	1.2	1.2	1.2	Wt% of Fresh Feed		6548.44	lb/hr H2S
NH3	9.90E-02	9.90E-02	9.90E-02	Wt% of Fresh Feed		540.25	lb/hr NH3
C1	6.40E-02	0.128	6.40E-02	Wt% of Fresh Feed		349.25	lb/hr C1
C2	8.70E-02	0.13	8.70E-02	Wt% of Fresh Feed		474.76	lb/hr C2
C3	0.126	0.214	0.126	Wt% of Fresh Feed		687.59	lb/hr C3
iC4	4.50E-02	8.20E-02	4.50E-02	Wt% of Fresh Feed		245.57	lb/hr iC4
nC4	9.00E-02	0.246	9.00E-02	Wt% of Fresh Feed		491.13	lb/hr nC4
Naphtha	1.742	3.277	1.742	Wt% of Fresh Feed		9506.15	lb/hr naphtha
Diesel	98.137	96.083	98.137	Wt% of Fresh Feed		535536.80	lb/hr diesel
H2 Consumption	1036.1	955.66	1036.1	scf H2/bbl fresh feed		41.44	MMSCFD H2
						-4550.29	lb mol/hr H2

User Inputs, Imported From Master Control

Reactor yield data from catalyst vendor

Calculated from straight line interpolation

Calculated and exported to individual streams in model

Figure 2 – Reactor Estimates Spreadsheet

Integrated System Simulation

Once all of the individual hydrotreater units were constructed within the overall model, the process team began to connect them as within the actual operating facility. Hydrogen producers were set as inlet streams of known composition and conditions. The integrated model showing all connected users and producers is shown in Figure 3.

Analysis of Operating Cases for a Complex Refinery Hydrogen System

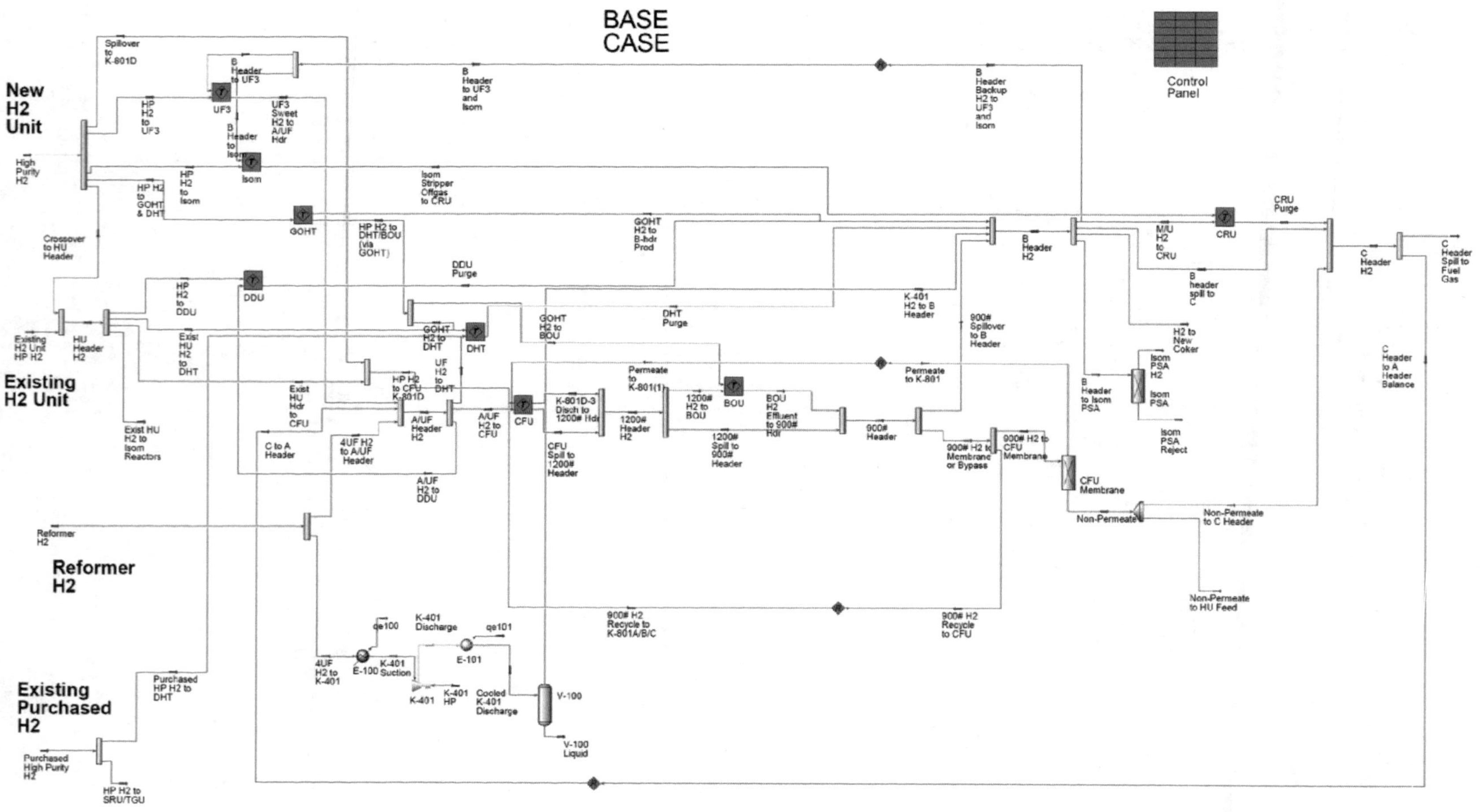

Figure 3 – Integrated Model with All Consumers and Producers

Control Panel

To accomplish the refinery's request for a simple-to-operate model, the process team decided on the use of a control panel within the simulation to facilitate making changes. The control panel allows units to be turned on or off, sets specific flows of oil and hydrogen within the model, and turns various compressors on and off. The control panel also allows the user to set any operating unit to any point of its run life from Start-of-Run (SOR) to End-of-Run (EOR), and changes the yields accordingly. The control panel also provides feedback on various key parameters calculated by the model. This allows the user to stay within the control panel view and not have to hunt through the model to find various data.

The control panel was set up using the spreadsheet operation available in HYSYS. The key parameters that a user would be most likely to change were set up as inputs in the simulation, and those inputs were in turn, moved within the spreadsheet control panel and exported from the spreadsheet to the appropriate stream or operation. Streams which can be lined up and switched to different destinations can be controlled through simple "If-then" logic exported to the appropriate tee branch. The control panel is illustrated in Figure 4.

Inputs available on the control panel are the oil flow, make-up gas flow, and recycle/quench gas flow for each unit. The user also inputs the point of run for each unit as a fraction of the End-of-Run. These values are exported from the control panel to the individual units and streams as appropriate. In some cases, hydrogen flows are controlled by various compressors. The user selects the number of compressors to run and inputs the loading percentage of the compressor selected from the loading steps available on the actual compressor. This then calculates the flow through the compressor and exports the value to the proper stream. Options are also available to direct certain streams among two possible destinations. The user inputs a 1 or a 2 corresponding to the destination listed in the control panel, and the value is then exported to control the affected tee branch.

The inputs for the hydrogen producers are somewhat different than the hydrogen consumers. For the existing high purity hydrogen producers, the user inputs the hydrogen production rate. The new hydrogen unit production is calculated by mass balance according to all the users connected to it. The reformer hydrogen production is based on the oil throughput through the unit input by the user. The spreadsheet simply multiplies the oil rate by the expected hydrogen production per barrel of oil to calculate the unit's hydrogen production. The composition of the reformer hydrogen is pre-set within the simulation.

Once the simulation is converged, it sends various selected values back into the control panel for easy viewing. The process team chose each unit's purge rate and actual treating ratio (scf H_2/bbl of oil) for viewing, but other parameters could be selected as desired. The simulation also calculates the hydrogen system's purge to fuel gas. Hydrogen purity and overall flows for each stream were generally displayed on the overall schematic as a stream table. These have been omitted from Figure 3 for clarity.

Analysis of Operating Cases for a Complex Refinery Hydrogen System

Unit	HTU-1	HTU-2	HTU-3	HTU-4	HTU-5	HTU-6	HTU-7
Oil Feed Flow, bpd	Input by User	Input by User	Input by User	Input by User	Input by User	Input by User	Input by User
Fraction EOR, 0(SOR) - 1(EOR)	0.5	0.5	0.5	0.5	N/A (Yields same)	0.5	0.5
M/U Gas Flow, MMSCFD	Input by User	Input by User	Input by User	See K-301 Control	Input by User	See Controls Below	See K-801 Control
Recommended M/U Gas Flow, MMSCFD	Calc'd by Simulation	Calc'd by Simulation	Calc'd by Simulation	Calc'd by Simulation	Calc'd by Simulation	Calc'd by Simulation	Calc'd by Simulation
Recycle + Quench Gas Flow, MMSCFD	N/A (once-through)	Input by User	N/A (once-through)	Input by User	N/A (once-through)	Input by User	Input by User
Recommended Recycle + Quench Gas Flow, MMSCFD	N/A (once-through)	Calc'd by Simulation	N/A (once-through)	Calc'd by Simulation	N/A (once-through)	Calc'd by Simulation	Calc'd by Simulation
Unit Purge Flow, MMSCFD	Calc'd by Simulation	Calc'd by Simulation	Calc'd by Simulation	Calc'd by Simulation	Calc'd by Simulation	Calc'd by Simulation	Calc'd by Simulation
Recommended Treat Ratio, scf H2/bbl	1775	3300 to 3200	644	1104 to 1072	280	1542 to 1564	2090 to 2005
Actual Treat Ratio, scf H2/bbl	1315	3250	644	1159	280	1550	2439
Total Treat H2, lbmol/hr	Calc'd by Simulation	Calc'd by Simulation	Calc'd by Simulation	Calc'd by Simulation	Calc'd by Simulation	Calc'd by Simulation	Calc'd by Simulation
Compressor Control	K-401			K-301A/B		K-601A	K-801A/B/C
No. Operating	0			1		1	2
Desired Loading, %	50			100		100	100
Available Load Steps, %	100, 90, 75, or 50			100, 87.5, 75, or 50		100 or 50	100, 85, or 55
Compressor Design Flow, MMSCFD	22.8			21.6		38.8	35
Compressor Actual Flow, MMSCFD	0			21.6		38.8	70
Compressor Control							K-801D
No. Operating							1
Desired Loading, %			Certain Data Omitted for Confidentiality				100
Available Load Steps, %							100, 90, 80, 60, or 50
Compressor Design Flow, MMSCFD							30
Compressor Actual Flow, MMSCFD							30
Auxiliary Control				A/UF Flow to HTU-4	M/U from B Hdr	H2 from A/UF Hdr	H2 from HU Hdr
Flow, MMSCFD				3	0	11.9	3
Normal Range, MMSCFD				0 to 5 MMSCFD	0	11.9	0 to 5 MMSCFD
Auxiliary Control	Import H2 fr. HTU-2					Import H2 fr. HU-2	Flow to Membrane/Bypass
Flow, MMSCFD	6					7.95	24
Normal Range, MMSCFD	0					9.61 to 6.28 MMSCFD	24 MMSCFD Max
Auxiliary Control		Purge Director					Membrane Bypass
Option Description		Enter 1 for HTU's					Enter 1 for Membrane
Option Description		Enter 2 for B Hdr					Enter 2 for Bypass to K-801A/B/C
Option Control		1					1
Auxiliary Control							K-801D Disch. Director
Option Description							Enter 1 for HTU-7
Option Description							Enter 2 for 1200# Hdr
Option Control							1

H2 PRODUCERS	Reformer
Oil Feed Flow, bpd	Input by User
H2, scf/bbl	678.5
Pure H2 Production, MMSCFD	Calc'd by Control Panel
H2 Purity	0.81
Total H2 Gas Production, MMSCFD	Calc'd by Control Panel
	Existing H2 Unit
H2 Production, MMSCFD	Input by User
	Existing Purchased H2
H2 Production, MMSCFD	Input by User
	New H2 Unit
H2 Production, MMSCFD	Calc'd by Simulation
Spill to K-801D, MMSCFD	6

Key
Input Value
Reference or Calculated Value
Imported from Simulation Calculation
Exported Value

Figure 4 – Simulation Control Panel Spreadsheet

As with any simulation, various errors are possible. Perhaps the most severe error associated with this simulation was the possibility of having negative flows and/or compositions. Since the simulator is ultimately only a mathematical model, such solutions are mathematically possible, but do not represent reality. Fortunately, the simulator will flag such problems. These issues usually pop up when insufficient hydrogen is being fed to a unit, or if too little hydrogen is fed to the overall system, resulting in a negative purge to fuel gas. These situations are best avoided by having an accurate assessment of the hydrogen that each unit needs.

Recycle loops deserve special mention for this simulation. The default convergence tolerances within HYSYS were not tight enough for this project and required resetting to smaller values. Generally, the overall flow and individual component flows for each recycle loop were tightened significantly. This is easy to do within HYSYS.

Once the model was developed, over a dozen individual operating scenarios were rigorously studied. The results of these studies were used to verify that the planned equipment and piping network will be adequate for all the cases. The study provided the data that the refinery operations and project team needed to have confidence in the design, and successfully complete the required operations and safety reviews of the system.

Although the use of the control panel required additional time to develop the model, the ease of manipulating the model was greatly improved. This more than offset the added development time. The use of a control panel operation is recommended for system simulations where frequent changes are expected or where the final users of the simulation are those with lesser simulation proficiency.

The authors:

Chip Allen is a principle process engineer with Fluor Corporation. He has over 20 years experience in process engineering for refinery and chemical projects. He holds a MS degree in chemical engineering from the University of Mississippi.

Virginia Brown is a process engineer who also works at Fluor Corporation. She graduated from the University of Texas at Austin with a BS in chemical engineering and has over 2 years of experience designing various utility and hydrocarbon systems for a major US refinery.

Jaime Robles is an associate process engineer with Fluor Corporation. He has over 2 years experience in the E&C industry and has worked on major US refinery modernization projects. He holds a BS degree in chemical engineering from Rice University.

Greg Willing is a Senior Operations and Maintenance Representative with BP. He has over 23 years experience for the BP Whiting Refinery. He holds a BS degree in chemical engineering from the University of Wisconsin – Madison.

The authors express their appreciation and gratitude to James Turner, Director of Process Technology and Engineering, Fluor Corporation, for his assistance and encouragement in the preparation of this paper.

Cascaded Hydrotreating Process Reduces Refiners' Costs

Authors:
James Turner, Jayce Mathews
Fluor Corporation, Sugar Land, Texas

Introduction

As refiners produce gasoline and diesel products to clean fuels specifications, severe hydrotreating is required of several streams. All diesel product streams require hydrotreating to meet ULSD specifications, which is a maximum of 15 ppmw sulfur in the United States. Most "off-road" U.S. diesel specifications are also transitioning to 15 ppmw maximum over the next few years. In order to meet gasoline clean fuels specifications, refiners operating Fluid Catalytic Cracker (FCC) units either need to hydrotreat the gas oil feed, hydrotreat the gasoline product, or both.

Many refineries have both gas oil (FCC feed) hydrotreaters, and diesel hydrotreaters. Future refinery expansions and modifications to handle heavier, higher sulfur crudes will also likely require additions of these types of units.

It is possible to design a hydrotreater reactor scheme to hydrotreat gas oil and diesel streams in series, in a single reactor loop. By using this cascaded hydrotreater scheme, capital costs and operating costs will be less than for two separate units processing these streams.

Fluor has been awarded a U.S. patent (#7,238,274, Combined Hydrotreating Process, Turner, issued July 3, 2007) for a processing scheme that hydrotreats a heavy hydrocarbon stream and a lighter hydrocarbon stream in series. The most common application for this process is for processing gas oil and diesel streams. While this process has been considered in refinery studies, no operating units have been constructed at this point in time.

There is a disadvantage to this combined hydrotreating scheme – when the unit needs to be taken out of service for maintenance, it causes a greater impact to the refinery operations, since both the gas oil and diesel hydrotreating capacity are lost at the same time. This disadvantage may be less severe for refiners that have other hydrotreater units that can mitigate this impact, or for refiners that plan to shut down many units for turnarounds at the same time.

This processing scheme can theoretically be applied to revamp an existing unit, although there may be many limitations to overcome with revamps that would not apply to new units. Therefore, the more common application for this scheme will be for new units.

Design Approach

In 2001, gasoline and diesel clean fuels requirements were announced by the U.S. EPA, with a phase in period for most refiners between 2004-2006. Many other countries followed suit with their own clean fuels requirements. Fluor commissioned an internal study to determine the optimum processing scheme for hydrotreating gas oil and diesel in one unit.

Most refinery FCC feed gas oil hydrotreaters require desulfurization down to between 500 and 3000 ppmw sulfur, depending on the overall refinery processing scheme. 500 to 800 ppmw sulfur is typically required if there is no gasoline post-treater on the FCC naphtha product. A higher sulfur limit may be acceptable if the refiner invests in an FCC naphtha post-treater. For any of these gas oil hydrotreaters, the diesel range material that comes out of the gas oil hydrotreater reactor will typically be 50-100 ppmw sulfur or more. Since the ultra low sulfur diesel specification is less than 15 ppmw, co-treating diesel and gas oil in the same reactor is not likely to be successful, and separate reactors for the gas oil and diesel will be required. In addition, care must be taken to keep diesel treated only in the gas oil reactor from comingling with the treated diesel product, as a small amount of this diesel can be a contaminant for the ultra low sulfur diesel product. The "trick" in the conceptual process design is to minimize equipment and simplify the process, without contaminating the diesel product.

Design Parameters

Many factors affect what combined processing scheme would be optimum for a given situation. The primary factors affecting this are:

- Relative capacity requirements
- Relative severity
- Desire for diesel recovery in Cat. Feed Hydrotreater
- Refinery Hydrogen System
- Risk tolerance/refinery impact for shutdown scenarios

Regardless of the specific situation, a combined unit approach can be designed that is likely to be significantly less expensive than two equivalent new units. However, the optimum configuration will vary depending on the above factors.

The overall concept of the scheme is to treat the gas oil first in the reactor loop, and then cool the effluent to about 500°F against the feed. The vapor and liquid are then separated in a hot separator. The vapor is mixed with the fresh diesel feed and preheated to enter the diesel reactor. The diesel effluent treating and fractionation scheme has several different alternatives depending on the feed content, hydrogen scheme, and other refinery specific factors.

Process Description

The scheme assumes the following:

- Similar throughput requirements for diesel and gas oil (more than a 2:1 ratio in either direction is probably not economical)
- Similar operating pressure requirements (the Cat. Feed operating pressure will be ~100 psi higher pressure than the diesel operating pressure).

Cascaded Hydrotreating Process Reduces Refiners' Costs

A PFD of the basic process is shown in Figure 1 below. Pumps, exchangers, and other process details have been eliminated from the drawing for clarity.

Figure 1 – Basic Cascaded Hydrotreating Process Scheme

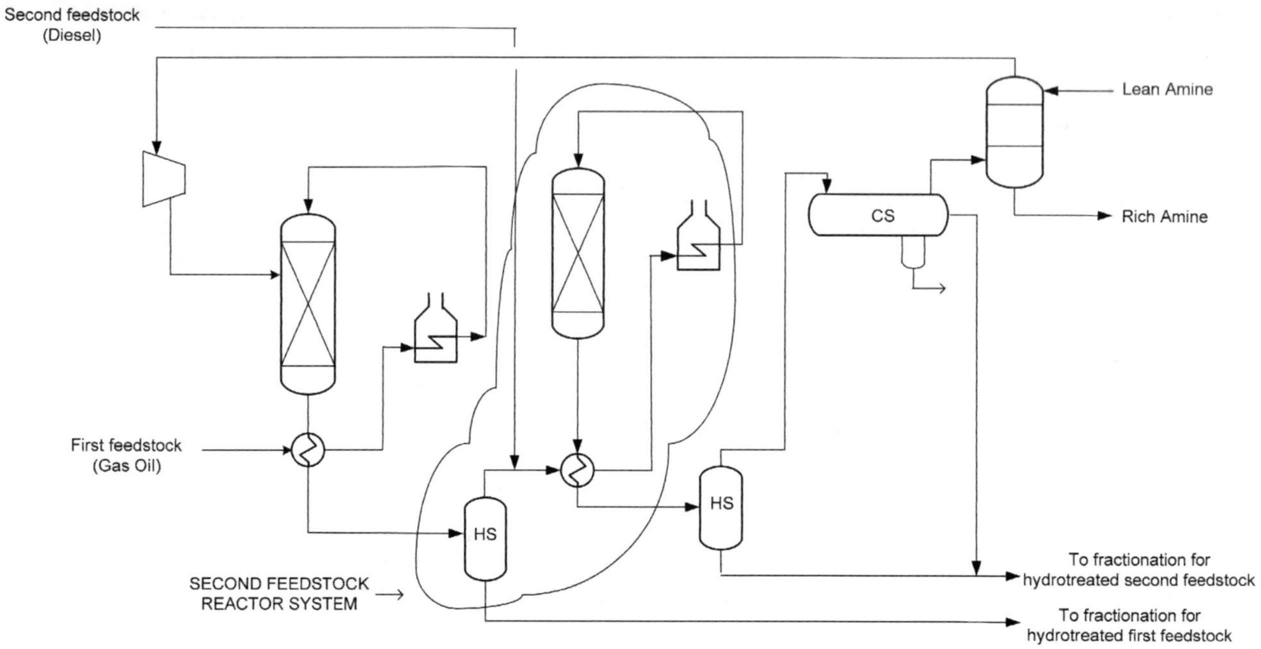

A brief description of the process, as applied to gas oil and diesel hydrotreating, is as follows:

The gas oil is pumped up to the required pressure, mixed with the treat gas, and heated against reactor effluent and in a fired heater, as in a typical Cat. Feed Hydrotreater. Following the reactor, the effluent is cooled to ~450-550°F against the feed before entering the gas oil hot separator. The hot separator temperature is optimized in the design to balance heat input for both reactors, and vaporize the right amount of diesel from the reactor effluent.

From the hot separator, instead of further cooling, the vapor is mixed with the diesel feed and preheated before entering the diesel reactor. An independent fired heater for the diesel reactor is shown in the scheme, but may be eliminated depending on the diesel and gas oil reactor heat releases and the desired operating temperatures. In this case, the diesel reactor feed would be preheated against the gas oil reactor effluent in addition to preheating from the diesel reactor effluent. Care should be taken to consider all possible operating scenarios in establishing the heat integration design.

From the diesel hot separator, the vapor is cooled in the reactor effluent air cooler, and enters the product separator. The vapor from the product separator is treated in an amine absorber before entering the recycle compressor.

Additional process optimization may be possible depending on the feed composition, product yields, refinery hydrogen scheme, and refinery economics for diesel recovery out of the FCC feed

The fractionation section has fewer opportunities for integration; however, some of the overhead systems can be combined, depending on the desired product recoveries. For example, if both the gas oil stripper and diesel stripper are designed for a naphtha overhead system, this system could be combined. However, care should be taken to determine the likelihood of cross-contamination of diesel range material from the gas oil stripper to the diesel sstripper. Another option is to draw diesel and naphtha overhead from the gas oil stripper, or draw diesel as a side product from the gas oil stripper, and feed it into the diesel reaction section along with the fresh diesel feed.

Cost Saving Opportunities

In the typical design, the following equipment items may be combined compared to two units:

- Reactor Effluent Air Cooler
- Product Separator
- Sour Water Flash Drum
- Amine Absorber
- Amine Absorber K.O. Drum
- Rich Amine Flash Drum
- Lean Amine Pumps
- Recycle Compressor
- Two Make-Up Compressors (total flow still the same as with two units)
- Recycle Compressor K.O. Drum
- Make-Up Compressor K.O. Drum
- Interstage Cooler
- Interstage K.O. Drum
- HP Stripper Overhead System (depends on chosen fractionation scheme)
- Water Break Tank
- Wash Water Pumps
- Fuel Gas K.O. Drum
- Flare K.O. Drum
- Piping associated with above

It has been estimated that a combined unit will cost 20-35% less than two independent units for the same processing conditions.

In addition to the above savings, the following opportunities may be present:

- If the throughputs are similar, a common spare charge pump can be used. Common spare product pumps and stripper reflux pumps may also be possible.
- Reactors can be built on a common structure.
- Heaters may be able to be combined in an efficient fashion, or the diesel charge heater may be eliminated depending on the processing requirements.
- Plot space can be reduced.

- Infrastructure and utilities are less than required for two units.

Conclusion

As refineries continue to change to meet changes in crude availability, product specifications, and product demands, there will continue to be a need for new hydrotreater units. Any time a refiner is looking to add two new hydrotreater services, building one combined unit can offer significant cost savings.

The authors:

James Turner is a Senior Director, Process Technology and Engineering, with Fluor Corporation. He has over 20 years experience in process design, mostly for refinery projects. He has expertise in hydrotreating, hydrogen management, process simulation, distillation, and clean fuels. He holds a BS degree in Chemical Engineering from Texas A&M University. He is active in AIChE, and has held positions of Director, Newsletter Editor, Treasurer, 2nd Vice Chair, 1st Vice Chair, and Chair of the Fuels and Petrochemicals Division.

Jayce Mathews is a Process/Specialty Engineer, Process Technology and Engineering, with Fluor Corporation. Prior to working for Fluor, he worked as a Process Specialist with Lurgi. He holds a MS degree in Chemical Engineering from Texas A&M University – Kingsville. He is a member of AIChE and is actively involved with the Houston local chapter, South Texas Section (STS) of AIChE where he is an officer and technical events coordinator for the Young Professionals (YP).

Cascaded Hydrotreating Process Reduces Refiners' Costs

Appendix I – Typical Hydrotreater Reactor Simulation Model

When building a simulation model of a hydrotreater unit, it is required to have a good reactor model. The description below is the approach typically used by the author to simulate hydrotreater reactors. The same general approach can be used regardless of which commercial simulator program is used, although the specific choice of simulation unit operations may be different depending on what options are available in the simulator.

Typically, yields are provided by a licensor or catalyst vendor. Yields are normally provided based on the fresh feed to the unit, and typically include H_2S, NH_3, hydrocarbon light ends, and heavier hydrocarbon cuts. An example yield structure for a gas oil hydrotreater is shown in Table 1. Separate distillation curve data is typically provided for the heavier hydrocarbon cuts, so that they can be properly specified in the simulation.

Table 1 – Example Reactor Yield Structure

	SOR	**EOR**
	SCF/BBL	SCF/BBL
Chemical H_2 Consumption	655 SCF/BBL 1.043 wt%	580 SCF/BBL 0.934 wt%
Net Yield	wt % FF	wt % FF
H_2S	2.541	2.541
NH_3	0.153	0.153
C_1	0.043	0.394
C_2	0.043	0.394
C_3	0.072	0.657
iC_4	0.020	0.171
nC_4	0.061	0.514
Naphtha	1.25	3.88
Diesel	12.15	15.51
Gas Oil	84.71	76.72
Total	101.043	100.934

Figure 2 depicts a simplified sketch of the typical reactor model.

The inputs to the model are the net reactor yield and hydrogen consumption, the fresh feed rate, and the recycle plus quench gas.

A stream is specified that has the same composition as the net reactor yields. Depending on which simulator is used, this may be either defined as a single stream, or as multiple streams combined in a mixer or balance type operation. This stream is specified using a mass balance module so that its total mass flow is equal to the mass flow of the fresh feed plus the mass flow of the hydrogen consumption.

Separately, the total treat/quench gas stream is processed in a module that removes the amount of hydrogen equivalent to the reactor hydrogen consumption.

Finally, the net reactor effluent stream is combined with the treat/quench minus the hydrogen consumption stream. This represents the reactor effluent.

Figure 2 – Simplified Reactor Model

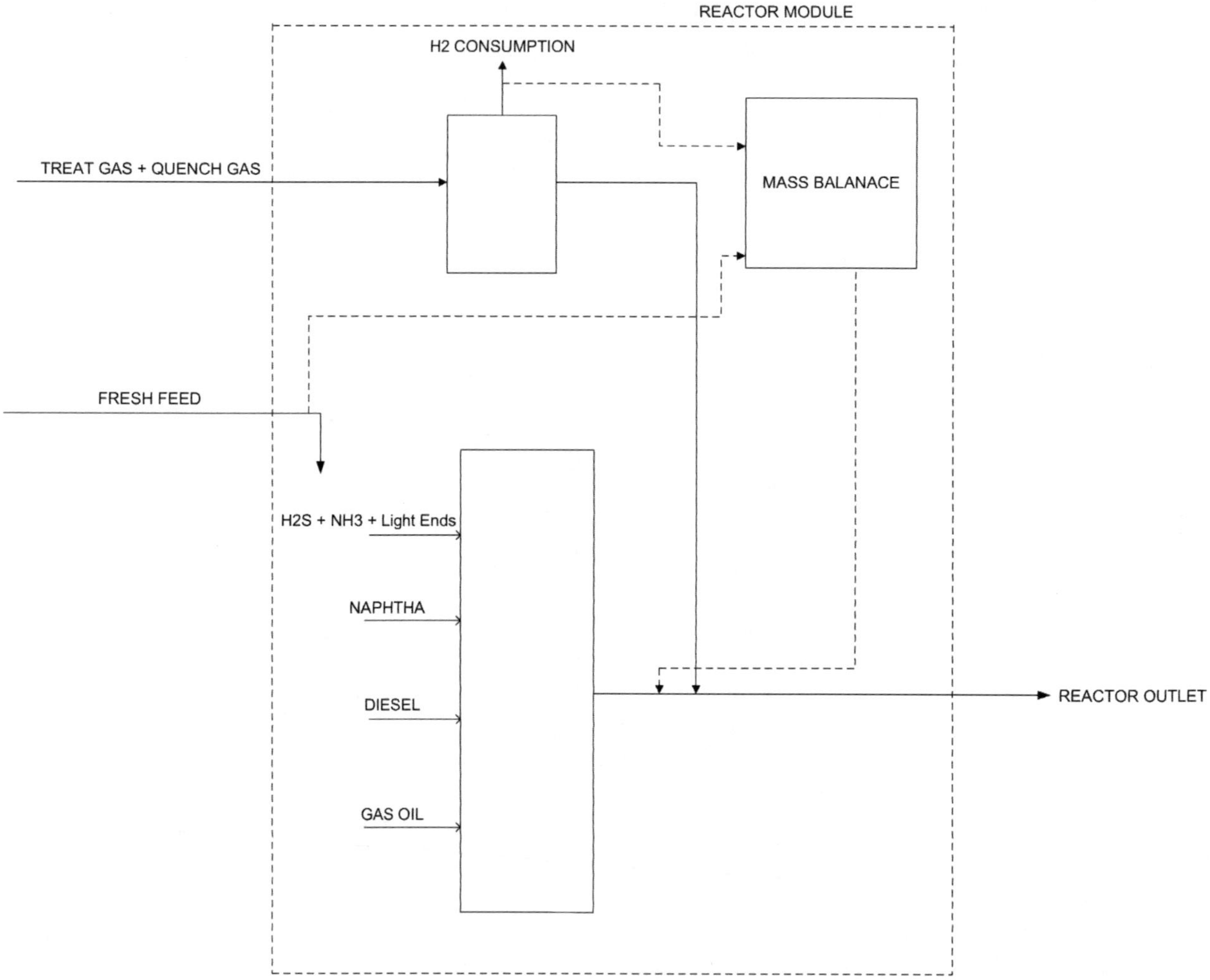

This general approach has been modified for various projects to divide the overall yields across multiple beds, and to calculate reactor and bed outlet temperatures based on bed heat release data and feed stream enthalpies.

87e Modeling and Simulation of Hydrotreating Reactors with the Effects of Vapor-Liquid Equilibrium

Neil Wang and Jinwen Chen. National Centre for Upgrading Technology (NCUT), 1 Oil Patch Drive, Devon, AB T9G 1A8, Canada

Hydrotreating (HDT) is a one of the most important processes used in refineries to improve the quality of distillate fractions by removing of sulfur, nitrogen, oxygen, and metals, and saturating olefins and aromatics. The HDT process is usually operated at temperatures of 350–400°C and pressures of 50–100 bar in a fixed bed. Under such conditions, the process typically involves gas, liquid, and solid phases (trickle bed) due to the presence of high-boiling-point fractions in the feedstocks.

In the trickle bed, hydrogen gas and liquid oil flow through a fixed bed of solid catalyst. There are complex couplings between the reactions, mass transfer, and heat transfer in this three-phase process. Although efforts to model trickle-bed reactors have achieved considerable progress in combining transport phenomena and heterogeneous reactions, variations in the flow rates and compositions of the individual phases due to the volatilization of liquid oil are often ignored because of the lack of vapor-liquid equilibrium (VLE) data for the HDT system.

It is generally assumed that the feed oil is non-volatile and the flow rates of both liquid and vapor phases therefore remain constant along the reactor. In fact, this assumption is not true for HDT reactors. Our previous work showed that large amounts of hydrocarbons are present in both phases, especially at higher temperatures and lower pressures. Accordingly, it is necessary to take VLE into account to ensure accurate modeling and simulation. Moreover, commercial HDT reactors are operated under non-isothermal conditions—temperature varies along the reactor. Since the vaporization of oil consumes significant energy, it is not possible to accurately predict reactor bed temperatures if this change of state is neglected.

In previous studies we conducted VLE flash experiments in a continuous-flow unit with hydrogen and various petroleum middle distillates under typical HDT conditions. The interaction coefficients between hydrogen and hydrocarbon boiling pseudo-components, required to perform VLE flash calculations, were estimated. Our previous work also demonstrated flow mapping of the desired operating regimes in a pilot plant trickle-bed reactor for hydrotreating two different middle distillate feeds. The present paper describes the development of a steady-state trickle-bed reactor model with VLE effects and gives predictions of commercial HDT reactor behavior. The detailed results are presented and discussed.

101a Fixed Bed Residue Hydroprocessing: Modeling Hydrotreating Performances and Catalyst Deactivation

Jan J. Verstraete[1], Isabelle Guibard[2], Damien Hudebine[1], and Hervé Toulhoat[3]. (1) Process Design and Modeling Division, IFP, PO Box 3, 69390 Vernaison, France, (2) Process Experimentation Division, IFP, PO Box 3, 69390 Vernaison, France, (3) Scientific Management, IFP, 1-4 avenue de Bois-Préau, 92852 Rueil-Malmaison

Crude oils contain large fractions of heavy products for which only few outlets exist. Refining processes converting heavy oils into more valuable products are therefore vital to a refinery. However, the heaviest cuts, termed residues, also contain the largest amounts of impurities, such as sulfur, nitrogen and metals. Fixed bed residue hydroprocessing is industrially carried out at high temperature and under a high hydrogen partial pressure. Under these conditions, the catalysts employed promote a complex network of parallel and consecutive reactions, among which are hydrodemetallization (HDM), hydrodesulfurization (HDS), hydrodenitrogenation (HDN), hydrocracking and hydrogenolysis. To enable all these chemical transformations, the residue feed successively passes over several fixed adiabatic beds containing different catalysts.

The reactivity of a residue feed depends both on the concentration and the nature of the various species. As it is not possible to fully separate and identify the various compounds in these mixtures, they are generally subdivided into several classes: saturates, aromatics, resins and asphaltenes (SARA). The latter are colloid structures whose radius of gyration typically lies between 10 Å and 500 Å, depending on the conditions. Asphaltenes can therefore only penetrate into the outer layers of porous catalysts. Their penetration is not only governed by molecular and Knudsen diffusion, but also by so-called configurational diffusion or surface diffusion. Hence, the performances of residue hydrotreating catalysts also strongly depend on the porous texture of the catalysts.

In this work, the performance of fixed bed residue hydrotreating units is predicted based on a model that accounts for the various physical and chemical phenomena. The porous structure of the catalyst is represented by Random Spheres, Random Needles and Random Coins models depending on the type of catalyst (Toulhoat et al., 2005). The hydrodynamic description accounts for transport of the species by convection and by Fickian diffusion inside the catalyst pellets. The molecular diffusion coefficients are calculated from the Stokes-Einstein equation and combined with Knudsen's law. Finally, as the molecular species can have a radius of gyration of the same order of magnitude as the pore size, configurational diffusion is also accounted for (Spry and Sawyer, 1975).

To describe the reactions, the hydrocarbons were first separated according to their boiling point into a low boiling (<520°C) and a high boiling (>520°C) fraction. Each fraction was subsequently separated by a SARA fractionation, which separates oil fractions into Saturates, Aromatics, Resins and Asphaltenes. To account for the purification reactions, the atomic composition (C, H, S, N, O, Ni and V) of each of these 7 lumps was also analysed and tracked in the model, leading to a total of 84 reactions.

The kinetic parameters were determined from experimental data obtained on a representative pilot unit operating on actual vacuum residues. From the simulations (Figure 1), it can be concluded that the model is able to correctly predict the unit operation. Indeed, a good agreement with the experimental data has been obtained, with regard not only to the concentration profiles of the various lumped families but also to the elemental composition of each family. As shown in Figure 1a, significant intraparticle profiles exist, even at the reactor outlet, illustrating the need to account for intraparticle diffusion.

In residue hydrotreating, catalyst aging is extremely rapid and run lengths are typically around 10 to 12 months. The build-up of deposited coke and metal sulfides over time strongly affects the catalyst accessibility and pore structure, and hence its activity. The use of Random Spheres, Random Needles and Random Coins models for the catalyst structure allows to advantageously represent catalyst aging. Indeed, the nucleation and growth of solid deposits inside the porosity can easily be accounted for by superimposing three random media (fresh catalyst, metal sulfides, coke), and calculate the residual porosity, surface area, volumes of coke deposit and of metal sulfides deposit, etc.

With this model, the simulations allow to predict not only the concentration profiles of the various chemical lumps, but also the evolution of elemental composition of each lump throughout the reactor. It is also shown that significant intraparticle profiles exist for asphaltenes and heavy resins, even at the reactor outlet, illustrating the need to account for intraparticle diffusion. The predicted intraparticle profiles also correspond well to the measured intraparticle profiles of the metal deposits (V and Ni).

Finally, the full-scale problem was also tackled with this model by simulating an industrial multi-bed adiabatic reactor with inter-reactor quenches. The various simulation outputs were then examined to assess the effects of the various operating variables.

References

1. Toulhoat, H., Hudebine D., Raybaud, P., Guillaume D. and Kressmann S.; A new model for combined simulation of operations and optimization of catalysts in residues hydroprocessing units, Catalysis Today, 109, p 135-153, 2005

2. Spry J. and Sawyer W., Paper presented at 68th Annual AIChE Meeting, Los Angeles, November 1975

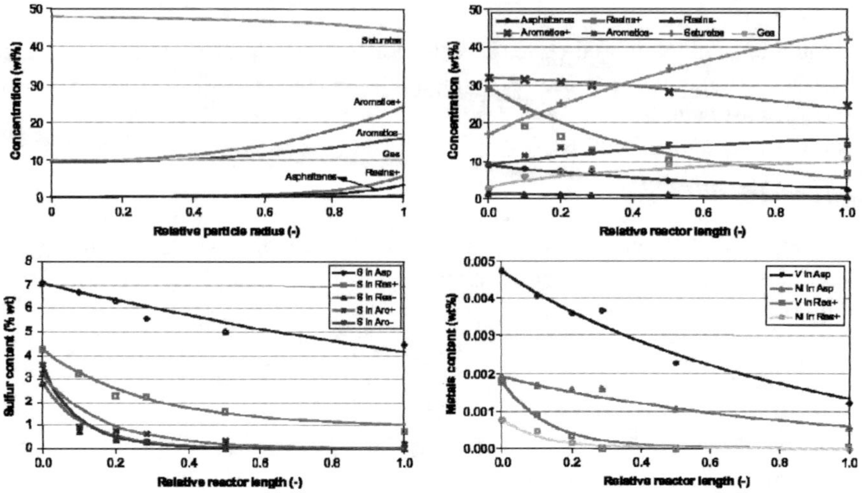

Figure 1: Simulation results (a: intraparticle profiles at the reactor outlet, b: axial reactor profiles for the various lumps, c: sulfur content of each lump, d: metal content of each lump)

UNDERSTANDING VAPOR-LIQUID EQUILIBRIUM IN BITUMEN-DERIVED HEAVY GAS OIL HYDROTREATERS

Jinwen Chen, Mugurel Munteanu, Neil Wang, Hena Farooqi, and Piotr Pieczykolan
National Centre for Upgrading Technology (NCUT)
One Oil Patch Drive
Devon, Alberta, Canada, T9G 1A8

Prepared for Presentation at the 2009 AIChE National Spring Meeting, T7005: Heavy Oil Chemistry and Bottom of the Barrel, April 26-30, 2009, Tampa, FL, USA

March 2009

Abstract

Vapor-liquid equilibrium (VLE) or flash experiments were conducted in a continuous-flow, bench-scale unit at temperatures, pressures, and gas/oil ratios typically used for heavy gas oil (HGO) hydrotreating or mild hydrocracking. Two HGO feeds with similar boiling ranges but different hydrocarbon type compositions were used in the experiments. Under similar operating conditions the feed that contained more aromatics showed a lower volatility than the other, which is partially attributed to the stronger interaction between hydrogen and the aromatics. The experimental data were used to verify the correlations, established with middle distillates hdyoprocessing systems in our previous studies, for estimating the interaction coefficients between hydrogen and hydrocarbon pseudo-components.

1. Introduction

Primary upgrading of bitumen (coking or hydroconversion) yields three distillate streams: naphtha, light gas oil (LGO), and heavy gas oil (HGO). These three streams have to undergo hydrotreating to remove most of the sulfur, nitrogen, and metals before they are blended into synthetic crude oil (SCO). The reactors used in these hydrotreating processes are operated under vapor-liquid equilibrium (VLE), with both vapor and liquid phases containing oil and hydrogen. Different operating conditions (temperature, pressure, gas/oil ratio, etc.) result in different flow dynamics and operating regimes in the hydrotreaters and, hence, differences in process performance and efficiency. It is always desirable to operate these reactors under ideal conditions (trickle flow, plug flow, fully wetted catalysts, etc.) in order to achieve optimal performance. Such an optimization is especially important in processing the HGO stream since the HGO hydrotreater accounts for about 30% of the total capital and operating cost in a bitumen upgrader.

Hydrotreating and/or mild hydrocracking of HGO involve oil and hydrogen flowing through trickle-bed reactors under high temperatures and pressures (350–420°C and 50–120 bar). Under such conditions, the reacting system (hydrogen and hydrocarbons) is close to thermodynamic VLE. In order to describe the flow dynamics precisely and to predict the operating regimes in HGO hydrotreating reactors under typical commercial operating conditions it is essential to understand the VLE behaviors and to conduct VLE calculations correctly. However, there are only limited number of studies reporting on this subject in the open literature (Akgerman et al., 1985; Smith and Satterfield, 1986; LaVopa and Satterfield, 1988; Jong, 1994; Tesser et al., 2003). There are even fewer studies on the effects of VLE on hydroprocessing (Kocis and Ho, 1986; Bellos and Papayannakos, 2003). Experimental data on VLE of hydrogen-petroleum fraction systems obtained under commercial hydroprocessing conditions are rarely published in the open literature due to the complexity of measurement and characterization (Laugier et al., 1983; Ramanujam et al., 1985; Wiegand et al., 1989). In the last several years, our research group has conducted studies on VLE measurement, calculation, and its effects on trickle-bed reactor hydrodynamics for middle distillates hydrotreating systems. The results have been reported elsewhere (Chen et al., 2005, 2006, 2009).

In the present study we extend our investigation of VLE behavior to HGO hydrotreating systems and examine its effects on reactor hydrodynamics. As a first step, this paper reports on the VLE experimental data and the verification of the correlations for estimating interaction coefficients between hydrogen and hydrocarbon pseudo-components.

2. Experimental

The pertinent details of the experimental setup were given in a previous paper (Chen et al., 2005). A similar experimental procedure was used in this study and is described only briefly here.

A bench-scale, continuous-flow VLE cell was used to conduct the VLE/flash measurements. The feed was pumped from a feed tank and then mixed with hydrogen (99.99%) under the desired pressure. The resulting mixture passed through a stainless steel preheating coil before entering the VLE cell where it was allowed to flash. Both the stainless steel coil and the VLE cell were housed in a temperature-controlled furnace to ensure a constant flash temperature. The VLE cell (I.D. of 2.54 cm and height of 30 cm) was arranged as a high-pressure, high-temperature phase separator with a two-point thermocouple positioned at the center to measure the temperatures of the liquid and vapor phase separately. The vapor phase from the top of the cell passed into a low-temperature, low-pressure phase separator where the hydrocarbons in the vapor were condensed to liquid while the uncondensed hydrogen went to a dry test meter (DTM) for measurement. The liquid phase from the bottom of the VLE cell entered another low-temperature, low-pressure phase separator where the hydrocarbons were condensed and the dissolved hydrogen was released and sent into a high-precision gas meter (Ruska 2331-801, Technel Engineering Inc.) for accurate volume measurement. Both of the condensed liquid samples were collected, weighed, and analyzed separately for density and simulated distillation (SimDis). The overall mass balance was calculated to ensure that the experiments were conducted properly.

Two HGO feeds were used in this study. Table 1 lists the key physical properties and chemical compositions of the two feeds. Feed 1 contained 61.2% aromatics while Feed 2 contained 36.3% aromatics. The two feeds had similar boiling ranges (IBPs and FBPs). However, Feed 2 had much lower boiling points from IBP to up to 40wt%, which is clearly shown by the two SimDis curves in Figure 1.

The temperatures and pressures used in this study ranged from 350°C to 430°C and 70 bar to 110 bar, respectively. The liquid feed and gas rates were fixed at 10 g/h and 8 NL/h, respectively, resulting in a gas/oil ratio of 800 NL/kg. Such conditions are commonly used in commercial HGO hydrotreating and mild hydrocracking.

Table 1. Physical properties and chemical compositions of the materials used in VLE experiments

	Feed 1	Feed 2
Density (15.6°C), g/mL (ASTM D4052)	0.9609	0.8866
Paraffins (wt%)	30.9	56.0
Aromatics (wt%)	61.2	36.3
Polars (wt%)	7.9	7.5
SimDis, °C (ASTM D2887)		
IBP (Initial boiling point)	201.8	201.2
10wt%	292.6	244.2
20wt%	329.2	259.8
30wt%	350.2	284.6
40wt%	367.4	353.6
50wt%	386.4	391.4
60wt%	412.4	420.0
70wt%	445.4	449.6
80wt%	484.6	486.0
90wt%	529.6	529.6
FBP (Final boiling point)	702.6	800.0

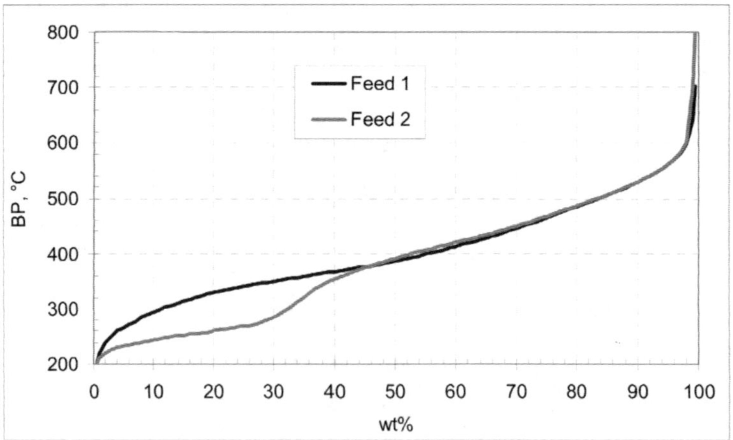

Figure 1. SimDis curves of Feed 1 and Feed 2

3. Results and Discussion

3.1. Comparison of VLE Data for Different Feeds

One of the important objectives of this study was to investigate the relative volatility of oils having differing hydrocarbon type compositions under similar operating conditions. It should be noted that it is difficult to find two (or more) feeds having different hydrocarbon type compositions for which all other bulk properties are the same or very similar. Changes in hydrocarbon type composition always lead to changes in density, SimDis, and other bulk properties, as is the case with the two feeds used in this study. Figures 2 and 3 show the weight percentages of oil in the vapor phase at different temperatures and pressures, respectively, for the two oils investigated in this study.

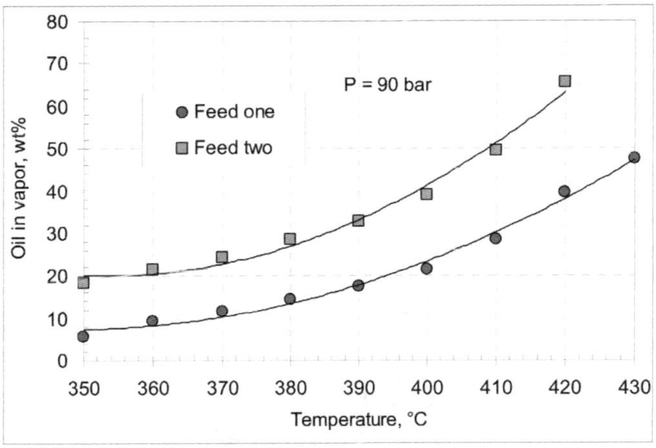

Figure 2. Wt% oil in vapor phase at different temperatures (P=90 bar)

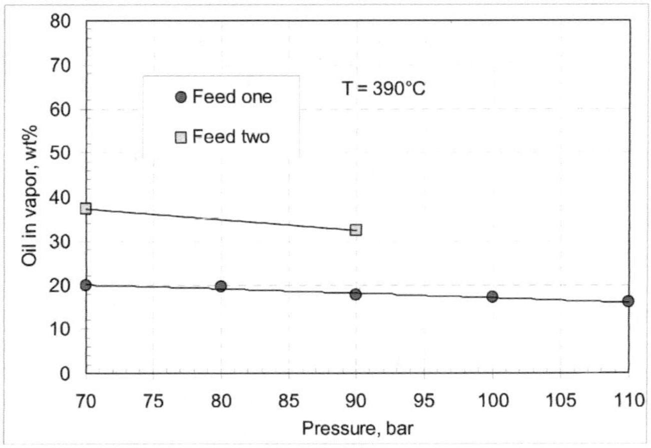

Figure 3. Wt% oil in vapor phase at different pressures (T=390°C)

As these figures show, the weight percentage of the individual oils in the vapor phase increase significantly with temperature. At 350°C (P=90 bar), only about 6wt% of the oil (for Feed 1) and 18wt% (for Feed 2) is vaporized. In this case, VLE would not have a significant effect on the flow dynamics in a reactor. However, at 420°C, about 40wt% of the oil (for Feed 1) and 65wt% (for Feed 2) are found in the vapor phase, which would dramatically affect the flow dynamics in a reactor. Over the pressure range from 70 to 110 bar investigated in this study, the change in the weight percentage of oil in the vapor phase with pressure is insignificant compared to that with temperature, from about 20wt% to about 16wt% at 390°C (for Feed 1). Feed 2, with only two points, seems to follow a similar trend (further measurements are under way).

As seen in Figures 1 and 2, within the range of operating conditions used in this study, for a given temperature and pressure, the weight percentage of oil in the vapor phase was always higher for Feed 2 than for Feed 1. There are two possible causes of this behavior: one is that Feed 2 contained more light material in the boiling range of IBP to 360°C than Feed 1, which is easier to vaporize. The other cause is that Feed 2 contained less aromatics than Feed 1, resulting in weaker hydrogen-hydrocarbon interaction.

3.2. Evaluation of Interaction Coefficients

The interaction coefficients between hydrogen and hydrocarbon pseudo-components are essential to the VLE/flash calculation when the equation of state (EOS) method is used. They must be known prior to the flash calculation. However, no methods or correlations are currently available in the open literature for predicting these coefficients, so they must be determined from experimental VLE/flash data. Assuming that the coefficients are correlated with the boiling points of the individual? pseudo-components and the aromatics content in the feed, and provided these measurements are available, it should be possible to obtain estimates using the correlation reported in our previously published papers (Chen et al., 2006, 2009) as follows:

$$d_{iH} = A + B \times BP_i \quad (1)$$

$$A = 0.3852 - 0.00241 \times C_{aromatics\%} \quad (2)$$

$$B = 2.245 \times 10^{-3} + 1.96 \times 10^{-5} \times C_{aromatics\%} \quad (3)$$

where d_{iH} represents the interaction coefficient between hydrogen and pseudo-component i and BP_i represents the boiling point of pseudo-component i. The same flash program as used previously, which is designed particularly for petroleum systems, was used to perform the flash calculation. Again the feed oil was divided into 29 boiling pseudo-components according to its SimDis data. Hydrogen was treated as a separate component. The aromatics contents of the two feeds are available as shown in Table 1. To use the program for flash calculation, the specific gravity, molecular weight, and boiling point of each pseudo-component must be known. All other properties are estimated using built-in correlations based on these three properties. The data for hydrogen all are available in the open literature.

To obtain the molecular weights and densities of the 29 hydrocarbon pseudo-components, distillations were performed to generate 8 fractions for each of the two feeds. Molecular weight, SimDis, and density/specific gravity were then measured for each fraction, resulting in separate correlations between molecular weight and boiling point, and between specific gravity and boiling point for each feedstock. These correlations, shown in Figures 4 and 5, were subsequently used to estimate the molecular weights and specific gravities of the 29 pseudo-components.

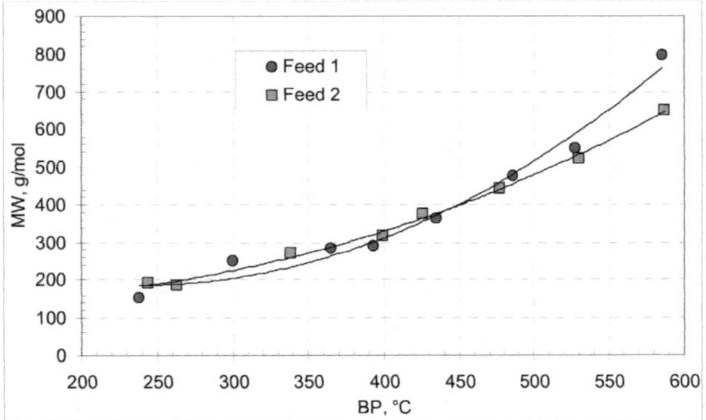

Figure 4. Molecular weight (MW) correlation with boiling point (BP)

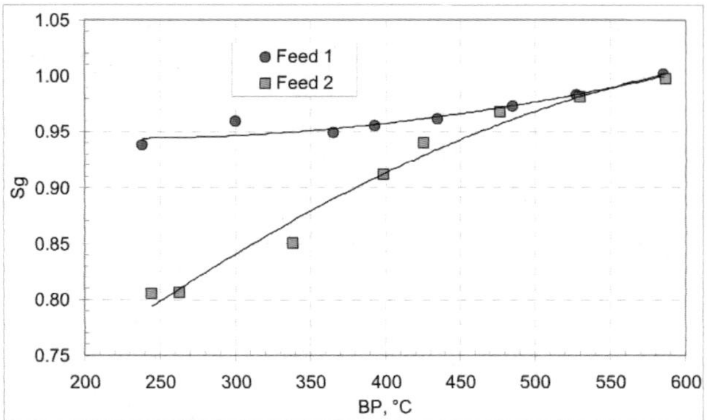

Figure 5. Specific gravity (Sg) correlation with boiling point (BP)

3.3. Discussion

The matches between the experimental and calculated data are shown in Figures 6–8. Figures 6(a) and 6(b) show the calculated and measured oil weight percentage in vapor phase at different temperatures for Feed 1 and Feed 2, respectively. Figures 7(a) and 7(b) show these quantities at different pressures. Figure 8(a) shows the calculated and measured SimDis data for the top and bottom liquid products of Run 01 for Feed 1, while Figure 8(b) shows these values of Run 15 for Feed 2. As Figures 6(a) and 6(b) show, the predictions of oil weight percentage in the vapor phase for different temperatures do not follow precisely the same trend as the experimental data: the predicted curve is flatter (less steep) than that for the experimental data. At the time of preparing this manuscript, the cause of this discrepancy was not clear and will be investigated further. However, the predictions of oil weight percentage in the vapor phase at different pressures match the experimental data quite well and also follow similar trends, as shown in Figures 7(a) and 7(b). It is worth noting that the flash calculations correctly predict the shapes of the SimDis curves of the liquid samples obtained from top and bottom of the VLE cell, as shown in Figures 8(a) and 8(b), which vary significantly among the two feeds, indicating a high predictive capability of the flash program.

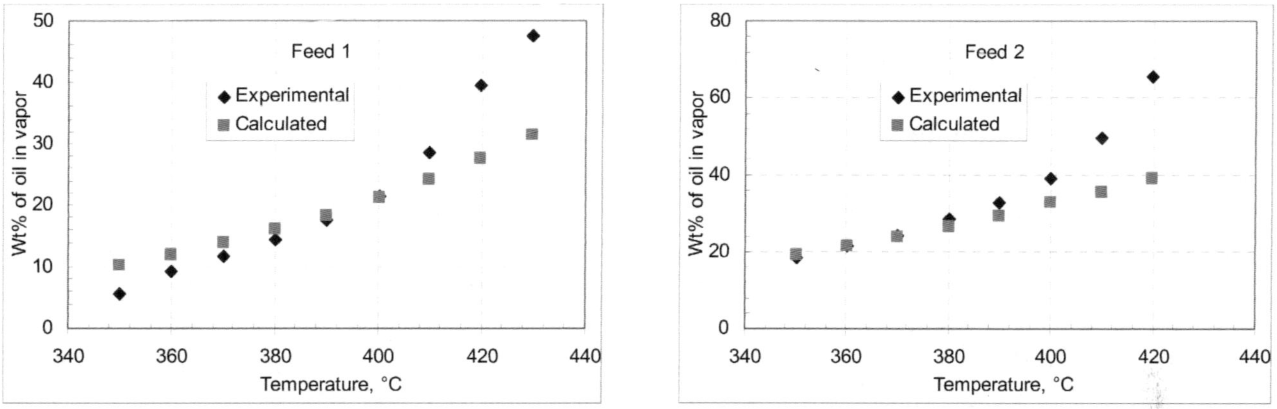

(a) Feed 1 (b) Feed 2

Figure 6. Calculated and measured wt% of oil in vapor phase at different temperatures

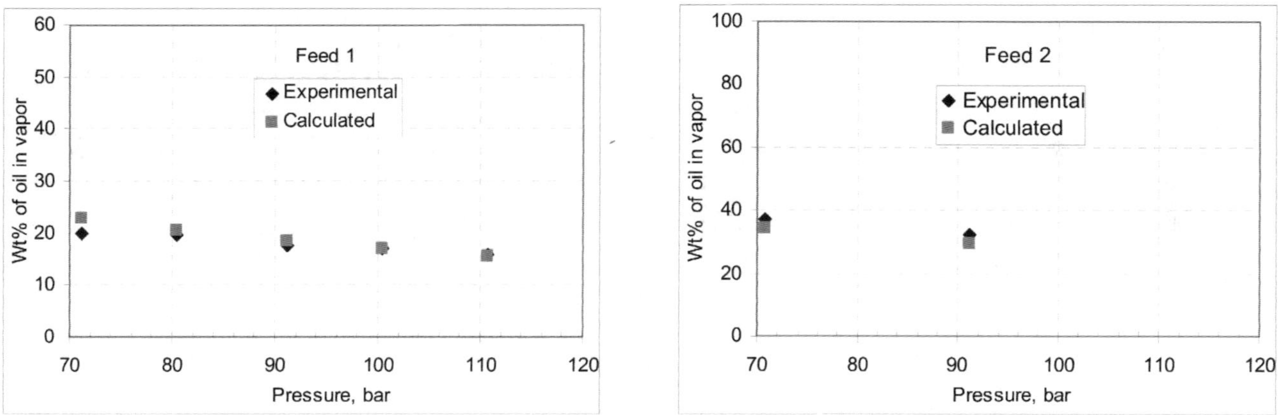

(a) Feed 1 (b) Feed 2

Figure 7. Calculated and measured wt% of oil in vapor phase at different pressures

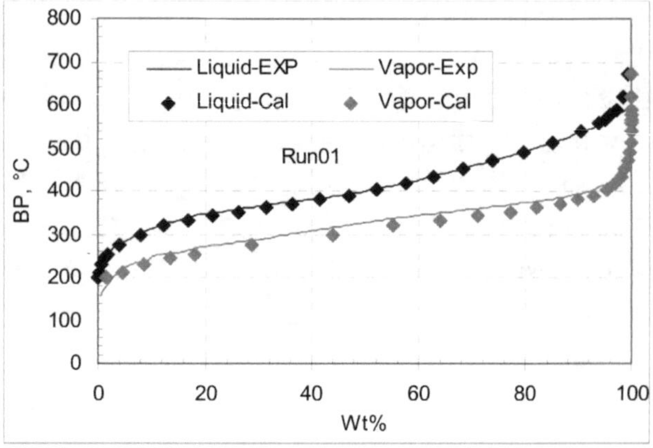

(a) Feed 1

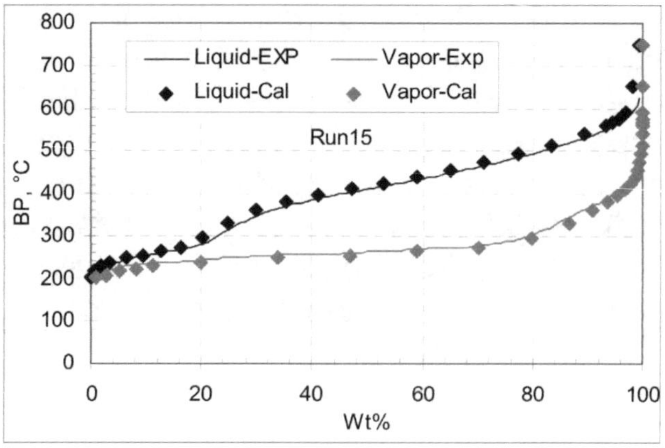

(b) Feed 2

Figure 8. Calculated and measured SimDis

It should be noted that the parameters and the correlations shown in Equations (1) to (3) can certainly be estimated by using the procedure described previously (Chen et al., 2006, 2009) to obtain improved agreement between model predictions and experimental data. This work will be conducted in the near future.

In order to perform the flash calculation for a particular feed, the SimDis data, density and molecular weight distributions with boiling points of the feed must be known. SimDis analysis is easy and can be conducted in any petroleum laboratory. However obtaining density and molecular weight distributions with boiling points normally requires real distillation followed by measurements of density and molecular weight for each individual fractions obtained from the distillation. These procedures are costly and time consuming. A methodology has been developed in our lab to measure the molecular weight distribution by boiling point using GC-FIMS (gas chromatography/field ionization mass spectrometry). The method utilizes the scanned mass of hydrocarbon ions and their abundance to calculate the molecular weight at different GC retention times, from which the boiling point of the ions can be calculated using pre-established calibration data between retention time and boiling point. The details on the development of this method have been reported elsewhere (Chen et al., 2007).

4. Conclusions

1) Comparative VLE experiments were conducted in a continuous-flow cell using hydrogen and two petroleum HGOs under typical hydroprocessing conditions used commercially.
2) The relative amount of oil in the vapor phase increased significantly with increasing temperature and decreasing pressure. Different volatilities were observed with the two feeds, which is attributed to the differences in aromatics contents and SimDis behaviors.
3) The experimental data were used to verify the correlations, established with middle distillates hdyoprocessing systems, for estimating the interaction coefficients between hydrogen and hydrocarbon pseudo-components. It showed that the correlations predicted the change in oil weight percentage in the vapor phase with pressure reasonably well. However, predicted changes in oil weight percentage in vapor phase with temperature were flatter than the experimental ones.

4) By using the correlations for estimating the interaction coefficients between hydrogen and hydrocarbon pseudo-components, the flash calculations correctly predicted the shapes of the SimDis curves of the liquid samples obtained from top and bottom of the VLE cell.

5. Acknowledgements

Partial funding for NCUT has been provided by the Canadian Program on Energy Research and Development (PERD), the Alberta Research Council (ARC), and the Alberta Energy Research Institute (AERI). The authors are grateful to staff of the NCUT analytical lab and pilot plant for their help and support in conducting the analyses and VLE experiments. The editing by Conrad Gietz and formatting by KJ Meharg of this paper are greatly appreciated.

6. References

1. Akgerman, A., Collins, G. M., Hook, B. D., "Effect of Feed Volatility on Conversion in Trickle-bed reactors", Ind. Engng. Chem. Fundam., 24, 398-401, 1985.
2. Bellos, G. D., and Papayannakos, N. G., "The Use of a Three Phase Microreactor to Investigate HDS Kinetics", Catal. Today, 79&80, 349-355, 2003.
3. Chen, J., Yang H., and Ring, Z., "Experimental Study on Vapour-Liquid Phase Equilibrium of LCO Hydrotreating Systems at High Temperatures and Pressures", Proceedings of 2005 AIChE National Spring Meeting, Topical A: 8th Topical Conference on Refinery Processing, Atlanta, GA, USA, April 10-14, 2005.
4. Chen, J., Jiang, W., Yang, H., Hawkins, R., and Ring, Z., "Comparative Study of Vapor-Liquid Equilibrium during Hydroprocessing of Different Petroleum Feedstocks", Proceedings of 2006 AIChE National Spring Meeting, Topical A: 9th Topical Conference on Refinery Processing, Orlando, FL, USA, April 23-27, 2006.
5. Chen, J., McLean, N., Briker, Y., Hager, D., and Ring, Z., "Molecular Weight and Its By-Boiling-Point Distribution of Middle Distillates for Hydroprocessing Modeling and Simulation", Proceedings of 2007 AIChE National Spring Meeting, Topical A: 10th Topical Conference on Refinery Processing, Houston, TX, USA, April 22-27, 2007.
6. Chen, J., Wang, N., Mederos, F., and Ancheyta, J., "Vapor-Liquid Equilibrium Study in Trickle-Bed Reactors", Ind. Eng. Chem. Res., 48, 1096–1106, 2009.
7. de Jong, K. P., "Effect of Vapour-Liquid Equilibrium on Coke Deposition in Trickle-Bed Reactors during Heavy Oil Processing. 2. Modeling", Ind. Engng. Chem. Res. 33, 3141-3145, 1994.
8. Kocis, G. R. and Ho, T. C., "Effects of Liquid Evaporation on the Performance of Trickle-Bed Reactors", Chem. Engng. Res. Des., 64, 288-291, 1986.
9. Laugier, S., Richon, D., and Renon, H., "Vapour-Liquid Equilibria of Hydrogen-Hydrocarbon Mixtures Up to 40 MPa, 400°C: Measurements and Representation", Chemical Engineering Thermodynamics, paper presented at the Thermodynamics Sessions of the 2nd World Congress on Chemical Engineering, Montreal, Que., Canada, 1983.
10. LaVopa, V., and Satterfield, C. N., "Some Effects of Vapour-Liquid Equilibria on Performance of a Trickle-Bed Reactor", Chem. Engng. Sci., 43, 2175-2180, 1988.

11. Ramanujam, S., Leipziger, S., and Well, S., "Vapour-Liquid Equilibrium for a Hydrogen/ Simulated Coal-Derived Liquid System", Ind. Engng. Chem. Proc. Des. Dev., 24, 364-368, 1985.
12. Smith, C. M. and Satterfield, C. N., "Some Effects of Vapour-Liquid Flow Ratio on Performance of a Trickle-bed reactor", Chem. Engng. Sci. 41, 839-843, 1986.
13. Tesser, R., Di Serio, M., and Santacesaria, E., "Influence of the Vapour-Liquid Equilibrium (VLE) on the Kinetics in Gas-Liquid and Gas-Liquid-Solid Systems", Catal. Today, 79&80, 323-331, 2003.
14. Wiegand, K., Strobel, B. and Hofmann, H., "Vapour-Liquid Equilibrium of Hydrogen/Coal Liquid and Methane/Coal Liquid Systems at Elevated Temperatures and Pressures", Chem. Engng. Technol., 12, 280-288, 1989.

101c Comparison of Thermochemical Conversion with Oilsand Bitumen in a Fluidized Bed Pyrolyzer and a Catalytic Pyrolyzer

See Hoon Lee[1], Jae Ho Kim[1], Kyung Seon Yoo[2], and Young Kwon Park[3]. (1) Gasification research center, KIER, 71-2 Jang-dong Yusung-gu, Daejeon, South Korea, (2) Kwangwoon University, Seoul, South Korea, (3) University of seoul, Seoul, South Korea

The thermochemical conversion of oils and bitumen were investigated in a fluidized bed pyrolyzer and a catalytic pyrolyzer. The fluidized bed pyrolyzer consisted of a MFC (MGF-Series, Rich Field), a gas pre-heater, a fluidized bed reactor (I.D. 0.076m and 0.8m high), a screw feeder, a quencher and an electrostatic precipitator. In the catalyst reactor, MMZ ZSM-5, ZSM-5, MESO C12, MCM-41 catalysts were used. At the temperature conditions of 400-550°C and Uo/Umf of 3.0, maximum oil yield was 70%. Oil composition over C20 is remained in the fluidized bed pyrolyzer. Also oil content from C5 to C9 is 0.6-9.5%. However oil content from C5 to C9 is 14-49% at the catalytic pyrolysis and oil content over C20 was not detected over 500oC.

101d Prediction of Heavy Oil Viscosity at Elevated Temperature

Adel Elsharkawy, Petroleum Engineering Department, Kuwait University, B.O. Box 5969, Safat 13060, Kuwait

prediction of heavy oil viscosity is important for calculations of cold oil production rate and planning for thermal means of oil recovery.

Several correlations, corresponding state models, and equation of state based models are used to calculate heavy oil viscosity when experimental data are unavailable or at temperatures higher than the formation temperature. These methods have serious limitations and have poor accuracy to predict heavy oil viscosity.

Rarely heavy oil viscosity is measured at temperature higher than the formation temperature. Thus productivity and recovery estimates are based on predictive tools that have poor accuracy of heavy oil viscosity. The prediction capability of these methods have never been tested at elevated temperature simply because of lacking the experimental data.

I this study viscosity data for several heavy oil samples having API gravity from 10 to 15 degree are measured at original formation temperature and at elevated temperature encountered in hot water, steam cycling and steam flooding. The measured data and other data collected from literatures are used to assess the accuracy of the existing methods used to predict heavy oil viscosity. The assessment indicate that the available model have serious accuracy problems. The data was used to develop a new compositional model to predict viscosity of heavy oil at formation temperature and at elevated temperature. The new model uses compositional data of crude oil (saturated, aromatic, asphaltenes, resins, and molecular weight), pressure and temperature to predict heavy oil viscosity. The proposed model was tested against unbiased experimental data. The comparison shows that the new model has better accuracy and valid to be used to predict heavy oil viscosity at elevated temperature in case these measurements are unavailable.

104a Transport and Kinetic Studies in Solid Acid Alkylation Processes

Subramanya Nayak, P. A Ramachandran, and M. P. Dudukovic. Chemical Engineering, Washington University in St Louis, St Louis, MO 63130

Commercial alkylation processes to produce desired heavier hydrocarbon (2,2,4 tri-methyl-pentane has research octane number of 100) use concentrated liquid sulfuric acid or anhydrous hydrogen fluoride as the catalysts at sub-ambient temperature with a mixture of butene highly diluted in isobutane. During last two decades, increasing safety and environmental concerns associated with use of these toxic acids and the disposal of their related acid soluble oil have led to alternative technologies based on solid acid catalysis.

Zeolites are active alkylation catalysts, giving product distribution similar to those characteristics of the liquid acids, but their unacceptably rapid deactivation was and still is the obstacle to commercialization (Feller and Lercher 2004). Zeolites are crystalline microporous materials mainly composed of silicon and oxygen atoms. How does organic molecule diffuse inside a zeolite and how does the intra-particle and intra-crystalline channel network of a zeolite influence the diffusion, adsorption/desorption and reaction pathway of organic molecules are necessarily important questions in solid acid alkylation processes.

The objectives of the current work are to understand and exploit different phenomena taking place during alkylation of iso-butane and n-butene on zeolites. Different particle sizes and pore morphologies of the zeolites are tested to enhance the understanding of the dynamics of the reactants and products in solid acid alkylation. The understanding gained should help in quantifying the key features needed for improved catalyst and regeneration and for selecting the optimal operation for solid acid alkylation processes.

References:

Feller A. and Lercher J. A., (2004). Chemistry and technology of isobuatne/ alkene alkylation catalyzed by liquid and solid acids. Adv. Catal. 48 229-295.

104b Intensification of Fluidized Bed Processes Using Novel Rotating Fluidized Bed Technologies

Juray De Wilde, Materials and Process Engineering Dept. (IMAP), Universite catholique de Louvain (UCL), Reaumur, Place Sainte Barbe, 2, Louvain-la-Neuve, B - 1348, Belgium

The operating conditions in conventional, that is, gravitational fluidized bed reactors are limited by the use of earth gravity. Internal mass and heat transfer limitations can be encountered, the van der Waals forces becoming too important when trying to fluidize smaller size particles. External mass and heat transfer limitations can be encountered and are related to limitations on the gas-solid slip velocities. With heterogeneous (catalytic) reactions, as well external mass or heat transfer limitations as internal mass transfer limitations may impose limitations on reaction rates and on the activity of the catalysts that can be used.

Fluidization in a centrifugal field may allow overcoming the above mentioned limitations and may, as such, allow fluidized bed process intensification and in particular the use of more active catalysts [1,2]. Current technology for fluidization in a centrifugal field is based on a rotating fluidization chamber [1,2] and its industrial application faces challenges with respect to sealing, mechanical vibrations and continuous operation. Two novel routes to fluidization in a centrifugal field are presented that facilitate industrial process application and offer additional specific advantages.

In rotating fluidized beds in a static geometry, the centrifugal field is generated by injecting the fluidization gas tangentially in the fluidization chamber via multiple gas inlet slots in its outer cylindrical wall [3,4,5]. A combined tangential-radial fluidization of the particle bed can be obtained by forcing the fluidization gas to leave the fluidization chamber via a central chimney. The fluidization gas flow rate influencing both the centrifugal force and the counteracting radial gas-solid drag force in a similar way renders rotating fluidized beds in a static geometry extremely flexible with respect to the fluidization gas flow rate and the gas-solid contact time. In particular, dense operation at high fluidization gas velocities is possible, allowing fast and highly endothermic or exothermic reactions to be carried out in a relatively small volume rotating fluidized bed in a static geometry.

Rotating fluidized beds in a static geometry furthermore allow angular multi-zone operation, the fluidization gases injected via successive gas inlet slots hardly being mixed in the rotating particle bed. Such multi-zone operation opens perspectives for carrying out catalytic reactions and catalyst regeneration within the same reactor vessel.

In rotating fluidized beds around a rotating chimney [6,7], the centrifugal field is generated by a rotating chimney consisting of multiple blades and positioned centrally in the fluidization chamber. The fluidization gas being forced to leave the fluidization chamber via the chimney, a radial gas-solid drag force counteracting the centrifugal force is generated. Furthermore, both the fluidization gas and the particles obtain a tangential velocity by the action of the rotating chimney. This results in a combined tangential-radial fluidization of the particle bed.

The two novel technologies for fluidization in a centrifugal field can eventually be combined to obtain extremely flexible fluidization technology. In such case, the rotating chimney allows to increase locally, that is, in the vicinity of the chimney, the centrifugal force. This allows to drastically reduce solids losses via the chimney [7]. This is particularly advantageous when fluidizing small, micro- or nano-scale particles. In such case, the rotating chimney also allows to increase the particle bed density, the rotating chimney acting like a compressor on the rotating particle bed.

Experimental and numerical data on the fluidization behavior with different types of particles, the pressure drop over the fluidization chamber and the rotating chimney, the gas-solid mass and heat transfer characteristics in the fluidization chamber and the solids losses via the chimney are presented and discussed. The advantages of fluidization in a centrifugal field using the novel technologies presented are illustrated with potential applications. Perspectives for using more active catalysts and facilitating catalyst regeneration are highlighted.

References

1. Qian, G.-H., Bagyi, I., Burdick, I.W., Pfeffer, R., Shaw, H., Stevens, J.G. AIChE J. 47(5), p. 1022-1034 (2001).

2. Watano, S., Nakamura, H., Hamada, K., Wakamatsu, Y., Tanabe, Y., Dave, R.N., Pfeffer, R. Powder Technol., 141, p. 172-176, (2004).

3. de Broqueville, Axel : Belgian Patent 2004/0186, Internat. Classif. : B01J C08F B01F; publication number : 1015976A3.

4. De Wilde, J., de Broqueville, A. AIChE J., 53(4), p. 793-810, (2007).

5. De Wilde, J., de Broqueville, A. Powder Technol., 183:(3), p. 426-435, (2008).

6. de Broqueville, A., De Wilde, J., April 23, 2007: PCT/EP2007/053941, European and United States of America patent, Pub. No. WO/2007/122211.

7. De Wilde, J., de Broqueville, A. AIChE J., 54:(8), p. 2029-2044, (2008).

Effectiveness of coatings to reduce fouling in Shell and Tube heat exchangers

Author: **Gabriel T. Fernandez**
Email: gabriel.fernandez@aramco.com

Co Author: **Talea A. Rothan**
Email: talea.rothan@aramco.com

Abstract

This paper studies the effectiveness of applying coating on the tube side to resist fouling in reboilers on crude oil stabilizers. In this process crude flows in the tubes of vertical reboilers heated by 60 psig steam to stabilize crude oil to meet True Vapor Pressure (TVP) and H_2S specifications. Due to the high temperature, reboilers are subject to fouling predominantly on the crude side, resulting in low heat transfer and shut down for cleaning. Based on vendor claims on the effectiveness of coatings to reduce fouling and increase the interval between cleanings, a study was undertaken to compare the operation of one set of reboilers on Column A coated on the crude side (inside) with a protective coating, with another set of similar reboilers and similar operating conditions with no coating, on an adjacent Column B. The study was conducted for an interval of about eleven (11) months and the results of this comparative study are presented in this paper.

Introduction

For the comparative study two clean shell & tube vertical reboilers E6/E6A were coated (inside walls of the tubes) and installed on Column A at the Stabilizer Plant. The operation of these two coated reboilers was monitored during the same period along with two uncoated reboilers E5/E5A installed on Column B. (For a Typical Crude Oil Stabilizer sketch see Figure 1.)

Both columns process Arab Extra Light (AXL) crude and are operated at almost the same conditions of feed throughput and H_2S ppm. Column A has 12 Glitsch valve trays with a maximum capacity of 280 MBOD, while column B has 14 Glitsch valve trays with a maximum capacity of 310 MBOD.

As shown in Figure 1 the sour crude is fed to the top of the column, passes down through glitch valve trays to the trap out tray located on the bottom tray. It is then fed to the vertical thermosiphon reboilers. The crude is stabilized to the desired true vapor pressure (TVP) and H_2S ppm by steam heating, steam injection to crude and by stripping in the column.

In the reboilers, the crude flows through tubes. The fouling is caused as a result of deposition of salts and impurities on the inside walls of the tubes during the process of heating (stabilization). The fouling results in resistance to heat transfer, thereby operational capabilities of the reboilers are reduced. This results in frequent water wash to remove the deposits from the tubes to maintain throughput. The consequence of this is reduced service factor.

Therefore, with the intention to reduce fouling and increase the service factor for the columns, plant operations and engineering decided to evaluate/test the technology that was recommended by the vendor, to coat the tube internals with polymer coating (Phenolic Epoxy Polymer). This coating would provide a special surface that would reduce the rate of scale deposits on the tubes surface. Thus, fouling rate would be controlled extending the interval between cleaning. It was decided to test the technology on Column A reboilers (E6/E6A), with intent that if proven successful all reboilers in stabilization service would be coated.

CASE STUDY-Highlights

- Two oil stabilization columns were selected and monitored. Reboilers (E6/E6A) on Column A were coated and the reboilers (E5/E5A) on the other Column B were left uncoated.
- The duration of the study was almost eleven months.
- Due to operations demand, both columns were operated at shallow stabilization (low bottoms temperature) for four months initially; then at deep stabilization (high bottoms temperature) for two months; and then back to the shallow stabilization for approximately five months.
- The following data was collected for both the columns:
    1. Crude oil feed flow, temperature and pressure.
    2. Reboiler crude outlet temperature
    3. Column crude outlet temperature
    4. Steam temperature and pressure
    5. Steam injection flow
    6. Condensate flow and temperature.
    7. Stabilization depth (H_2S ppm).
- Both columns, each with two reboilers, were simulated on Hysys using the data collected from the field. The simulations were calibrated to match field data.
- Hysys cases were imported into STX (shell & tube Exchanger software suite), to calculate h (film coefficients), U_c (clean coefficient) and U_d (dirty overall heat transfer rate).

(Note: The flows through the reboilers were obtained from simulations).

BASIS & ASSUMPTIONS

- The clean overall heat transfer rate was calculated as follows:

$$\frac{1}{U_c} = \frac{1}{h_{io}} + \frac{1}{h_o}$$

where,

$$h_{io} = h_i \times \frac{d_i}{d_o}$$

- The crude was heated by 60 psig steam on the shell side, the outside film coefficient h_o was assumed to be 1500 BTU/h.ft².F for steam condensation. Since we were mainly interested in the fouling trend, such an assumption does not affect the results.
- Fouling Factor of the reboiler was calculated by the following equation:

$$r_d = \frac{1}{U_d} - \frac{1}{U_c}$$

- E6 & E6A "coated reboilers" on Column A were assumed to perform identically.
- Similarly, E5 & E5A "uncoated reboilers" on Column B were assumed to perform identically.
- Both E6/E6A "coated reboilers" and E5/E5A "uncoated reboilers" are identical in construction, so the data sheet of the reboilers E5/E5A was used to evaluate both (reboiler data sheet for E6/6A was not available).
- The performance of E6/E6A "coated reboiler" was compared with E5/E5A "uncoated reboilers based on:
    1. Fouling Factor.
    2. Condensate temperature trend.
    3. Total production from column A and column B.

OBSERVATIONS & DISCUSSION

- For the first five months of the study the plant was operating at shallow stabilization (low bottoms temperature). The operations then switched to deep stabilization (high bottoms temperature) for almost two months; and for the rest of the testing period it operated at shallow stabilization. The reason for the changes in stabilization depth was due to operations commitment to meet the business requirements. This change gave us an opportunity to see the performance for a short duration under deep stabilization conditions and this is also documented here.

- **Chart 1** (attached) shows the Fouling factor trend for both E6/6A 'coated reboilers' and E5/5A uncoated reboilers. As shown from the chart, fouling resistance in E6/6A coated reboilers was mostly higher than in E5/5A uncoated reboilers, this can be attributed to the additional resistance from the coating. The trend is similar at the different operating conditions between shallow stabilization to deep stabilization operation.

- **Chart 2** (attached) shows the condensate temperature trend for both coated and uncoated reboilers. From the chart, the following observations can be made:
    1. In the first five months (Jul 3, 03 till Nov 15, 04), condensate temperature in E6/6A coated reboilers was higher than in E5/5A uncoated reboilers, which supports the previous observation of higher fouling resistance in coated reboilers.
    At the deep stabilization operation, E6/6A coated reboilers condensate temperature jumped up to more than 200°F (considerably high) while in E5/5A it did not exceed 180°F. Similarly the observations of the condensate outlet temperatures at near design throughput of 280 MBOD was higher for the coated reboilers; E6/6A coated reboilers condensate temperature reached about 170°F, while for E5/5A uncoated reboilers the maximum reading was 144°F. This indicates that the coating introduces inherent additional fouling leading to reduced subcooling, higher steam flow thereby higher condensate outlet temperature.

- The total production from Columns A & B during the duration of the study was compared; Column B (with uncoated reboilers) produced 5.6 million barrels of stabilized crude more than column A during the same period (11 months). The reason for the difference is that, throughput through column A had to be restricted due to higher condensate outlet temperature to avoid uncondensed steam flow into the condensate header. This restriction, though not intended, was a consequence of inherent fouling introduced by the coating.

Conclusion

1. Based on the above observations from the study, we can conclude that coated reboilers (E6/6A) have an inherent higher resistance for heat transfer compared to uncoated reboilers (E5/5A). This is because the coating material adds more resistance to the heat transfer and thus reduces heat transfer in the reboilers.
2. The coating did not help to reduce the fouling over time as was claimed by the technology vendor.
3. The condensate outlet temperature on the coated reboilers was higher compared to uncoated reboilers, therefore directionally it reduced maximum throughput to the column with coated reboilers and would get worse with fouling.

Glossary of Terms:

MBOD-Thousand Barrels per Operating Day: (M=1000)
STX: Shell & Tube Exchanger design software from Hyprotech (ASPEN Tech)
H_2S: Hydrogen Sulfide
ppm: parts per million

References:

Kern, D. Q: Process Heat Transfer

Figure 1.

Typical Crude Oil Stabilizer

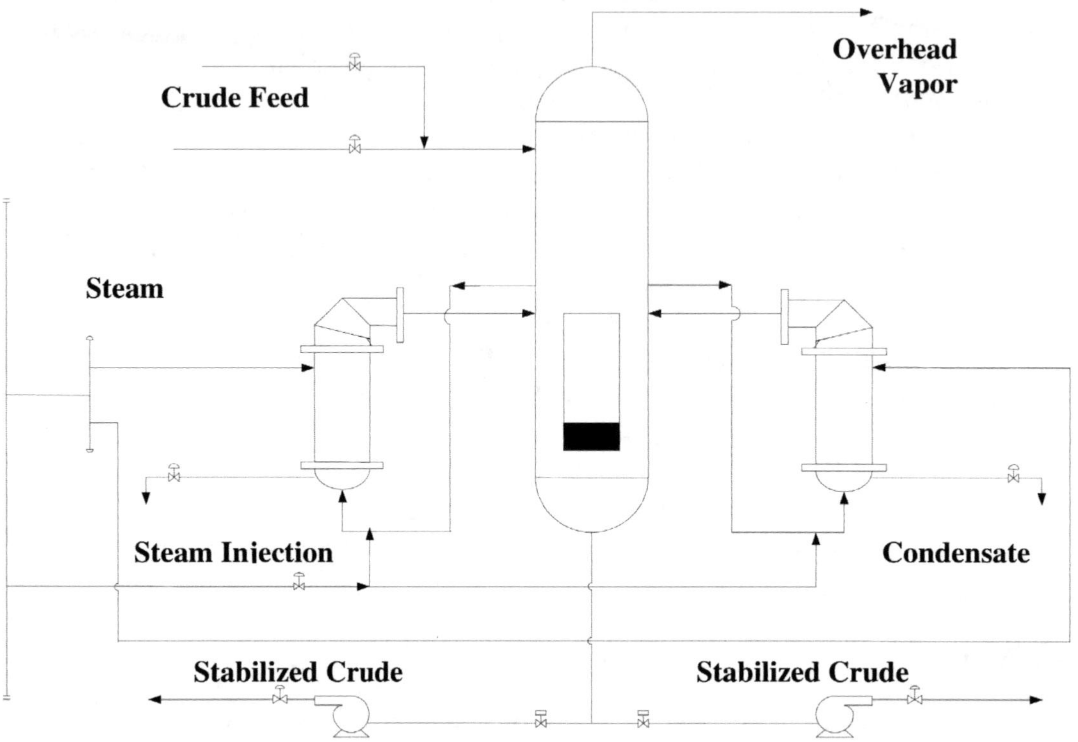

Chart 1: Fouling Factor Trend

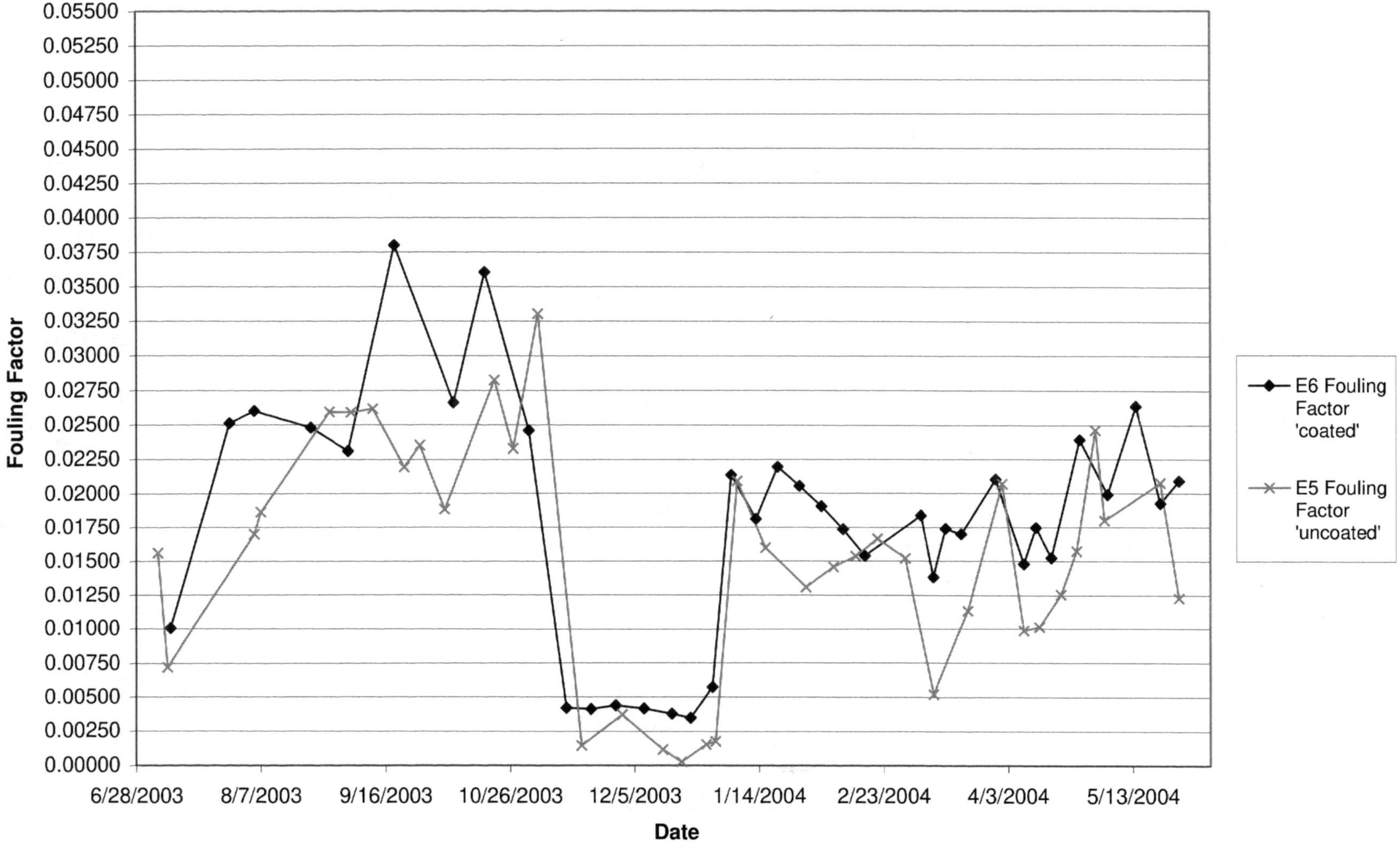

Chart 2: Condensate Temp. Trend

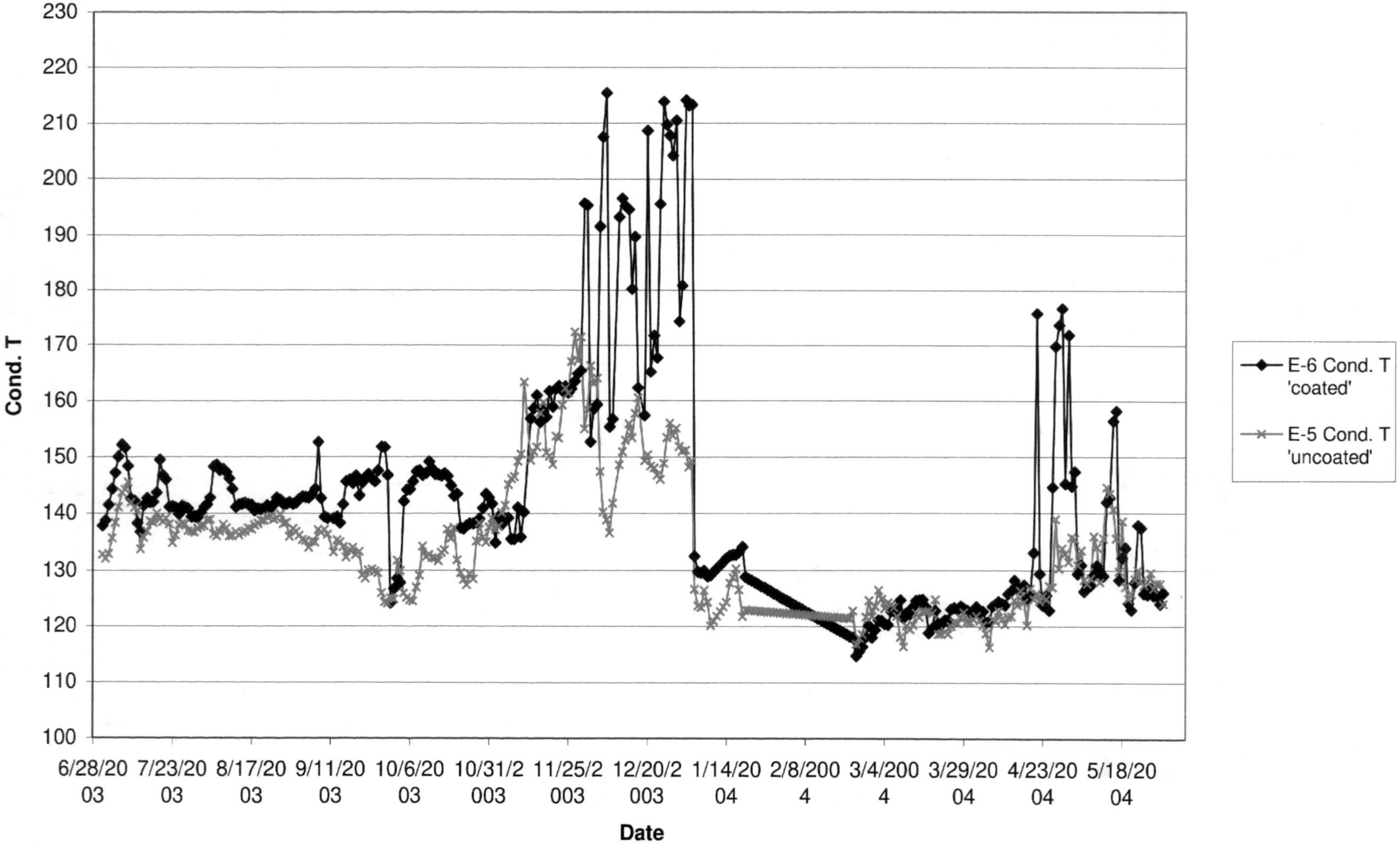

Selective Oxidation of a hydrotreated LCO to increase the cetane number

Roberto Galiasso Tailleur[1]

Department of Thermodynamics Simon Bolivar University Baruta Miranda Venezuela, galiassa@cantv.net

Introduction

Recent studies demonstrated the importance of adding oxygenated compounds to reduce the diesel emission of diesel [1]. However, this solution is expensive because these compounds must fulfill diesel specifications, such as diesel storage stability, lower toxicity, viscosity, density, and a boiling point compatible with the diesel flash point.

Another solution is a direct oxidation of the diesel, either by a nonselective auto-thermal oxidation or via a catalytic oxidation process. The author [2] developed a process that hydrogenates the diesel to some extent to produce naphtho-aromatics compounds, and then these partially hydrogenated molecules are oxidized into ketones types, raising the cetane number and reducing emissions [2,3]. This paper deals with the application of this technology to upgrade a low-value light cycle oil (LCO) stream. This fuel was deeply hydrogenated to remove sulfur to specification (~15 ppm) and to convert most of the alkyl-poly-aromatics in alkyl-naphtho-aromatics for further oxidation.

Experimental

A deep hydrotreated LCO to ultra-low sulfur (LSLCO) was oxidized using air and a CuCr/P(4-VP) catalyst to selectively incorporate oxygen in the alkyl-naphto-aromatics (ANAr) molecules. The hydrotreated product was characterized using HPLC with three different solvents to separate the naphto-aromatics from the other aromatics and naphthenes. The naphto-aromatics fractions were then separated by families using a PDVB column coupled with a GC-MS and by HNMR analysis of the fractions. The analysis of the naphto-aromatics fraction was particularly targeted to measure the potential oxidized compounds before and after reaction. The oxidation was carried out at three temperatures and six space velocities in a continuous CSTR device [2,3] operating at 1.2 MPa of total pressure. A total of 45 experiments were performed using the same catalyst at different temperature and residence time. The oxygenated products were characterized by the same technique to identify the type of compounds formed [2,3]. In addition, their acid and cetane numbers were also measured using ASTM methods. In particular, an in-house-developed storage stability test was used to control this important fuel property. This test involves storing 1 liter of sample in a steel tank for 45 days at 50°C under 0.1 Pa of air partial pressure. Samples were taken at three levels in the vessel every 3 days. Color, solids formation, oxygen uptake, and acid number analysis were performed on each sample to assess the degree of fuel deterioration. This method was compared to the storage of a ULSD (a control fuel that meets the requirements of US Federal Specification VV-F-800a).

Results and discussions

Hydrotreated product: A large sample of LCO was hydrogenated in the pilot plant operating at 350°C, 0.2 h^{-1} 6 MPa of total pressure and H_2//HC of 6 to obtain on-spec sulfur fuel (~15ppm LSLCO). A WNiPd/$TiO_2Al_2O_3$ catalyst was used for that [2]. We observed that the cetane number decreases with molecular weight in agreement with the negative contribution of the poly-aromatics compounds that prevail over the positive effect of the higher naphthenes and paraffins content in the LSLCO. The mono/diring-aromatics ratio also increases, while that of mono/diring-naphthenes decreases as a function of the boiling point [2]. Table 1 show the average value of the LSLCO properties.

Table 1 Properties of LSLCO

Boiling range °C	180-360
Mass wt%	100.0
Di-ring-aromatics	5.82
Mono-ring-aromatics	7.52
Tri ring aromatics	2.08
Naphtho-aromatics	14.1
Dicycloparaffins	20.21
Monocycloparaffins	18.34
Paraffins	31
Density kg/m3	823.0
Viscosity 40°C (cSt)	1.24
Cetane number	42.5

[1] Current address: Oklahoma University, Chem. Biol. & Mat Engineering School, Norman OK 73071 (405)7950432

It is important to determine the amount of naphto-aromatics that represent the potential "oxidable" compounds [2]. The HPLC-GC-MS analysis allowed us to recognize the (semi-quantitative) distribution of some families of compounds depicted in Table 2. The poly-alkyl-tetralin, poly-alkyl-indane, and poly-alkyl-fluorene families are the most abundant. The prevalence of the first family increases with boiling point, while the other two follow an opposite trend. The others represent more than 20 different types of compounds in minor proportion. The calibration of the HPLC techniques by blending model molecules with the feed and oxidation product indicated a 96% recovery of the total naphtho-aromatics compound present in the LSLCO. Table 2 shows the average value for the whole families.

Table 2 Alkyl-naphtho-aromatics (ANA) families in LSLCO

Cuts °C	180-360
Tetralin	1.896
Indane	1.764
Fluorene	1.464
TH Pentalene	0.288
DH and TH antracene	0.252
Others (difference)	2.136

LSLCO—Thermal oxidation. Thermal oxidation of the LSLCO was performed at 80°C, 90°C, and 100°C at five space velocities. The product after reaction was filtered using the methodology of ASTM 2274, and the liquids and solids analyzed. The results shown that the rate of oxygen uptake is initially very high (measured in liquid phase by elemental analysis), but by increasing the contact time it slows. The high rate of oxygen uptake is mostly associated with peroxide formation and then transformed into ketones and ketols in equal proportions. Gradually, the amount of peroxide is reduced, while that of acid and insoluble material starts to increase very rapidly when the contact time increases. Soon after this reduction in oxygen uptake, the quality of the fuel is totally deteriorated, the cetane number could not be measured in the diesel engine due to filter plugging, and the families of oxygenates could not be separated by HPLC due to fuel instability. The thermal studies of pure tetralin shown very poor selectivity toward ketones formation [2].

LSLCO—Catalytic oxidation of LSLCO

The effect of residence time in the activity and selectivity of the oxidation reaction in presence of CuCr/(4-PVP) catalyst is reported in Figure 1 for 90°C as an example. The right axis represents the wt% of oxygen uptake, measured by elemental analysis of the LSLCO and the wt% acid formed by titration with a standardized base (ASTM D664). The left ordinate depicts the cetane number for the OLSLCO products determined by testing the samples in a diesel engine (D613). For all temperatures studied the hydroperoxide initially increases, then decreases as a function of contact time.

We founded that in presence of CuCr/P(4-VP) catalyst, most of the hydroperoxides are still initially produced by autothermal reactions, but in a smaller proportion than in the absence of catalyst, for the same operating conditions. The autothermal hydroperoxide activates the catalyst [2-4], which starts to form its own "on-surface" peroxide intermediaries of reaction. The shape of the oxygen uptake curve is mostly the result of catalytic reactions. These catalytic reactions

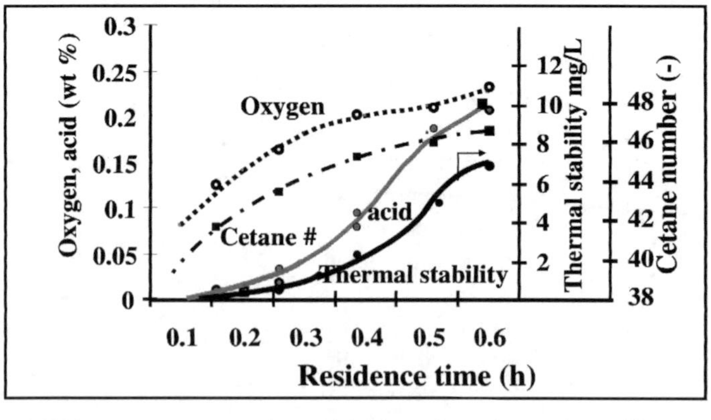

Figure 1 effect of residence time (90°C) on oxygen uptake, acid formation, thermal stability and cetane number

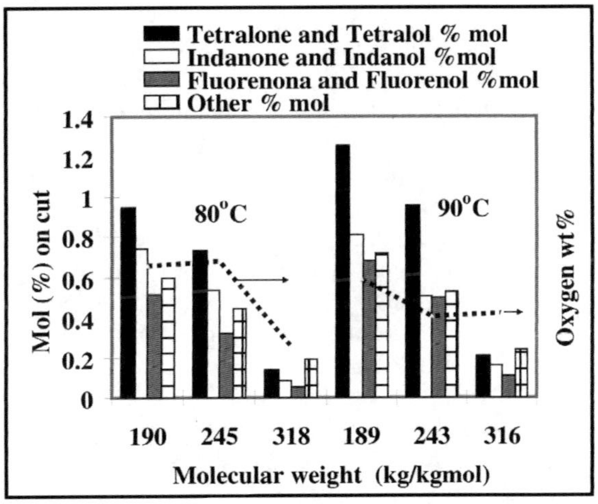

Figure 2 Main oxygenates families and total oxygen content distribution along the molecular weight (left OLSLCO1 - 80°C, and right OLSLCO2 - 90°C).

contribute to ketones (1)), ketols (reaction (2)), and acid formation (reaction (3)) by converting the naphto-aromatics families. The cetane number increases proportionally less as a function of oxygen content of the product.. Within the naphtho-aromatics families, oxidation is initiated in a naphthenic carbon (alpha and beta position in the molecules) which might be electronically-promoted or sterically depressed by paraffinic branches located in the neighbor carbons. As the oxygen uptake progresses, the reactions rates decreased because the already oxidized compounds and aromatics molecules (less reactive than the naphto-aromatics compounds are involved in the oxidation. Some of these secondary reactions lead to poly-oxygenates and carboxylic acid. The polymerization of unconverted peroxide and poly-oxygenated compounds produce gums and insoluble carboides. It can be see that the higher the residence time, the higher the amount of acid formed and the higher the product instability. For commercial application the oxidation can done until the product reaches the specification for acid number and/or product stability specifications recommended for diesel fuel. We observed that this limit is reduced by increasing the temperature for the same residence time. Further on this limit, the reaction became unfeasible because no commercial fuel could be derived. Even when the temperature favored the incorporation of oxygen and increased the cetane number, the higher the temperature, the higher is the amount of acid and poly-oxygenated compounds present in the OLSCO.

Figure 3 shows the oxygenate families distribution along the boiling point (three cuts) as well as the profile of oxygen distribution. The oxygenated-tetralin is more abundant in the lighter part while oxygenated indane is in the heavier fractions. Clearly the two product for near the same amount of total oxygen (~0.45 wt %) has completely different families and oxygen distribution. Acid and poly-oxygenated compounds (others in figure 2) accumulated in the middle and heavy part of the diesel and are responsible for gum formation (see equation (2)).

$$ANAr + O_2 \xrightarrow{k2Cat} ANArO \quad (1)$$

$$ANAr + O_2 \xrightarrow{k3CAT} ANArOH \xrightarrow{k5CAT} ANArOOH \rightarrow gum \quad (2)$$

$$ANAr + O_2 \xrightarrow{k3Cat} HOOANAr \quad (3)$$

Thermal stability of oxidized product. Next is the analysis of the effect of temperature on diesel stability for the same level of the total oxygen content and after washing all samples with caustic to extract the acid and peroxides formed during the oxidation. Table 3 shows the results of the amount of solid and acid formed during the storage of OLSLCO and LSLCO samples for 3 and 45 days. This test predicts well the one year diesel storage stability of ULSD.

Table 3 Storage stability (acid and solid formation)

Fuel	LSLCO		OLSLCO1 80°C		OLSLCO2 90°C	
Days	3	45	3	45	3	45
Acid*	0.17	0.24	0.10	0.14	0.13	0.14
Solid**	13	16	8	12	9	13

*Units: mg KOH/g ; ** mg / 100 cm^3 similar to ASTM 2274)

Table 3 shows that the LSLCO is the most unstable of the three fuels at beginning of the storage due to the loss of some oxidation inhibitor, mostly nitrogen compounds, eliminated during the deep hydrotreating of LCO. The reduction in the values of acids and solids formed in initial storage of OLSLCO1 and OLSLCO2 (after three days) are attributed to the selective formation of more stable oxidized compounds (catalytic oxidation of naphto-aromatics present in the LSLCO) and to the elimination of others gum-precursor. A commercial sample of ultralow-sulphur-diesel obtained from the market produce 40% higher acid and solid formation than the LSLCO during 45 days storage.

After 45 days in storage, the low-temperature, and more selectively oxidized fuel (OLSLCO1), produces less solid and acid compounds than the high-temperature oxidized one (OLSLCO2), and the hydrogenate fuel (LSLCO); therefore the storage results indicated that the previous selective (controlled) oxidation produced more storage-stable compounds than the original fuel which is an important benefits in addition of its higher cetane number. It was demonstrated that by storing during 45 days 100% tetralone, or 100% of tetralol at the same condition as indicated above that tetralone is not oxidized further, while tetralol is converted into poly-oxygenate compounds and then into some gum (insoluble) at high rate. In addition, the catalyst stability was studied by carrying out successive oxidation of fresh feed with the same catalyst. The results confirm our previous study finding [5]: lower the selectivity to ketones higher the catalyst deactivation, which represent another drawback of having a non-selective oxidation. At low temperature, the selectivity and stability of the catalyst is better than at high temperature. LSLCO oxidation at low temperature produces mainly ketones (reaction (1) compounds that are more stable to the additional oxidation than the original hydrocarbon. Therefore by increasing the oxidation temperature or the residence time the catalyst start to produce more unwanted compounds (reactions (2) and (3)) that are not removed by a caustic and can be reoxidized during the storage. Only a selective oxidation (1) generates a stable-high cetane number fuel. The values of the activation energies obtained in a detailed kinetic study [3] with model molecules and the same catalyst confirmed that reactions (2) and (3) have higher activation energies than reaction (1).

Effect of adding tetralol and tetralone to the LSLCO. The LSLCO fuel was blend with the tetralol and tetralone to have the same amount of oxygen than OLSLCO1. Table 4 compares the results in cetane number and storage stability at similar oxygen content.

Table 4 Effect of adding Tetralol (TOH) and tetralone (TO) to have 0.5 wt % of oxygen

Fuel	TO	TOH	OLSLCO1
Cetane number	46	45	47
Acid (45 days)mg/g	0.2	0.44	0.14
Solid (45 days)microg/cm^3	16	56	12

Table 4 shows that the oxidation of the naphto-aromatics produces better storage properties than blending the fuel with tetralone or tetralin, a model molecules that represent only one of the oxidable families. The difference is associated to the fact that the blend contain alkyl-naphto-aromatics with low storage stability while in the oxidized fuel most of then had been already converted into ketones, stable molecules for further oxidation.

Conclusions

The oxidation of the hydrotreated LCO demonstrated that there is a set of optimal operating conditions (temperature and residence time) for catalytic oxidation with CuCr/P(4-VP) catalyst that allows selective incorporation of oxygen and the production of a commercial fuel. The selective production of ketone type molecules by oxidation of alkyl-naphto-aromatics increase the cetane number, improve the fuel storage stability and reduce the catalyst deactivation. The present study demonstrated that by increasing the temperature the rate of oxygen uptake increase but the selectivity is reduced due to higher activation energies of the unwanted reactions. The accelerate storage test demonstrated that tetralol incorporation in the fuel dramatically reduce the fuel stability while the tetralone do not affect it.

References

(1) Gonzalez M.A., Liney E., Piel W., Natarajan M., Asmas T., Naegeli D.W., Yost D., Frame E.A., Clark W., Wallace J.P, Garback J., *SAE Tech. Pap. Ser.*2001, 01-3632, (2001).
(2) Galiasso Tailleur R., and Casanova Caris P.; *IJCRE*, 2007, 25.
(3) Galiasso Tailleur R., *Fuel & Energy* , 23(2), (2009) 799
(4) Galiasso Tailleur R., and Gomez C., *J. of Catal.*, (2007) 250, 1, 110.
(5) Casanova O., Galiasso Tailleur. R., Corma A., "Effect of tetralol and tetralone in the tetralin oxidation with copper and chromium heterogeneous catalysts", *preprint* of the ECCE Congress, Copenhagen, Sept. (2007).

AUTHOR INDEX

AUTHOR INDEX

American Institute of Chemical Engineers
3 Park Avenue
New York, New York 10016-5991

CURRAN ASSOCIATES INC
proceedings
.com